STUDENT SOLUTIONS MANUAL

to accompany

Physics

Eighth Edition

John D. Cutnell
Kenneth W. Johnson

Southern Illinois University at Carbondale

WILEY

John Wiley & Sons, Inc.

PREFACE

This volume contains the complete solutions to those Problems in the text that are marked with an $\boxed{\textbf{SSM}}$ icon. There are about 600 such Problems, and they are found at the end of each chapter in the text.

The solutions are worked out with great care, and all the steps are included. Most of the solutions are comprised of two parts, a REASONING part followed by a SOLUTION part. In the REASONING part we explain what motivates our procedure for solving the problem, before any algebraic or numerical work is done. During the SOLUTION part, numerical calculations are performed, and the answer to the problem is obtained.

We welcome any suggestions that you may have for improving the usefulness of these solutions. Please feel free to write us care of Physics Editor, Higher Education Division, John Wiley & Sons, Inc., 111 River Street, Hoboken, NJ 07030 or contact us at **www.wiley.com/college/cutnell**

CONTENTS

CHAPTER 1 | *INTRODUCTION AND MATHEMATICAL CONCEPTS*

3. ***REASONING*** We use the facts that 1 mi = 5280 ft, 1 m = 3.281 ft, and 1 yd = 3 ft. With these facts we construct three conversion factors: (5280 ft)/(1 mi) = 1, (1 m)/(3.281 ft) = 1, and (3 ft)/(1 yd) = 1.

    SOLUTION By multiplying by the given distance d of the fall by the appropriate conversion factors we find that

    $$d = \left(6 \ \text{mi}\right)\left(\frac{5280 \ \text{ft}}{1 \ \text{mi}}\right)\left(\frac{1 \ \text{m}}{3.281 \ \text{ft}}\right) + \left(551 \ \text{yd}\right)\left(\frac{3 \ \text{ft}}{1 \ \text{yd}}\right)\left(\frac{1 \ \text{m}}{3.281 \ \text{ft}}\right) = \boxed{10 \ 159 \ \text{m}}$$

7. ***REASONING*** This problem involves using unit conversions to determine the number of magnums in one jeroboam. The necessary relationships are

    $$1.0 \ \text{magnum} = 1.5 \ \text{liters}$$
    $$1.0 \ \text{jeroboam} = 0.792 \ \text{U. S. gallons}$$
    $$1.00 \ \text{U. S. gallon} = 3.785 \times 10^{-3} \ \text{m}^3 = 3.785 \ \text{liters}$$

    These relationships may be used to construct the appropriate conversion factors.

    SOLUTION By multiplying one jeroboam by the appropriate conversion factors we can determine the number of magnums in a jeroboam as shown below:

    $$\left(1.0 \ \text{jeroboam}\right)\left(\frac{0.792 \ \text{gallons}}{1.0 \ \text{jeroboam}}\right)\left(\frac{3.785 \ \text{liters}}{1.0 \ \text{gallon}}\right)\left(\frac{1.0 \ \text{magnum}}{1.5 \ \text{liters}}\right) = \boxed{2.0 \ \text{magnums}}$$

11. ***REASONING*** The dimension of the spring constant k can be determined by first solving the equation $T = 2\pi\sqrt{m/k}$ for k in terms of the time T and the mass m. Then, the dimensions of T and m can be substituted into this expression to yield the dimension of k.

    SOLUTION Algebraically solving the expression above for k gives $k = 4\pi^2 m/T^2$. The term $4\pi^2$ is a numerical factor that does not have a dimension, so it can be ignored in this analysis. Since the dimension for mass is [M] and that for time is [T], the dimension of k is

$$\text{Dimension of } k = \left|\frac{[M]}{[T]^2}\right|$$

13. **REASONING** The shortest distance between the two towns is along the line that joins them. This distance, h, is the hypotenuse of a right triangle whose other sides are $h_o = 35.0$ km and $h_a = 72.0$ km, as shown in the figure below.

SOLUTION The angle θ is given by $\tan\theta = h_o / h_a$ so that

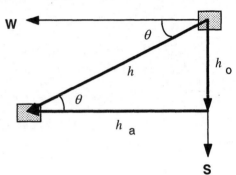

$$\theta = \tan^{-1}\left(\frac{35.0 \text{ km}}{72.0 \text{ km}}\right) = \boxed{25.9° \text{ S of W}}$$

We can then use the Pythagorean theorem to find h.

$$h = \sqrt{h_o^2 + h_a^2} = \sqrt{(35.0 \text{ km})^2 + (72.0 \text{ km})^2} = \boxed{80.1 \text{ km}}$$

21. **REASONING** The drawing at the right shows the location of each deer A, B, and C. From the problem statement it follows that

$$b = 62 \text{ m}$$

$$c = 95 \text{ m}$$

$$\gamma = 180° - 51° - 77° = 52°$$

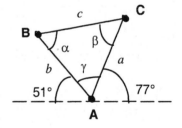

Applying the law of cosines (given in Appendix E) to the geometry in the figure, we have

$$a^2 - 2ab \cos\gamma + (b^2 - c^2) = 0$$

which is an expression that is quadratic in a. It can be simplified to $Aa^2 + Ba + C = 0$, with

$$A = 1$$

$$B = -2b \cos\gamma = -2(62 \text{ m}) \cos 52° = -76 \text{ m}$$

$$C = (b^2 - c^2) = (62 \text{ m})^2 - (95 \text{ m})^2 = -5181 \text{ m}^2$$

This quadratic equation can be solved for the desired quantity a.

SOLUTION Suppressing units, we obtain from the quadratic formula

$$a = \frac{-(-76) \pm \sqrt{(-76)^2 - 4(1)(-5181)}}{2(1)} = 1.2 \times 10^2 \text{ m} \quad \text{and} \quad -43 \text{ m}$$

Discarding the negative root, which has no physical significance, we conclude that the distance between deer A and C is $\boxed{1.2 \times 10^2 \text{ m}}$.

---

23. *REASONING AND SOLUTION* A single rope must supply the resultant of the two forces. Since the forces are perpendicular, the magnitude of the resultant can be found from the Pythagorean theorem.

a. Applying the Pythagorean theorem,

$$F = \sqrt{(475 \text{ N})^2 + (315 \text{ N})^2} = \boxed{5.70 \times 10^2 \text{ N}}$$

b. The angle θ that the resultant makes with the westward direction is

$$\theta = \tan^{-1}\left(\frac{315 \text{ N}}{475 \text{ N}}\right) = 33.6°$$

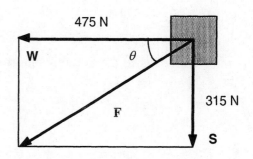

Thus, the rope must make an angle of $\boxed{33.6° \text{ south of west}}$.

---

25. *REASONING*

a. Since the two force vectors **A** and **B** have directions due west and due north, they are perpendicular. Therefore, the resultant vector **F** = **A** + **B** has a magnitude given by the Pythagorean theorem: $F^2 = A^2 + B^2$. Knowing the magnitudes of **A** and **B**, we can calculate the magnitude of **F**. The direction of the resultant can be obtained using trigonometry.

b. For the vector **F′** = **A** − **B** we note that the subtraction can be regarded as an addition in the following sense: **F′** = **A** + (−**B**). The vector −**B** points due south, opposite the vector **B**, so the two vectors are once again perpendicular and the magnitude of **F′** again is given by the Pythagorean theorem. The direction again can be obtained using trigonometry.

SOLUTION   a. The drawing shows the two vectors and the resultant vector. According to the Pythagorean theorem, we have

$$F^2 = A^2 + B^2$$

$$F = \sqrt{A^2 + B^2}$$

$$F = \sqrt{(445 \text{ N})^2 + (325 \text{ N})^2}$$

$$= \boxed{551 \text{ N}}$$

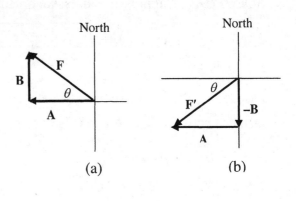

(a)   (b)

Using trigonometry, we can see that the direction of the resultant is

$$\tan\theta = \frac{B}{A} \quad \text{or} \quad \theta = \tan^{-1}\left(\frac{325 \text{ N}}{445 \text{ N}}\right) = \boxed{36.1° \text{ north of west}}$$

b. Referring to the drawing and following the same procedure as in part a, we find

$$F'^2 = A^2 + (-B)^2 \quad \text{or} \quad F' = \sqrt{A^2 + (-B)^2} = \sqrt{(445 \text{ N})^2 + (-325 \text{ N})^2} = \boxed{551 \text{ N}}$$

$$\tan\theta = \frac{B}{A} \quad \text{or} \quad \theta = \tan^{-1}\left(\frac{325 \text{ N}}{445 \text{ N}}\right) = \boxed{36.1° \text{ south of west}}$$

---

31. *REASONING AND SOLUTION*   The single force needed to produce the same effect is equal to the resultant of the forces provided by the two ropes. The figure below shows the force vectors drawn to scale and arranged tail to head. The magnitude and direction of the resultant can be found by direct measurement using the scale factor shown in the figure.

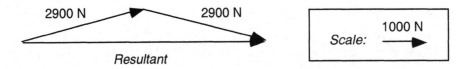

a. From the figure, the magnitude of the resultant is $\boxed{5600 \text{ N}}$.

b. The single rope should be directed $\boxed{\text{along the dashed line}}$ in the text drawing.

---

35. **REASONING** The ostrich's velocity vector **v** and the desired components are shown in the figure at the right. The components of the velocity in the directions due west and due north are v_W and v_N, respectively. The sine and cosine functions can be used to find the components.

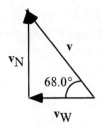

SOLUTION
a. According to the definition of the sine function, we have for the vectors in the figure

$$\sin \theta = \frac{v_N}{v} \quad \text{or} \quad v_N = v \sin \theta = (17.0 \text{ m/s}) \sin 68° = \boxed{15.8 \text{ m/s}}$$

b. Similarly,

$$\cos \theta = \frac{v_W}{v} \quad \text{or} \quad v_W = v \cos \theta = (17.0 \text{ m/s}) \cos 68.0° = \boxed{6.37 \text{ m/s}}$$

---

39. **REASONING**   The x and y components of **r** are mutually perpendicular; therefore, the magnitude of **r** can be found using the Pythagorean theorem. The direction of **r** can be found using the definition of the tangent function.

SOLUTION   According to the Pythagorean theorem, we have

$$r = \sqrt{x^2 + y^2} = \sqrt{(-125 \text{ m})^2 + (-184 \text{ m})^2} = \boxed{222 \text{ m}}$$

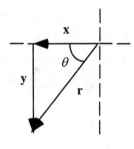

The angle θ is
$$\theta = \tan^{-1}\left(\frac{184 \text{ m}}{125 \text{ m}}\right) = \boxed{55.8°}$$

---

43. **REASONING AND SOLUTION**   We take due north to be the direction of the $+y$ axis. Vectors **A** and **B** are the components of the resultant, **C**. The angle that **C** makes with the x axis is then $\theta = \tan^{-1}(B/A)$. The symbol u denotes the units of the vectors.

a. Solving for B gives

$$B = A \tan \theta = (6.00 \text{ u}) \tan 60.0° = \boxed{10.4 \text{ u}}$$

b. The magnitude of **C** is

$$C = \sqrt{A^2 + B^2} = \sqrt{(6.00 \text{ u})^2 + (10.4 \text{ u})^2} = \boxed{12.0 \text{ u}}$$

45. ***REASONING*** The individual displacements of the golf ball, **A**, **B**, and **C** are shown in the figure. Their resultant, **R,** is the displacement that would have been needed to "hole the ball" on the very first putt. We will use the component method to find **R**.

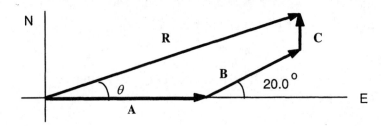

SOLUTION The components of each displacement vector are given in the table below.

Vector	x Components	y Components
A	(5.0 m) cos 0° = 5.0 m	(5.0 m) sin 0° = 0
B	(2.1 m) cos 20.0° = 2.0 m	(2.1 m) sin 20.0° = 0.72 m
C	(0.50 m) cos 90.0° = 0	(0.50 m) sin 90.0° = 0.50 m
R = A + B + C	7.0 m	1.22 m

The resultant vector **R** has magnitude

$$R = \sqrt{(7.0 \text{ m})^2 + (1.22 \text{ m})^2} = \boxed{7.1 \text{ m}}$$

and the angle θ is

$$\theta = \tan^{-1}\left(\frac{1.22 \text{ m}}{7.0 \text{ m}}\right) = 9.9°$$

Thus, the required direction is $\boxed{9.9° \text{ north of east}}$.

---

53. ***REASONING*** Since the finish line is coincident with the starting line, the net displacement of the sailboat is zero. Hence the sum of the components of the displacement vectors of the individual legs must be zero. In the drawing in the text, the directions to the right and upward are taken as positive.

SOLUTION In the horizontal direction $R_h = A_h + B_h + C_h + D_h = 0$

$$R_h = (3.20 \text{ km}) \cos 40.0° - (5.10 \text{ km}) \cos 35.0° - (4.80 \text{ km}) \cos 23.0° + D \cos \theta = 0$$

$$D \cos \theta = 6.14 \text{ km} \qquad\qquad (1)$$

In the vertical direction $R_V = A_V + B_V + C_V + D_V = 0$.

$R_V = (3.20 \text{ km}) \sin 40.0° + (5.10 \text{ km}) \sin 35.0° - (4.80 \text{ km}) \sin 23.0° - D \sin \theta = 0$.

$$D \sin \theta = 3.11 \text{ km} \qquad\qquad (2)$$

Dividing (2) by (1) gives

$$\tan \theta = (3.11 \text{ km})/(6.14 \text{ km}) \quad \text{or} \quad \theta = \boxed{26.9°}$$

Solving (1) gives

$$D = (6.14 \text{ km})/\cos 26.9° = \boxed{6.88 \text{ km}}$$

---

59. **REASONING AND SOLUTION** In order to determine which vector has the largest x and y components, we calculate the magnitude of the x and y components explicitly and compare them. In the calculations, the symbol u denotes the units of the vectors.

$A_x = (100.0 \text{ u}) \cos 90.0° = 0.00 \text{ u}$ $\qquad$ $A_y = (100.0 \text{ u}) \sin 90.0° = 1.00 \times 10^2 \text{ u}$.

$B_x = (200.0 \text{ u}) \cos 60.0° = 1.00 \times 10^2 \text{ u}$ $\qquad$ $B_y = (200.0 \text{ u}) \sin 60.0° = 173 \text{ u}$

$C_x = (150.0 \text{ u}) \cos 0.00° = 150.0 \text{ u}$ $\qquad$ $C_y = (150.0 \text{ u}) \sin 0.00° = 0.00 \text{ u}$

a. $\boxed{\text{C has the largest } x \text{ component.}}$

b. $\boxed{\text{B has the largest } y \text{ component.}}$

---

63. **REASONING** The performer walks out on the wire a distance d, and the vertical distance to the net is h. Since these two distances are perpendicular, the magnitude of the displacement is given by the Pythagorean theorem as $s = \sqrt{d^2 + h^2}$. Values for s and h are given, so we can solve this expression for the distance d. The angle that the performer's displacement makes below the horizontal can be found using trigonometry.

SOLUTION
a. Using the Pythagorean theorem, we find that

$$s = \sqrt{d^2 + h^2} \quad \text{or} \quad d = \sqrt{s^2 - h^2} = \sqrt{(26.7 \text{ ft})^2 - (25.0 \text{ ft})^2} = \boxed{9.4 \text{ ft}}$$

b. The angle θ that the performer's displacement makes below the horizontal is given by

$$\tan \theta = \frac{h}{d} \quad \text{or} \quad \theta = \tan^{-1}\left(\frac{h}{d}\right) = \tan^{-1}\left(\frac{25.0 \text{ ft}}{9.4 \text{ ft}}\right) = \boxed{69°}$$

65. ***REASONING*** The force **F** and its two components form a right triangle. The hypotenuse is 82.3 newtons, and the side parallel to the +x axis is F_x = 74.6 newtons. Therefore, we can use the trigonometric cosine and sine functions to determine the angle of **F** relative to the +x axis and the component F_y of **F** along the +y axis.

SOLUTION

a. The direction of **F** relative to the +x axis is specified by the angle θ as

$$\theta = \cos^{-1}\left(\frac{74.6 \text{ newtons}}{82.3 \text{ newtons}}\right) = \boxed{25.0°} \tag{1.5}$$

b. The component of **F** along the +y axis is

$$F_y = F \sin 25.0° = (82.3 \text{ newtons})\sin 25.0° = \boxed{34.8 \text{ newtons}} \tag{1.4}$$

CHAPTER 2 | KINEMATICS IN ONE DIMENSION

1. **REASONING AND SOLUTION**
   a. The total distance traveled is found by adding the distances traveled during each segment of the trip.

$$6.9 \text{ km} + 1.8 \text{ km} + 3.7 \text{ km} = \boxed{12.4 \text{ km}}$$

   b. All three segments of the trip lie along the east-west line. Taking east as the positive direction, the individual displacements can then be added to yield the resultant displacement.

$$6.9 \text{ km} + (-1.8 \text{ km}) + 3.7 \text{ km} = +8.8 \text{ km}$$

The displacement is positive, indicating that it points due east. Therefore,

$$\text{Displacement of the whale} = \boxed{8.8 \text{ km, due east}}$$

3. **REASONING** The average speed is the distance traveled divided by the elapsed time (Equation 2.1). Since the average speed and distance are known, we can use this relation to find the time.

SOLUTION The time it takes for the continents to drift apart by 1500 m is

$$\text{Elapsed time} = \frac{\text{Distance}}{\text{Average speed}} = \frac{1500 \text{ m}}{\left(3 \dfrac{\text{cm}}{\text{yr}}\right)\left(\dfrac{1 \text{ m}}{100 \text{ cm}}\right)} = \boxed{5 \times 10^4 \text{ yr}}$$

11. **REASONING** Since the woman runs for a known distance at a known constant speed, we can find the time it takes for her to reach the water from Equation 2.1. We can then use Equation 2.1 to determine the total distance traveled by the dog in this time.

SOLUTION The time required for the woman to reach the water is

$$\text{Elapsed time} = \frac{d_{\text{woman}}}{v_{\text{woman}}} = \left(\frac{4.0 \text{ km}}{2.5 \text{ m/s}}\right)\left(\frac{1000 \text{ m}}{1.0 \text{ km}}\right) = 1600 \text{ s}$$

In 1600 s, the dog travels a total distance of

$$d_{dog} = v_{dog}t = (4.5 \text{ m/s})(1600 \text{ s}) = \boxed{7.2 \times 10^3 \text{ m}}$$

---

17. **REASONING**   The average acceleration is defined by Equation 2.4 as the change in velocity divided by the elapsed time. We can find the elapsed time from this relation because the acceleration and the change in velocity are given. Since the acceleration of the spacecraft is constant, it is equal to the average acceleration.

SOLUTION
a.   The time Δt that it takes for the spacecraft to change its velocity by an amount $\Delta v = +2700$ m/s is

$$\Delta t = \frac{\Delta v}{a} = \frac{+2700 \text{ m/s}}{+9.0 \dfrac{\text{m/s}}{\text{day}}} = \boxed{3.0 \times 10^2 \text{ days}}$$

b. Since 24 hr = 1 day and 3600 s = 1 hr, the acceleration of the spacecraft (in m/s^2) is

$$a = \frac{\Delta v}{t} = \frac{+9.0 \text{ m/s}}{(1 \text{ day})\left(\dfrac{24 \text{ hr}}{1 \text{ day}}\right)\left(\dfrac{3600 \text{ s}}{1 \text{ hr}}\right)} = \boxed{+1.04 \times 10^{-4} \text{ m/s}^2}$$

---

21. **REASONING AND SOLUTION**   The velocity of the automobile for each stage is given by Equation 2.4: $v = v_0 + at$. Therefore,

$$v_1 = v_0 + a_1 t = 0 \text{ m/s} + a_1 t \quad \text{and} \quad v_2 = v_1 + a_2 t$$

Since the magnitude of the car's velocity at the end of stage 2 is 2.5 times greater than it is at the end of stage 1, $v_2 = 2.5v_1$. Thus, rearranging the result for v_2, we find

$$a_2 = \frac{v_2 - v_1}{t} = \frac{2.5v_1 - v_1}{t} = \frac{1.5v_1}{t} = \frac{1.5(a_1 t)}{t} = 1.5a_1 = 1.5(3.0 \text{ m/s}^2) = \boxed{4.5 \text{ m/s}^2}$$

---

25. **REASONING AND SOLUTION**   The average acceleration of the plane can be found by solving Equation 2.9 $\left(v^2 = v_0^2 + 2ax\right)$ for a. Taking the direction of motion as positive, we have

$$a = \frac{v^2 - v_0^2}{2x} = \frac{(+6.1 \text{ m/s})^2 - (+69 \text{ m/s})^2}{2(+750 \text{ m})} = \boxed{-3.1 \text{ m/s}^2}$$

The minus sign indicates that the direction of the acceleration is opposite to the direction of motion, and the plane is slowing down.

---

29. **REASONING AND SOLUTION**

a. The magnitude of the acceleration can be found from Equation 2.4 ($v = v_0 + at$) as

$$a = \frac{v - v_0}{t} = \frac{3.0 \text{ m/s} - 0 \text{ m/s}}{2.0 \text{ s}} = \boxed{1.5 \text{ m/s}^2}$$

b. Similarly the magnitude of the acceleration of the car is

$$a = \frac{v - v_0}{t} = \frac{41.0 \text{ m/s} - 38.0 \text{ m/s}}{2.0 \text{ s}} = \boxed{1.5 \text{ m/s}^2}$$

c. Assuming that the acceleration is constant, the displacement covered by the car can be found from Equation 2.9 ($v^2 = v_0^2 + 2ax$):

$$x = \frac{v^2 - v_0^2}{2a} = \frac{(41.0 \text{ m/s})^2 - (38.0 \text{ m/s})^2}{2(1.5 \text{ m/s}^2)} = 79 \text{ m}$$

Similarly, the displacement traveled by the jogger is

$$x = \frac{v^2 - v_0^2}{2a} = \frac{(3.0 \text{ m/s})^2 - (0 \text{ m/s})^2}{2(1.5 \text{ m/s}^2)} = 3.0 \text{ m}$$

Therefore, the car travels 79 m – 3.0 m = $\boxed{76 \text{ m}}$ further than the jogger.

---

31. **REASONING** Since the belt is moving with constant velocity, the displacement ($x_0 = 0$ m) covered by the belt in a time t_{belt} is giving by Equation 2.2 (with x_0 assumed to be zero) as

$$x = v_{belt} t_{belt} \tag{1}$$

Since Clifford moves with constant acceleration, the displacement covered by Clifford in a time t_{Cliff} is, from Equation 2.8,

$$x = v_0 t_{Cliff} + \tfrac{1}{2} a t_{Cliff}^2 = \tfrac{1}{2} a t_{Cliff}^2 \tag{2}$$

The speed v_{belt} with which the belt of the ramp is moving can be found by eliminating x between Equations (1) and (2).

SOLUTION   Equating the right hand sides of Equations (1) and (2), and noting that $t_{Cliff} = \frac{1}{4} t_{belt}$, we have

$$v_{belt} t_{belt} = \frac{1}{2} a \left(\frac{1}{4} t_{belt} \right)^2$$

$$v_{belt} = \frac{1}{32} a t_{belt} = \frac{1}{32} (0.37 \text{ m/s}^2)(64 \text{ s}) = \boxed{0.74 \text{ m/s}}$$

---

41. **REASONING**   As the train passes through the crossing, its motion is described by Equations 2.4 ($v = v_0 + at$) and 2.7 $\left[x = \frac{1}{2}(v+v_0)t \right]$, which can be rearranged to give

$$v - v_0 = at \quad \text{and} \quad v + v_0 = \frac{2x}{t}$$

These can be solved simultaneously to obtain the speed v when the train reaches the end of the crossing. Once v is known, Equation 2.4 can be used to find the time required for the train to reach a speed of 32 m/s.

SOLUTION   Adding the above equations and solving for v, we obtain

$$v = \frac{1}{2} \left(at + \frac{2x}{t} \right) = \frac{1}{2} \left[(1.6 \text{ m/s}^2)(2.4 \text{ s}) + \frac{2(20.0 \text{ m})}{2.4 \text{ s}} \right] = 1.0 \times 10^1 \text{ m/s}$$

The motion from the end of the crossing until the locomotive reaches a speed of 32 m/s requires a time

$$t = \frac{v - v_0}{a} = \frac{32 \text{ m/s} - 1.0 \times 10^1 \text{ m/s}}{1.6 \text{ m/s}^2} = \boxed{14 \text{ s}}$$

---

43. **REASONING AND SOLUTION**
a. Once the pebble has left the slingshot, it is subject only to the acceleration due to gravity. Since the downward direction is negative, the acceleration of the pebble is $\boxed{-9.80 \text{ m/s}^2}$.
The pebble is not decelerating. Since its velocity and acceleration both point downward, the magnitude of the pebble's velocity is increasing, not decreasing.

b. The displacement y traveled by the pebble as a function of the time t can be found from Equation 2.8. Using Equation 2.8, we have

$$y = v_0 t + \tfrac{1}{2} a_y t^2 = (-9.0 \text{ m/s})(0.50 \text{ s}) + \tfrac{1}{2}\left[(-9.80 \text{ m/s}^2)(0.50 \text{ s})^2\right] = -5.7 \text{ m}$$

Thus, after 0.50 s, the pebble is $\boxed{5.7 \text{ m}}$ beneath the cliff-top.

---

49. ***REASONING AND SOLUTION*** Equation 2.8 can be used to determine the displacement that the ball covers as it falls halfway to the ground. Since the ball falls from rest, its initial velocity is zero. Taking down to be the negative direction, we have

$$y = v_0 t + \tfrac{1}{2} a t^2 = \tfrac{1}{2} a t^2 = \tfrac{1}{2}(-9.80 \text{ m/s}^2)(1.2 \text{ s})^2 = -7.1 \text{ m}$$

In falling all the way to the ground, the ball has a displacement of $y = -14.2$ m. Solving Equation 2.8 with this displacement then yields the time

$$t = \sqrt{\frac{2y}{a}} = \sqrt{\frac{2(-14.2 \text{ m})}{-9.80 \text{ m/s}^2}} = \boxed{1.7 \text{ s}}$$

---

53. ***REASONING AND SOLUTION*** Since the balloon is released from rest, its initial velocity is zero. The time required to fall through a vertical displacement y can be found from Equation 2.8 $\left(y = v_0 t + \tfrac{1}{2} a t^2\right)$ with $v_0 = 0$ m/s. Assuming upward to be the positive direction, we find

$$t = \sqrt{\frac{2y}{a}} = \sqrt{\frac{2(-6.0 \text{ m})}{-9.80 \text{ m/s}^2}} = \boxed{1.1 \text{ s}}$$

---

59. ***REASONING AND SOLUTION*** The stone requires a time, t_1, to reach the bottom of the hole, a distance y below the ground. Assuming downward to be the positive direction, the variables are related by Equation 2.8 with $v_0 = 0$ m/s:

$$y = \tfrac{1}{2} a t_1^2 \qquad (1)$$

The sound travels the distance y from the bottom to the top of the hole in a time t_2. Since the sound does not experience any acceleration, the variables y and t_2 are related by Equation 2.8 with $a = 0$ m/s^2 and v_{sound} denoting the speed of sound:

$$y = v_{\text{sound}} t_2 \qquad (2)$$

Equating the right hand sides of Equations (1) and (2) and using the fact that the total elapsed time is $t = t_1 + t_2$, we have

$$\tfrac{1}{2}at_1^2 = v_{\text{sound}}t_2 \qquad \text{or} \qquad \tfrac{1}{2}at_1^2 = v_{\text{sound}}(t - t_1)$$

Rearranging gives

$$\tfrac{1}{2}at_1^2 + v_{\text{sound}}t_1 - v_{\text{sound}}t = 0$$

Substituting values and suppressing units for brevity, we obtain the following quadratic equation for t_1:

$$4.90t_1^2 + 343t_1 - 514 = 0$$

From the quadratic formula, we obtain

$$t_1 = \frac{-343 \pm \sqrt{(343)^2 - 4(4.90)(-514)}}{2(4.90)} = 1.47\text{ s} \quad \text{or} \quad -71.5\text{ s}$$

The negative time corresponds to a nonphysical result and is rejected. The depth of the hole is then found using Equation 2.8 with the value of t_1 obtained above:

$$y = v_0 t_1 + \tfrac{1}{2}at_1^2 = (0\text{ m/s})(1.47\text{ s}) + \tfrac{1}{2}(9.80\text{ m/s}^2)(1.47\text{ s})^2 = \boxed{10.6\text{ m}}$$

---

61. ***REASONING***   Once the man sees the block, the man must get out of the way in the time it takes for the block to fall through an additional 12.0 m. The velocity of the block at the instant that the man looks up can be determined from Equation 2.9. Once the velocity is known at that instant, Equation 2.8 can be used to find the time required for the block to fall through the additional distance.

SOLUTION   When the man first notices the block, it is 14.0 m above the ground and its displacement from the starting point is $y = 14.0\text{ m} - 53.0\text{ m}$. Its velocity is given by Equation 2.9 $\left(v^2 = v_0^2 + 2ay\right)$. Since the block is moving down, its velocity has a negative value,

$$v = -\sqrt{v_0 + 2ay} = -\sqrt{(0\text{ m/s})^2 + 2(-9.80\text{ m/s}^2)(14.0\text{ m} - 53.0\text{ m})} = -27.7\text{ m/s}$$

The block then falls the additional 12.0 m to the level of the man's head in a time t which satisfies Equation 2.8:

$$y = v_0 t + \tfrac{1}{2}at^2$$

where $y = -12.0$ m and $v_0 = -27.7$ m/s. Thus, t is the solution to the quadratic equation

$$4.90t^2 + 27.7t - 12.0 = 0$$

where the units have been suppressed for brevity. From the quadratic formula, we obtain

$$t = \frac{-27.7 \pm \sqrt{(27.7)^2 - 4(4.90)(-12.0)}}{2(4.90)} = 0.40 \text{ s} \quad \text{or} \quad -6.1 \text{ s}$$

The negative solution can be rejected as nonphysical, and the time it takes for the block to reach the level of the man is $\boxed{0.40 \text{ s}}$.

---

67. **REASONING** The slope of a straight-line segment in a position-versus-time graph is the average velocity. The algebraic sign of the average velocity, therefore, corresponds to the sign of the slope.

SOLUTION
a. The slope, and hence the average velocity, is *positive* for segments A and C, *negative* for segment B, and *zero* for segment D.

b. In the given position-versus-time graph, we find the slopes of the four straight-line segments to be

$$v_A = \frac{1.25 \text{ km} - 0 \text{ km}}{0.20 \text{ h} - 0 \text{ h}} = \boxed{+6.3 \text{ km/h}}$$

$$v_B = \frac{0.50 \text{ km} - 1.25 \text{ km}}{0.40 \text{ h} - 0.20 \text{ h}} = \boxed{-3.8 \text{ km/h}}$$

$$v_C = \frac{0.75 \text{ km} - 0.50 \text{ km}}{0.80 \text{ h} - 0.40 \text{ h}} = \boxed{+0.63 \text{ km/h}}$$

$$v_D = \frac{0.75 \text{ km} - 0.75 \text{ km}}{1.00 \text{ h} - 0.80 \text{ h}} = \boxed{0 \text{ km/h}}$$

---

71. **REASONING** The two runners start one hundred meters apart and run toward each other. Each runs ten meters during the first second and, during each second thereafter, each runner runs ninety percent of the distance he ran in the previous second. While the velocity of each runner changes from second to second, it remains constant during any one second.

SOLUTION The following table shows the distance covered during each second for one of the runners, and the position at the end of each second (assuming that he begins at the origin) for the first eight seconds.

Time t (s)	Distance covered (m)	Position x (m)
0.00		0.00
1.00	10.00	10.00
2.00	9.00	19.00
3.00	8.10	27.10
4.00	7.29	34.39
5.00	6.56	40.95
6.00	5.90	46.85
7.00	5.31	52.16
8.00	4.78	56.94

The following graph is the position-time graph constructed from the data in the table above.

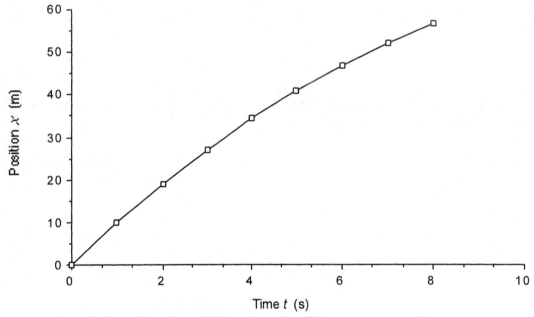

a. Since the two runners are running toward each other in exactly the same way, they will meet halfway between their respective starting points. That is, they will meet at $x = 50.0$ m. According to the graph, therefore, this position corresponds to a time of $\boxed{6.6 \text{ s}}$.

b. Since the runners collide during the seventh second, the speed at the instant of collision can be found by taking the slope of the position-time graph for the seventh second. The speed of either runner in the interval from $t = 6.00$ s to $t = 7.00$ s is

$$v = \frac{\Delta x}{\Delta t} = \frac{52.16 \text{ m} - 46.85 \text{ m}}{7.00 \text{ s} - 6.00 \text{ s}} = 5.3 \text{ m/s}$$

Therefore, at the moment of collision, the speed of either runner is $\boxed{5.3 \text{ m/s}}$.

73. **REASONING AND SOLUTION** When air resistance is neglected, free fall conditions are applicable. The final speed can be found from Equation 2.9;

$$v^2 = v_0^2 + 2ay$$

where v_0 is zero since the stunt man falls from rest. If the origin is chosen at the top of the hotel and the upward direction is positive, then the displacement is $y = -99.4$ m. Solving for v, we have

$$v = -\sqrt{2ay} = -\sqrt{2(-9.80 \text{ m/s}^2)(-99.4 \text{ m})} = -44.1 \text{ m/s}$$

The speed at impact is the magnitude of this result or $\boxed{44.1 \text{ m/s}}$.

75. **REASONING** The cart has an initial velocity of $v_0 = +5.0$ m/s, so initially it is moving to the right, which is the positive direction. It eventually reaches a point where the displacement is $x = +12.5$ m, and it begins to move to the left. This must mean that the cart comes to a momentary halt at this point (final velocity is $v = 0$ m/s), before beginning to move to the left. In other words, the cart is decelerating, and its acceleration must point opposite to the velocity, or to the left. Thus, the acceleration is negative. Since the initial velocity, the final velocity, and the displacement are known, Equation 2.9 $\left(v^2 = v_0^2 + 2ax\right)$ can be used to determine the acceleration.

SOLUTION Solving Equation 2.9 for the acceleration a shows that

$$a = \frac{v^2 - v_0^2}{2x} = \frac{(0 \text{ m/s})^2 - (+5.0 \text{ m/s})^2}{2(+12.5 \text{ m})} = \boxed{-1.0 \text{ m/s}^2}$$

81. **REASONING AND SOLUTION**
a. The total displacement traveled by the bicyclist for the entire trip is equal to the sum of the displacements traveled during each part of the trip. The displacement traveled during each part of the trip is given by Equation 2.2: $\Delta x = \bar{v}\Delta t$. Therefore,

$$\Delta x_1 = (7.2 \text{ m/s})(22 \text{ min})\left(\frac{60 \text{ s}}{1 \text{ min}}\right) = 9500 \text{ m}$$

$$\Delta x_2 = (5.1 \text{ m/s})(36 \text{ min})\left(\frac{60 \text{ s}}{1 \text{ min}}\right) = 11\ 000 \text{ m}$$

$$\Delta x_3 = (13 \text{ m/s})(8.0 \text{ min})\left(\frac{60 \text{ s}}{1 \text{ min}}\right) = 6200 \text{ m}$$

The total displacement traveled by the bicyclist during the entire trip is then

$$\Delta x = 9500 \text{ m} + 11\ 000 \text{ m} + 6200 \text{ m} = \boxed{2.67 \times 10^4 \text{ m}}$$

b. The average velocity can be found from Equation 2.2.

$$\overline{v} = \frac{\Delta x}{\Delta t} = \frac{2.67 \times 10^4 \text{ m}}{(22 \text{ min} + 36 \text{ min} + 8.0 \text{ min})}\left(\frac{1 \text{ min}}{60 \text{ s}}\right) = \boxed{6.74 \text{ m/s, due north}}$$

---

83. **REASONING AND SOLUTION** The stone will reach the water (and hence the log) after falling for a time t, where t can be determined from Equation 2.8: $y = v_0 t + \frac{1}{2} a t^2$. Since the stone is dropped from rest, $v_0 = 0$ m/s. Assuming that downward is positive and solving for t, we have

$$t = \sqrt{\frac{2y}{a}} = \sqrt{\frac{2(75 \text{ m})}{9.80 \text{ m/s}^2}} = 3.9 \text{ s}$$

During that time, the displacement of the log can be found from Equation 2.8. Since the log moves with constant velocity, $a = 0$ m/s^2, and v_0 is equal to the velocity of the log.

$$x = v_0 t = (5.0 \text{ m/s})(3.9 \text{ s}) = 2.0 \times 10^1 \text{ m}$$

Therefore, the horizontal distance between the log and the bridge when the stone is released is $\boxed{2.0 \times 10^1 \text{ m}}$.

---

87. **REASONING** Since the car is moving with a constant velocity, the displacement of the car in a time t can be found from Equation 2.8 with $a = 0$ m/s^2 and v_0 equal to the velocity of the car: $x_{car} = v_{car} t$. Since the train starts from rest with a constant acceleration, the displacement of the train in a time t is given by Equation 2.8 with $v_0 = 0$ m/s:

$$x_{train} = \tfrac{1}{2} a_{train} t^2$$

At a time t_1, when the car just reaches the front of the train, $x_{car} = L_{train} + x_{train}$, where L_{train} is the length of the train. Thus, at time t_1,

$$v_{car} t_1 = L_{train} + \tfrac{1}{2} a_{train} t_1^2 \tag{1}$$

At a time t_2, when the car is again at the rear of the train, $x_{car} = x_{train}$. Thus, at time t_2

$$v_{car} t_2 = \tfrac{1}{2} a_{train} t_2^2 \tag{2}$$

Equations (1) and (2) can be solved simultaneously for the speed of the car v_{car} and the acceleration of the train a_{train}.

SOLUTION
a. Solving Equation (2) for a_{train} we have

$$a_{train} = \frac{2 v_{car}}{t_2} \tag{3}$$

Substituting this expression for a_{train} into Equation (1) and solving for v_{car}, we have

$$v_{car} = \frac{L_{train}}{t_1 \left(1 - \dfrac{t_1}{t_2}\right)} = \frac{92 \text{ m}}{(14 \text{ s})\left(1 - \dfrac{14 \text{ s}}{28 \text{ s}}\right)} = \boxed{13 \text{ m/s}}$$

b. Direct substitution into Equation (3) gives the acceleration of the train:

$$a_{train} = \frac{2 v_{car}}{t_2} = \frac{2 (13 \text{ m/s})}{28 \text{ s}} = \boxed{0.93 \text{ m/s}^2}$$

CHAPTER 3 | *KINEMATICS IN TWO DIMENSIONS*

1. ***REASONING*** The displacement is a vector drawn from the initial position to the final position. The magnitude of the displacement is the shortest distance between the positions. Note that it is only the initial and final positions that determine the displacement. The fact that the squirrel jumps to an intermediate position before reaching his final position is not important. The trees are perfectly straight and both growing perpendicular to the flat horizontal ground beneath them. Thus, the distance between the trees and the length of the trunk of the second tree below the squirrel's final landing spot form the two perpendicular sides of a right triangle, as the drawing shows. To this triangle, we can apply the Pythagorean theorem and determine the magnitude A of the displacement vector **A**.

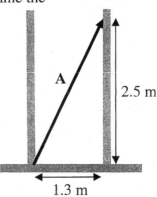

SOLUTION According to the Pythagorean theorem, we have

$$A = \sqrt{(1.3 \text{ m})^2 + (2.5 \text{ m})^2} = \boxed{2.8 \text{ m}}$$

5. ***REASONING AND SOLUTION***

$$x = r \cos \theta = (162 \text{ km}) \cos 62.3° = \boxed{75.3 \text{ km}}$$

$$y = r \sin \theta = (162 \text{ km}) \sin 62.3° = \boxed{143 \text{ km}}$$

7. ***REASONING*** The displacement of the elephant seal has two components; 460 m due east and 750 m downward. These components are mutually perpendicular; hence, the Pythagorean theorem can be used to determine their resultant.

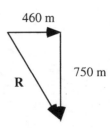

SOLUTION From the Pythagorean theorem,

$$R^2 = (460 \text{ m})^2 + (750 \text{ m})^2$$

Therefore,

$$R = \sqrt{(460 \text{ m})^2 + (750 \text{ m})^2} = \boxed{8.8 \times 10^2 \text{ m}}$$

9. REASONING

a. We designate the direction down and parallel to the ramp as the $+x$ direction, and the table shows the variables that are known. Since three of the five kinematic variables have values, one of the equations of kinematics can be employed to find the acceleration a_x.

x-Direction Data

x	a_x	v_x	v_{0x}	t
+12.0 m	?	+7.70 m/s	0 m/s	

b. The acceleration vector points down and parallel to the ramp, and the angle of the ramp is 25.0° relative to the ground (see the drawing). Therefore, trigonometry can be used to determine the component $a_{parallel}$ of the acceleration that is parallel to the ground.

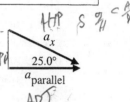

SOLUTION

a. Equation 3.6a $\left(v_x^2 = v_{0x}^2 + 2a_x x\right)$ can be used to find the acceleration in terms of the three known variables. Solving this equation for a_x gives

$$a_x = \frac{v_x^2 - v_{0x}^2}{2x} = \frac{(+7.70 \text{ m/s})^2 - (0 \text{ m/s})^2}{2(+12.0 \text{ m})} = \boxed{2.47 \text{ m/s}^2}$$

b. The drawing shows that the acceleration vector is oriented 25.0° relative to the ground. The component $a_{parallel}$ of the acceleration that is parallel to the ground is

$$a_{parallel} = a_x \cos 25.0° = (2.47 \text{ m/s}^2) \cos 25.0° = \boxed{2.24 \text{ m/s}^2}$$

---

13. REASONING AND SOLUTION

As shown in Example 3, the time required for the package to hit the ground is given by $t = \sqrt{2y/a_y}$ and is independent of the plane's horizontal velocity. Thus, the time needed for the package to hit the ground is still $\boxed{14.6 \text{ s}}$.

---

17. REASONING

The upward direction is chosen as positive. Since the ballast bag is released from rest relative to the balloon, its initial velocity relative to the ground is equal to the velocity of the balloon relative to the ground, so that $v_{0y} = 3.0 \text{ m/s}$. Time required for the ballast to reach the ground can be found by solving Equation 3.5b for t.

SOLUTION Using Equation 3.5b, we have

$$\tfrac{1}{2}a_y t^2 + v_{0y}t - y = 0 \quad \text{or} \quad \tfrac{1}{2}(-9.80 \text{ m/s}^2)t^2 + (3.0 \text{ m/s})t - (-9.5 \text{ m}) = 0$$

This equation is quadratic in t, and t may be found from the quadratic formula. Using the quadratic formula, suppressing the units, and discarding the negative root, we find

$$t = \frac{-3.0 \pm \sqrt{(3.0)^2 - 4(-4.90)(9.5)}}{2(-4.90)} = \boxed{1.7 \text{ s}}$$

---

21. *REASONING* The time that the ball spends in the air is determined by its vertical motion. The time required for the ball to reach the lake can be found by solving Equation 3.5b for t. The motion of the golf ball is characterized by constant velocity in the x direction and accelerated motion (due to gravity) in the y direction. Thus, the x component of the velocity of the golf ball is constant, and the y component of the velocity at any time t can be found from Equation 3.3b. Once the x and y components of the velocity are known for a particular time t, the speed can be obtained from $v = \sqrt{v_x^2 + v_y^2}$.

SOLUTION
a. Since the ball rolls off the cliff horizontally, $v_{0y} = 0$. If the origin is chosen at top of the cliff and upward is assumed to be the positive direction, then the vertical component of the ball's displacement is $y = -15.5$ m. Thus, Equation 3.5b gives

$$t = \sqrt{\frac{2y}{a_y}} = \sqrt{\frac{2(-15.5 \text{ m})}{(-9.80 \text{ m/s}^2)}} = \boxed{1.78 \text{ s}}$$

b. Since there is no acceleration in the x direction, $v_x = v_{0x} = 11.4$ m/s . The y component of the velocity of the ball just before it strikes the water is, according to Equation 3.3b,

$$v_y = v_{0y} + a_y t = \left[0 + (-9.80 \text{ m/s}^2)(1.78 \text{ s}) \right] = -17.4 \text{ m/s}$$

The speed of the ball just before it strikes the water is, therefore,

$$v = \sqrt{v_x^2 + v_y^2} = \sqrt{(11.4 \text{ m/s})^2 + (-17.4 \text{ m/s})^2} = \boxed{20.8 \text{ m/s}}$$

---

23. *REASONING* Since the magnitude of the velocity of the fuel tank is given by $v = \sqrt{v_x^2 + v_y^2}$, it is necessary to know the velocity components v_x and v_y just before impact. At the instant of release, the empty fuel tank has the same velocity as that of the plane. Therefore, the magnitudes of the initial velocity components of the fuel tank are given

by $v_{0x} = v_0 \cos\theta$ and $v_{0y} = v_0 \sin\theta$, where v_0 is the speed of the plane at the instant of release. Since the x motion has zero acceleration, the x component of the velocity of the plane remains equal to v_{0x} for all later times while the tank is airborne. The y component of the velocity of the tank after it has undergone a vertical displacement y is given by Equation 3.6b.

SOLUTION

a. Taking up as the positive direction, the velocity components of the fuel tank just before it hits the ground are

$$v_x = v_{0x} = v\cos\theta = (135 \text{ m/s}) \cos 15° = 1.30 \times 10^2 \text{ m/s}$$

From Equation 3.6b, we have

$$v_y = -\sqrt{v_{0y}^2 + 2a_y y} = -\sqrt{(v_0 \sin\theta)^2 + 2a_y y}$$

$$= -\sqrt{\left[(135 \text{ m/s}) \sin 15.0°\right]^2 + \left[2(-9.80 \text{ m/s}^2)(-2.00 \times 10^3 \text{ m})\right]} = -201 \text{ m/s}$$

Therefore, the magnitude of the velocity of the fuel tank just before impact is

$$v = \sqrt{v_x^2 + v_y^2} = \sqrt{(1.30 \times 10^2 \text{ m/s})^2 + (201 \text{ m/s})^2} = \boxed{239 \text{ m/s}}$$

The velocity vector just before impact is inclined at an angle ϕ above the horizontal. This angle is

$$\phi = \tan^{-1}\left(\frac{201 \text{ m/s}}{1.30 \times 10^2 \text{ m/s}}\right) = \boxed{57.1°}$$

b. As shown in Conceptual Example 10, once the fuel tank in part a rises and falls to the same altitude at which it was released, its motion is identical to the fuel tank in part b. Therefore, the velocity of the fuel tank in part b just before impact is $\boxed{239 \text{ m/s at an angle of } 57.1° \text{ above the horizontal}}$.

---

27. **REASONING AND SOLUTION** The water exhibits projectile motion. The x component of the motion has zero acceleration while the y component is subject to the acceleration due to gravity. In order to reach the highest possible fire, the displacement of the hose from the building is x, where, according to Equation 3.5a (with $a_x = 0 \text{ m/s}^2$),

$$x = v_{0x}t = (v_0 \cos\theta)t$$

with t equal to the time required for the water the reach its maximum vertical displacement. The time t can be found by considering the vertical motion. From Equation 3.3b,

$$v_y = v_{0y} + a_y t$$

When the water has reached its maximum vertical displacement, $v_y = 0$ m/s. Taking up and to the right as the positive directions, we find that

$$t = \frac{-v_{0y}}{a_y} = \frac{-v_0 \sin \theta}{a_y}$$

and

$$x = (v_0 \cos \theta)\left(\frac{-v_0 \sin \theta}{a_y}\right)$$

Therefore, we have

$$x = -\frac{v_0^2 \cos \theta \, \sin \theta}{a_y} = -\frac{(25.0 \text{ m/s})^2 \cos 35.0° \sin 35.0°}{-9.80 \text{ m/s}^2} = \boxed{30.0 \text{ m}}$$

---

31. ***REASONING*** Once the diver is airborne, he moves in the x direction with constant velocity while his motion in the y direction is accelerated (at the acceleration due to gravity). Therefore, the magnitude of the x component of his velocity remains constant at 1.20 m/s for all times t. The magnitude of the y component of the diver's velocity after he has fallen through a vertical displacement y can be determined from Equation 3.6b: $v_y^2 = v_{0y}^2 + 2a_y y$. Since the diver runs off the platform horizontally, $v_{0y} = 0$ m/s. Once the x and y components of the velocity are known for a particular vertical displacement y, the speed of the diver can be obtained from $v = \sqrt{v_x^2 + v_y^2}$.

SOLUTION For convenience, we will take downward as the positive y direction. After the diver has fallen 10.0 m, the y component of his velocity is, from Equation 3.6b,

$$v_y = \sqrt{v_{0y}^2 + 2a_y y} = \sqrt{0^2 + 2(9.80 \text{ m/s}^2)(10.0 \text{ m})} = 14.0 \text{ m/s}$$

Therefore,

$$v = \sqrt{v_x^2 + v_y^2} = \sqrt{(1.20 \text{ m/s})^2 + (14.0 \text{ m/s})^2} = \boxed{14.1 \text{ m/s}}$$

---

37. **REASONING**

a. The drawing shows the initial velocity v_0 of the package when it is released. The initial speed of the package is 97.5 m/s. The component of its displacement along the ground is labeled as x. The data for the x direction are indicated in the data table below.

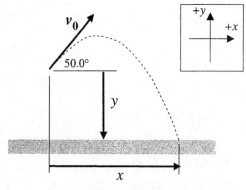

x-Direction Data

x	a_x	v_x	v_{0x}	t
?	0 m/s^2		+(97.5 m/s) cos 50.0° = +62.7 m/s	

Since only two variables are known, it is not possible to determine x from the data in this table. A value for a third variable is needed. We know that the time of flight t is the same for both the x and y motions, so let's now look at the data in the y direction.

y-Direction Data

y	a_y	v_y	v_{0y}	t
−732 m	-9.80 m/s^2		+(97.5 m/s) sin 50.0° = +74.7 m/s	?

Note that the displacement y of the package points from its initial position toward the ground, so its value is negative, i.e., $y = -732$ m. The data in this table, along with the appropriate equation of kinematics, can be used to find the time of flight t. This value for t can, in turn, be used in conjunction with the x-direction data to determine x.

b. The drawing at the right shows the velocity of the package just before impact. The angle that the velocity makes with respect to the ground can be found from the inverse tangent function as $\theta = \tan^{-1}\left(v_y / v_x\right)$. Once the time has been found in part (a), the values of v_y and v_x can be determined from the data in the tables and the appropriate equations of kinematics.

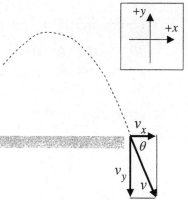

SOLUTION

a. To determine the time that the package is in the air, we will use Equation 3.5b $\left(y = v_{0y}t + \frac{1}{2}a_y t^2\right)$ and the data in the y-direction data table. Solving this quadratic equation for the time yields

$$t = \frac{-v_{0y} \pm \sqrt{v_{0y}^2 - 4\left(\frac{1}{2}a_y\right)(-y)}}{2\left(\frac{1}{2}a_y\right)}$$

$$t = \frac{-(74.7 \text{ m/s}) \pm \sqrt{(74.7 \text{ m/s})^2 - 4\left(\frac{1}{2}\right)(-9.80 \text{ m/s}^2)(732 \text{ m})}}{2\left(\frac{1}{2}\right)(-9.80 \text{ m/s}^2)} = -6.78 \text{ s} \quad \text{and} \quad 22.0 \text{ s}$$

We discard the first solution, since it is a negative value and, hence, unrealistic. The displacement x can be found using $t = 22.0$ s, the data in the x-direction data table, and Equation 3.5a:

$$x = v_{0x}t + \tfrac{1}{2}a_x t^2 = (+62.7 \text{ m/s})(22.0 \text{ s}) + \underbrace{\tfrac{1}{2}(0 \text{ m/s}^2)(22.0 \text{ s})^2}_{=0} = \boxed{+1380 \text{ m}}$$

b. The angle θ that the velocity of the package makes with respect to the ground is given by $\theta = \tan^{-1}(v_y/v_x)$. Since there is no acceleration in the x direction ($a_x = 0 \text{ m/s}^2$), v_x is the same as v_{0x}, so that $v_x = v_{0x} = +62.7$ m/s. Equation 3.3b can be employed with the y-direction data to find v_y:

$$v_y = v_{0y} + a_y t = +74.7 \text{ m/s} + (-9.80 \text{ m/s}^2)(22.0 \text{ s}) = -141 \text{ m/s}$$

Therefore,

$$\theta = \tan^{-1}\left(\frac{v_y}{v_x}\right) = \tan^{-1}\left(\frac{-141 \text{ m/s}}{+62.7 \text{ m/s}}\right) = -66.0°$$

where the minus sign indicates that the angle is $\boxed{66.0° \text{ below the horizontal}}$.

---

43. **REASONING**   The horizontal distance covered by stone **1** is equal to the distance covered by stone **2** after it passes point **P** in the following diagram. Thus, the distance Δx between the points where the stones strike the ground is equal to x_2, the horizontal distance covered by stone **2** when it reaches **P**. In the diagram, we assume up and to the right are positive.

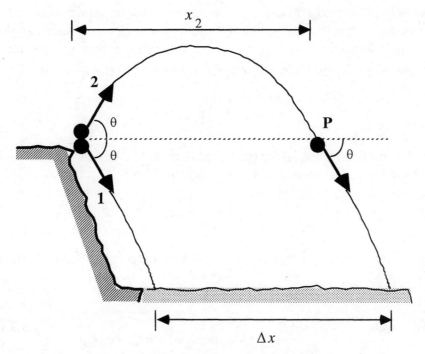

SOLUTION   If t_P is the time required for stone **2** to reach **P**, then

$$x_2 = v_{0x}t_P = (v_0 \cos\theta)t_P$$

For the vertical motion of stone **2**, $v_y = v_0 \sin\theta + a_y t$.   Solving for t gives

$$t = \frac{v_y - v_0 \sin\theta}{a_y}$$

When stone **2** reaches **P**, $v_y = -v_0 \sin\theta$, so the time required to reach **P** is

$$t_P = \frac{-2v_0 \sin\theta}{a_y}$$

Then,

$$x_2 = v_{0x}t_P = (v_0 \cos\theta)\left(\frac{-2v_0 \sin\theta}{a_y}\right)$$

$$x_2 = \frac{-2v_0^2 \sin\theta \ \cos\theta}{a_y} = \frac{-2(13.0 \text{ m/s})^2 \sin 30.0° \cos 30.0°}{-9.80 \text{ m/s}^2} = \boxed{14.9 \text{ m}}$$

47. ***REASONING AND SOLUTION*** In the absence of air resistance, the bullet exhibits projectile motion. The x component of the motion has zero acceleration while the y component of the motion is subject to the acceleration due to gravity. The horizontal distance traveled by the bullet is given by Equation 3.5a (with $a_x = 0$ m/s^2):

$$x = v_{0x}t = (v_0 \cos\theta)t$$

with t equal to the time required for the bullet to reach the target. The time t can be found by considering the vertical motion. From Equation 3.3b,

$$v_y = v_{0y} + a_y t$$

When the bullet reaches the target, $v_y = -v_{0y}$. Assuming that up and to the right are the positive directions, we have

$$t = \frac{-2v_{0y}}{a_y} = \frac{-2v_0 \sin\theta}{a_y} \qquad \text{and} \qquad x = (v_0 \cos\theta)\left(\frac{-2v_0 \sin\theta}{a_y}\right)$$

Using the fact that $2\sin\theta \cos\theta = \sin 2\theta$, we have

$$x = -\frac{2v_0^2 \cos\theta \sin\theta}{a_y} = -\frac{v_0^2 \sin 2\theta}{a_y}$$

Thus, we find that

$$\sin 2\theta = -\frac{x\, a_y}{v_0^2} = -\frac{(91.4 \text{ m})\,(-9.80 \text{ m/s}^2)}{(427 \text{ m/s})^2} = 4.91 \times 10^{-3}$$

and

$$2\theta = 0.281° \quad \text{or} \quad 2\theta = 180.000° - 0.281° = 179.719°$$

Therefore,

$$\theta = \boxed{0.141° \text{ and } 89.860°}$$

---

49. ***REASONING*** Since the horizontal motion is not accelerated, we know that the x component of the velocity remains constant at 340 m/s. Thus, we can use Equation 3.5a (with $a_x = 0$ m/s^2) to determine the time that the bullet spends in the building before it is embedded in the wall. Since we know the vertical displacement of the bullet after it enters the building, we can use the flight time in the building and Equation 3.5b to find the y component of the velocity of the bullet as it enters the window. Then, Equation 3.6b can be

used (with $v_{0y} = 0$ m/s) to determine the vertical displacement y of the bullet as it passes between the buildings. We can determine the distance H by adding the magnitude of y to the vertical distance of 0.50 m within the building.

Once we know the vertical displacement of the bullet as it passes between the buildings, we can determine the time t_1 required for the bullet to reach the window using Equation 3.4b. Since the motion in the x direction is not accelerated, the distance D can then be found from $D = v_{0x}t_1$.

SOLUTION Assuming that the direction to the right is positive, we find that the time that the bullet spends in the building is (according to Equation 3.5a)

$$t = \frac{x}{v_{0x}} = \frac{6.9 \text{ m}}{340 \text{ m/s}} = 0.0203 \text{ s}$$

The vertical displacement of the bullet after it enters the building is, taking down as the negative direction, equal to –0.50 m. Therefore, the vertical component of the velocity of the bullet as it passes through the window is, from Equation 3.5b,

$$v_{0y(\text{window})} = \frac{y - \frac{1}{2}a_y t^2}{t} = \frac{y}{t} - \frac{1}{2}a_y t = \frac{-0.50 \text{ m}}{0.0203 \text{ s}} - \frac{1}{2}(-9.80 \text{ m/s}^2)(0.0203 \text{ s}) = -24.5 \text{ m/s}$$

The vertical displacement of the bullet as it travels between the buildings is (according to Equation 3.6b with $v_{0y} = 0$ m/s)

$$y = \frac{v_y^2}{2a_y} = \frac{(-24.5 \text{ m/s})^2}{2(-9.80 \text{ m/s}^2)} = -30.6 \text{ m}$$

Therefore, the distance H is

$$H = 30.6 \text{ m} + 0.50 \text{ m} = \boxed{31 \text{ m}}$$

The time for the bullet to reach the window, according to Equation 3.4b, is

$$t_1 = \frac{2y}{v_{0y} + v_y} = \frac{2y}{v_y} = \frac{2(-30.6 \text{ m})}{(-24.5 \text{ m/s})} = 2.50 \text{ s}$$

Hence, the distance D is given by

$$D = v_{0x}t_1 = (340 \text{ m/s})(2.50 \text{ s}) = \boxed{850 \text{ m}}$$

---

53. **REASONING** The velocity $\mathbf{v}_{SG}$ of the swimmer relative to the ground is the vector sum of the velocity $\mathbf{v}_{SW}$ of the swimmer relative to the water and the velocity $\mathbf{v}_{WG}$ of the water relative to the ground as shown at the right: $\mathbf{v}_{SG} = \mathbf{v}_{SW} + \mathbf{v}_{WG}$.

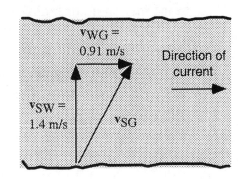

The component of $\mathbf{v}_{SG}$ that is parallel to the width of the river determines how fast the swimmer is moving across the river; this parallel component is v_{SW}. The time for the swimmer to cross the river is equal to the width of the river divided by the magnitude of this velocity component.

The component of $\mathbf{v}_{SG}$ that is parallel to the direction of the current determines how far the swimmer is carried down stream; this component is v_{WG}. Since the motion occurs with constant velocity, the distance that the swimmer is carried downstream while crossing the river is equal to the magnitude of v_{WG} multiplied by the time it takes for the swimmer to cross the river.

SOLUTION
a. The time t for the swimmer to cross the river is

$$t = \frac{\text{width}}{v_{SW}} = \frac{2.8 \times 10^3 \text{ m}}{1.4 \text{ m/s}} = \boxed{2.0 \times 10^3 \text{ s}}$$

b. The distance x that the swimmer is carried downstream while crossing the river is

$$x = v_{WG}t = (0.91 \text{ m/s})(2.0 \times 10^3 \text{ s}) = \boxed{1.8 \times 10^3 \text{ m}}$$

---

57. **REASONING** The velocity $\mathbf{v}_{AB}$ of train **A** relative to train **B** is the vector sum of the velocity $\mathbf{v}_{AG}$ of train **A** relative to the ground and the velocity $\mathbf{v}_{GB}$ of the ground relative to train **B**, as indicated by Equation 3.7: $\mathbf{v}_{AB} = \mathbf{v}_{AG} + \mathbf{v}_{GB}$. The values of $\mathbf{v}_{AG}$ and $\mathbf{v}_{BG}$ are given in the statement of the problem. We must also make use of the fact that $\mathbf{v}_{GB} = -\mathbf{v}_{BG}$.

SOLUTION

a. Taking east as the positive direction, the velocity of **A** relative to **B** is, according to Equation 3.7,

$$\mathbf{v}_{AB} = \mathbf{v}_{AG} + \mathbf{v}_{GB} = \mathbf{v}_{AG} - \mathbf{v}_{BG} = (+13 \text{ m/s}) - (-28 \text{ m/s}) = \boxed{+41 \text{ m/s}}$$

The positive sign indicates that the direction of $\mathbf{v}_{AB}$ is $\boxed{\text{due east}}$.

b. Similarly, the velocity of **B** relative to **A** is

$$\mathbf{v}_{BA} = \mathbf{v}_{BG} + \mathbf{v}_{GA} = \mathbf{v}_{BG} - \mathbf{v}_{AG} = (-28 \text{ m/s}) - (+13 \text{ m/s}) = \boxed{-41 \text{ m/s}}$$

The negative sign indicates that the direction of $\mathbf{v}_{BA}$ is $\boxed{\text{due west}}$.

---

63. **REASONING** The velocity $\mathbf{v}_{PM}$ of the puck relative to Mario is the vector sum of the velocity $\mathbf{v}_{PI}$ of the puck relative to the ice and the velocity $\mathbf{v}_{IM}$ of the ice relative to Mario as indicated by Equation 3.7: $\mathbf{v}_{PM} = \mathbf{v}_{PI} + \mathbf{v}_{IM}$. The values of $\mathbf{v}_{MI}$ and $\mathbf{v}_{PI}$ are given in the statement of the problem. In order to use the data, we must make use of the fact that $\mathbf{v}_{IM} = -\mathbf{v}_{MI}$, with the result that $\mathbf{v}_{PM} = \mathbf{v}_{PI} - \mathbf{v}_{MI}$.

SOLUTION The first two rows of the following table give the east/west and north/south components of the vectors $\mathbf{v}_{PI}$ and $-\mathbf{v}_{MI}$. The third row gives the components of their resultant $\mathbf{v}_{PM} = \mathbf{v}_{PI} - \mathbf{v}_{MI}$. Due east and due north have been taken as positive.

Vector	East/West Component	North/South Component
$\mathbf{v}_{PI}$	$-(11.0 \text{ m/s}) \sin 22° = -4.1 \text{ m/s}$	$-(11.0 \text{ m/s}) \cos 22° = -10.2 \text{ m/s}$
$-\mathbf{v}_{MI}$	0	$+7.0 \text{ m/s}$
$\mathbf{v}_{PM} = \mathbf{v}_{PI} - \mathbf{v}_{MI}$	-4.1 m/s	-3.2 m/s

Now that the components of v_{PM} are known, the Pythagorean theorem can be used to find the magnitude.

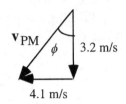

$$v_{PM} = \sqrt{(-4.1 \text{ m/s})^2 + (-3.2 \text{ m/s})^2} = \boxed{5.2 \text{ m/s}}$$

The direction of v_{PM} is found from

$$\phi = \tan^{-1}\left(\frac{4.1 \text{ m/s}}{3.2 \text{ m/s}}\right) = \boxed{52° \text{ west of south}}$$

---

65. **REASONING**   The velocity v_{OW} of the object relative to the water is the vector sum of the velocity v_{OS} of the object relative to the ship and the velocity v_{SW} of the ship relative to the water, as indicated by Equation 3.7: $v_{OW} = v_{OS} + v_{SW}$. The value of v_{SW} is given in the statement of the problem.   We can find the value of v_{OS} from the fact that we know the position of the object relative to the ship at two different times.   The initial position is r_{OS1}, and the final position is r_{OS2}.   Since the object moves with constant velocity,

$$v_{OS} = \frac{\Delta r_{OS}}{\Delta t} = \frac{r_{OS2} - r_{OS1}}{\Delta t} \qquad (1)$$

SOLUTION   The first two rows of the following table give the east/west and north/south components of the vectors r_{OS2} and $-r_{OS1}$. The third row of the table gives the components of $\Delta r_{OS} = r_{OS2} - r_{OS1}$. Due east and due north have been taken as positive.

Vector	*East/West Component*	*North/South Component*
r_{OS2}	$-(1120 \text{ m}) \cos 57.0°$ $= -6.10 \times 10^2 \text{ m}$	$-(1120 \text{ m}) \sin 57.0°$ $= -9.39 \times 10^2 \text{ m}$
$-r_{OS1}$	$-(2310 \text{ m}) \cos 32.0°$ $= -1.96 \times 10^3 \text{ m}$	$+(2310 \text{ m}) \sin 32.0°$ $= 1.22 \times 10^3 \text{ m}$
$\Delta r_{OS} = r_{OS2} - r_{OS1}$	$-2.57 \times 10^3 \text{ m}$	$2.81 \times 10^2 \text{ m}$

Now that the components of $\Delta \mathbf{r}_{OS}$ are known, the Pythagorean theorem can be used to find the magnitude.

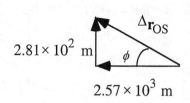

$$\Delta r_{OS} = \sqrt{(-2.57 \times 10^3 \text{ m})^2 + (2.81 \times 10^2 \text{ m})^2} = 2.59 \times 10^3 \text{ m}$$

The direction of $\Delta \mathbf{r}_{OS}$ is found from

$$\phi = \tan^{-1}\left(\frac{2.81 \times 10^2 \text{ m}}{2.57 \times 10^3 \text{ m}}\right) = 6.24°$$

Therefore, from Equation (1),

$$\mathbf{v}_{OS} = \frac{\Delta \mathbf{r}_{OS}}{\Delta t} = \frac{\mathbf{r}_{OS2} - \mathbf{r}_{OS1}}{\Delta t} = \frac{2.59 \times 10^3 \text{ m}}{360 \text{ s}} = 7.19 \text{ m/s}, \ 6.24° \text{ north of west}$$

Now that $\mathbf{v}_{OS}$ is known, we can find $\mathbf{v}_{OW}$, as indicated by Equation 3.7: $\mathbf{v}_{OW} = \mathbf{v}_{OS} + \mathbf{v}_{SW}$. The following table summarizes the vector addition:

Vector	East/West Component	North/South Component
$\mathbf{v}_{OS}$	$-(7.19 \text{ m/s}) \cos 6.24° = -7.15 \text{ m/s}$	$(7.19 \text{ m/s}) \sin 6.24° = 0.782 \text{ m/s}$
$\mathbf{v}_{SW}$	$+4.20 \text{ m/s}$	0 m/s
$\mathbf{v}_{OW} = \mathbf{v}_{OS} + \mathbf{v}_{SW}$	-2.95 m/s	0.782 m/s

Now that the components of $\mathbf{v}_{OW}$ are known, the Pythagorean theorem can be used to find the magnitude.

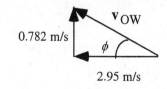

$$v_{OW} = \sqrt{(-2.95 \text{ m/s})^2 + (0.782 \text{ m/s})^2} = \boxed{3.05 \text{ m/s}}$$

The direction of $\mathbf{v}_{OW}$ is found from

$$\phi = \tan^{-1}\left(\frac{0.782 \text{ m/s}}{2.95 \text{ m/s}}\right) = \boxed{14.8° \text{ north of west}}$$

71. **REASONING** The speed of the fish at any time t is given by $v = \sqrt{v_x^2 + v_y^2}$, where v_x and v_y are the x and y components of the velocity at that instant. Since the horizontal motion of the fish has zero acceleration, $v_x = v_{0x}$ for all times t. Since the fish is dropped by the eagle, v_{0x} is equal to the horizontal speed of the eagle and $v_{0y} = 0$. The y component of the velocity of the fish for any time t is given by Equation 3.3b with $v_{0y} = 0$. Thus, the speed at any time t is given by $v = \sqrt{v_{0x}^2 + (a_y t)^2}$.

SOLUTION

a. The initial speed of the fish is $v_0 = \sqrt{v_{0x}^2 + v_{0y}^2} = \sqrt{v_{0x}^2 + 0^2} = v_{0x}$. When the fish's speed doubles, $v = 2v_{0x}$. Therefore,

$$2v_{0x} = \sqrt{v_{0x}^2 + (a_y t)^2} \qquad \text{or} \qquad 4v_{0x}^2 = v_{0x}^2 + (a_y t)^2$$

Assuming that downward is positive and solving for t, we have

$$t = \sqrt{3}\,\frac{v_{0x}}{a_y} = \sqrt{3}\left(\frac{6.0 \text{ m/s}}{9.80 \text{ m/s}^2}\right) = \boxed{1.1 \text{ s}}$$

b. When the fish's speed doubles again, $v = 4v_{0x}$. Therefore,

$$4v_{0x} = \sqrt{v_{0x}^2 + (a_y t)^2} \qquad \text{or} \qquad 16v_{0x}^2 = v_{0x}^2 + (a_y t)^2$$

Solving for t, we have

$$t = \sqrt{15}\,\frac{v_{0x}}{a_y} = \sqrt{15}\left(\frac{6.0 \text{ m/s}}{9.80 \text{ m/s}^2}\right) = 2.37 \text{ s}$$

Therefore, the additional time for the speed to double again is $(2.4 \text{ s}) - (1.1 \text{ s}) = \boxed{1.3 \text{ s}}$.

---

75. **REASONING** The angle θ can be found from

$$\theta = \tan^{-1}\left(\frac{2400 \text{ m}}{x}\right) \qquad\qquad (1)$$

where x is the horizontal displacement of the flare. Since $a_x = 0 \text{ m/s}^2$, it follows that $x = (v_0 \cos 30.0°)t$. The flight time t is determined by the vertical motion. In particular, the time t can be found from Equation 3.5b. Once the time is known, x can be calculated.

SOLUTION From Equation 3.5b, assuming upward is the positive direction, we have

$$y = -(v_0 \sin 30.0°)t + \tfrac{1}{2}a_y t^2$$

which can be rearranged to give the following equation that is quadratic in t:

$$\tfrac{1}{2}a_y t^2 - (v_0 \sin 30.0°)t - y = 0$$

Using $y = -2400$ m and $a_y = -9.80$ m/s^2 and suppressing the units, we obtain the quadratic equation

$$4.9t^2 + 120t - 2400 = 0$$

Using the quadratic formula, we obtain $t = 13$ s. Therefore, we find that

$$x = (v_0 \cos 30.0°)t = (240 \text{ m/s})(\cos 30.0°)(13 \text{ s}) = 2700 \text{ m}$$

Equation (1) then gives

$$\theta = \tan^{-1}\left(\frac{2400 \text{ m}}{2700 \text{ m}}\right) = \boxed{42°}$$

CHAPTER 4 | FORCES AND NEWTON'S LAWS OF MOTION

5. **REASONING** The net force acting on the ball can be calculated using Newton's second law. Before we can use Newton's second law, however, we must use Equation 2.9 from the equations of kinematics to determine the acceleration of the ball.

SOLUTION According to Equation 2.9, the acceleration of the ball is given by

$$a = \frac{v^2 - v_0^2}{2x}$$

Thus, the magnitude of the net force on the ball is given by

$$\Sigma F = ma = m\left(\frac{v^2 - v_0^2}{2x}\right) = (0.058 \text{ kg})\left[\frac{(45 \text{ m/s})^2 - (0 \text{ m/s})^2}{2(0.44 \text{ m})}\right] = \boxed{130 \text{ N}}$$

7. **REASONING AND SOLUTION** The acceleration required is

$$a = \frac{v^2 - v_0^2}{2x} = \frac{-(15.0 \text{ m/s})^2}{2(50.0 \text{ m})} = -2.25 \text{ m/s}^2$$

Newton's second law then gives the magnitude of the net force as

$$F = ma = (1580 \text{ kg})(2.25 \text{ m/s}^2) = \boxed{3560 \text{ N}}$$

9. **REASONING** Let due east be chosen as the positive direction. Then, when both forces point due east, Newton's second law gives

$$\underbrace{F_A + F_B}_{\Sigma F} = ma_1 \tag{1}$$

where $a_1 = 0.50 \text{ m/s}^2$. When F_A points due east and F_B points due west, Newton's second law gives

$$\underbrace{F_A - F_B}_{\Sigma F} = ma_2 \tag{2}$$

where $a_2 = 0.40$ m/s^2. These two equations can be used to find the magnitude of each force.

SOLUTION

a. Adding Equations 1 and 2 gives

$$F_A = \frac{m(a_1 + a_2)}{2} = \frac{(8.0 \text{ kg})(0.50 \text{ m/s}^2 + 0.40 \text{ m/s}^2)}{2} = \boxed{3.6 \text{ N}}$$

b. Subtracting Equation 2 from Equation 1 gives

$$F_B = \frac{m(a_1 - a_2)}{2} = \frac{(8.0 \text{ kg})(0.50 \text{ m/s}^2 - 0.40 \text{ m/s}^2)}{2} = \boxed{0.40 \text{ N}}$$

---

13. **REASONING** To determine the acceleration we will use Newton's second law $\Sigma\mathbf{F} = m\mathbf{a}$. Two forces act on the rocket, the thrust T and the rocket's weight W, which is $mg = (4.50 \times 10^5 \text{ kg})(9.80 \text{ m/s}^2) = 4.41 \times 10^6$ N. Both of these forces must be considered when determining the net force $\Sigma\mathbf{F}$. The direction of the acceleration is the same as the direction of the net force.

SOLUTION In constructing the free-body diagram for the rocket we choose upward and to the right as the positive directions. The free-body diagram is as follows:
The x component of the net force is

$$\Sigma F_x = T\cos 55.0°$$
$$= (7.50 \times 10^6 \text{ N})\cos 55.0° = 4.30 \times 10^6 \text{ N}$$

The y component of the net force is

$$\Sigma F_y = T\sin 55.0° - W = (7.50 \times 10^6 \text{ N})\sin 55.0° - 4.41 \times 10^6 \text{ N} = 1.73 \times 10^6 \text{ N}$$

The magnitudes of the net force and of the acceleration are

$$\Sigma F = \sqrt{(\Sigma F_x)^2 + (\Sigma F_y)^2}$$

$$a = \frac{\sqrt{(\Sigma F_x)^2 + (\Sigma F_y)^2}}{m} = \frac{\sqrt{(4.30 \times 10^6 \text{ N})^2 + (1.73 \times 10^6 \text{ N})^2}}{4.50 \times 10^5 \text{ kg}} = \boxed{10.3 \text{ m/s}^2}$$

The direction of the acceleration is the same as the direction of the net force. Thus, it is directed above the horizontal at an angle of

$$\theta = \tan^{-1}\left(\frac{\Sigma F_y}{\Sigma F_x}\right) = \tan^{-1}\left(\frac{1.73\times10^6 \text{ N}}{4.30\times10^6 \text{ N}}\right) = \boxed{21.9°}$$

---

19. ***REASONING*** We first determine the acceleration of the boat. Then, using Newton's second law, we can find the net force $\Sigma \mathbf{F}$ that acts on the boat. Since two of the three forces are known, we can solve for the unknown force $\mathbf{F}_W$ once the net force $\Sigma \mathbf{F}$ is known.

SOLUTION   Let the direction due east be the positive x direction and the direction due north be the positive y direction. The x and y components of the initial velocity of the boat are then

$$v_{0x} = (2.00 \text{ m/s}) \cos 15.0° = 1.93 \text{ m/s}$$

$$v_{0y} = (2.00 \text{ m/s}) \sin 15.0° = 0.518 \text{ m/s}$$

Thirty seconds later, the x and y velocity components of the boat are

$$v_x = (4.00 \text{ m/s}) \cos 35.0° = 3.28 \text{ m/s}$$

$$v_y = (4.00 \text{ m/s}) \sin 35.0° = 2.29 \text{ m/s}$$

Therefore, according to Equations 3.3a and 3.3b, the x and y components of the acceleration of the boat are

$$a_x = \frac{v_x - v_{0x}}{t} = \frac{3.28 \text{ m/s} - 1.93 \text{ m/s}}{30.0 \text{ s}} = 4.50\times10^{-2} \text{ m/s}^2$$

$$a_y = \frac{v_y - v_{0y}}{t} = \frac{2.29 \text{ m/s} - 0.518 \text{ m/s}}{30.0 \text{ s}} = 5.91\times10^{-2} \text{ m/s}^2$$

Thus, the x and y components of the net force that act on the boat are

$$\Sigma F_x = ma_x = (325 \text{ kg})(4.50\times10^{-2} \text{ m/s}^2) = 14.6 \text{ N}$$

$$\Sigma F_y = ma_y = (325 \text{ kg})(5.91\times10^{-2} \text{ m/s}^2) = 19.2 \text{ N}$$

The following table gives the x and y components of the net force $\sum \mathbf{F}$ and the two known forces that act on the boat.   The fourth row of that table gives the components of the unknown force $\mathbf{F}_W$.

Force	x-Component	y-Component
$\sum \mathbf{F}$	14.6 N	19.2 N
$\mathbf{F}_1$	(31.0 N) cos 15.0° = 29.9 N	(31.0 N) sin 15.0° = 8.02 N
$\mathbf{F}_2$	−(23.0 N) cos 15.0° = −22.2 N	−(23.0 N) sin 15.0° = −5.95 N

$\mathbf{F}_W = \sum \mathbf{F} - \mathbf{F}_1 - \mathbf{F}_2$ 14.6 N − 29.9 N + 22.2 N = 6.9 N   19.2 N − 8.02 N + 5.95 N = 17.1 N

The magnitude of $\mathbf{F}_W$ is given by the Pythagorean theorem as

$$F_W = \sqrt{(6.9 \text{ N})^2 + (17.1 \text{ N})^2} = \boxed{18.4 \text{ N}}$$

The angle θ that $\mathbf{F}_W$ makes with the x axis is

$$\theta = \tan^{-1}\left(\frac{17.1 \text{ N}}{6.9 \text{ N}}\right) = 68°$$

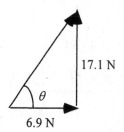

17.1 N

6.9 N

Therefore, the direction of $\mathbf{F}_W$ is $\boxed{68°, \text{ north of east}}$.

---

25. **REASONING AND SOLUTION**
a.   Combining Equations 4.4 and 4.5, we see that the acceleration due to gravity on the surface of Saturn can be calculated as follows:

$$g_{\text{Saturn}} = G \frac{M_{\text{Saturn}}}{r_{\text{Saturn}}^2} = \left(6.67 \times 10^{-11} \text{ N} \cdot \text{m}^2/\text{kg}^2\right) \frac{\left(5.67 \times 10^{26} \text{ kg}\right)}{\left(6.00 \times 10^7 \text{ m}\right)^2} = \boxed{10.5 \text{ m/s}^2}$$

b.   The ratio of the person's weight on Saturn to that on earth is

$$\frac{W_{\text{Saturn}}}{W_{\text{earth}}} = \frac{mg_{\text{Saturn}}}{mg_{\text{earth}}} = \frac{g_{\text{Saturn}}}{g_{\text{earth}}} = \frac{10.5 \text{ m/s}^2}{9.80 \text{ m/s}^2} = \boxed{1.07}$$

27. **_REASONING AND SOLUTION_**   According to Equations 4.4 and 4.5, the weight of an object of mass m at a distance r from the *center* of the earth is

$$mg = \frac{GM_E m}{r^2}$$

In a circular orbit that is 3.59×10^7 m above the surface of the earth (radius $= 6.38 \times 10^6$ m, mass $= 5.98 \times 10^{24}$ kg), the total distance from the center of the earth is $r = 3.59 \times 10^7$ m $+ 6.38 \times 10^6$ m. Thus the acceleration g due to gravity is

$$g = \frac{GM_E}{r^2} = \frac{(6.67 \times 10^{-11} \text{N} \cdot \text{m}^2/\text{kg}^2)(5.98 \times 10^{24} \text{kg})}{(3.59 \times 10^7 \text{m} + 6.38 \times 10^6 \text{m})^2} = \boxed{0.223 \text{ m/s}^2}$$

31. **_REASONING_**   According to Equation 4.4, the weights of an object of mass m on the surfaces of planet A (mass $= M_A$, radius $= R$) and planet B (mass $= M_B$, radius $= R$) are

$$W_A = \frac{GM_A m}{R^2} \quad \text{and} \quad W_B = \frac{GM_B m}{R^2}$$

The difference between these weights is given in the problem.

SOLUTION   The difference in weights is

$$W_A - W_B = \frac{GM_A m}{R^2} - \frac{GM_B m}{R^2} = \frac{Gm}{R^2}\left(M_A - M_B\right)$$

Rearranging this result, we find

$$M_A - M_B = \frac{\left(W_A - W_B\right)R^2}{Gm} = \frac{(3620 \text{ N})\left(1.33 \times 10^7 \text{ m}\right)^2}{(6.67 \times 10^{-11} \text{N} \cdot \text{m}^2/\text{kg}^2)(5450 \text{ kg})} = \boxed{1.76 \times 10^{24} \text{ kg}}$$

33. **_REASONING AND SOLUTION_**   There are two forces that act on the balloon; they are, the combined weight of the balloon and its load, Mg, and the upward buoyant force F_B. If we take upward as the positive direction, then, initially when the balloon is motionless, Newton's second law gives $F_B - Mg = 0$. If an amount of mass m is dropped overboard so that the balloon has an upward acceleration, Newton's second law for this situation is

$$F_B - (M - m)g = (M - m)a$$

But $F_B = mg$, so that

$$Mg - (M - m)g = mg = (M - m)a$$

Solving for the mass m that should be dropped overboard, we obtain

$$m = \frac{Ma}{g+a} = \frac{(310 \text{ kg})(0.15 \text{ m/s}^2)}{9.80 \text{ m/s}^2 + 0.15 \text{ m/s}^2} = \boxed{4.7 \text{ kg}}$$

---

39. **REASONING** The book is kept from falling as long as the total static frictional force balances the weight of the book. The forces that act on the book are shown in the following free-body diagram, where P is the pressing force applied by each hand.

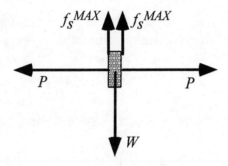

In this diagram, note that there are two pressing forces, one from each hand. Each hand also applies a static frictional force, and, therefore, two static frictional forces are shown. The maximum static frictional force is related in the usual way to a normal force F_N, but in this problem the normal force is provided by the pressing force, so that $F_N = P$.

SOLUTION Since the frictional forces balance the weight, we have

$$2 f_s^{MAX} = 2(\mu_s F_N) = 2(\mu_s P) = W$$

Solving for P, we find that

$$P = \frac{W}{2\mu_s} = \frac{31 \text{ N}}{2(0.40)} = \boxed{39 \text{ N}}$$

---

45. **REASONING AND SOLUTION** Four forces act on the sled. They are the pulling force **P**, the force of kinetic friction $\mathbf{f}_k$, the weight $m\mathbf{g}$ of the sled, and the normal force $\mathbf{F}_N$ exerted on the sled by the surface on which it slides. The following figures show free-body diagrams for the sled. In the diagram on the right, the forces have been resolved into their x and y components.

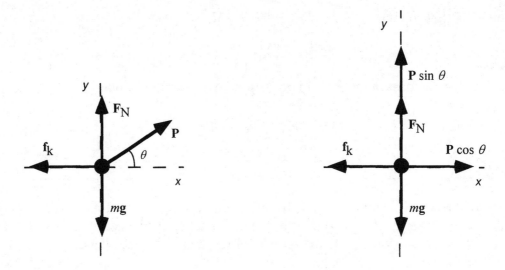

Since the sled is pulled at constant velocity, its acceleration is zero, and Newton's second law in the direction of motion is (with right chosen as the positive direction)

$$\sum F_x = P\cos\theta - f_k = ma_x = 0$$

From Equation 4.8, we know that $f_k = \mu_k F_N$, so that the above expression becomes

$$P\cos\theta - \mu_k F_N = 0 \qquad (1)$$

In the vertical direction,

$$\sum F_y = P\sin\theta + F_N - mg = ma_y = 0 \qquad (2)$$

Solving Equation (2) for the normal force, and substituting into Equation (1), we obtain

$$P\cos\theta - \mu_k\left(mg - P\sin\theta\right) = 0$$

Solving for μ_k, the coefficient of kinetic friction, we find

$$\mu_k = \frac{P\cos\theta}{mg - P\sin\theta} = \frac{(80.0\text{ N})\cos 30.0^\circ}{(20.0\text{ kg})(9.80\text{ m/s}^2) - (80.0\text{ N})\sin 30.0^\circ} = \boxed{0.444}$$

---

49. ***REASONING*** Let us assume that the skater is moving horizontally along the $+x$ axis. The time t it takes for the skater to reduce her velocity to $v_x = +2.8$ m/s from $v_{0x} = +6.3$ m/s can be obtained from one of the equations of kinematics:

$$v_x = v_{0x} + a_x t \qquad (3.3a)$$

The initial and final velocities are known, but the acceleration is not. We can obtain the acceleration from Newton's second law $\left(\Sigma F_x = ma_x, \text{ Equation 4.2a} \right)$ in the following manner. The kinetic frictional force is the only horizontal force that acts on the skater, and, since it is a resistive force, it acts opposite to the direction of the motion. Thus, the net force in the x direction is $\Sigma F_x = -f_k$, where f_k is the magnitude of the kinetic frictional force. Therefore, the acceleration of the skater is $a_x = \Sigma F_x / m = -f_k / m$.

The magnitude of the frictional force is $f_k = \mu_k F_N$ (Equation 4.8), where μ_k is the coefficient of kinetic friction between the ice and the skate blades and F_N is the magnitude of the normal force. There are two vertical forces acting on the skater: the upward-acting normal force $\mathbf{F_N}$ and the downward pull of gravity (her weight) $m\mathbf{g}$. Since the skater has no vertical acceleration, Newton's second law in the vertical direction gives (taking upward as the positive direction) $\Sigma F_y = F_N - mg = 0$. Therefore, the magnitude of the normal force is $F_N = mg$ and the magnitude of the acceleration is

$$a_x = \frac{-f_k}{m} = \frac{-\mu_k F_N}{m} = \frac{-\mu_k \cancel{m} g}{\cancel{m}} = -\mu_k g$$

SOLUTION
Solving the equation $v_x = v_{0x} + a_x t$ for the time and substituting the expression above for the acceleration yields

$$t = \frac{v_x - v_{0x}}{a_x} = \frac{v_x - v_{0x}}{-\mu_k g} = \frac{2.8 \text{ m/s} - 6.3 \text{ m/s}}{-(0.081)(9.80 \text{ m/s}^2)} = \boxed{4.4 \text{ s}}$$

---

53. ***REASONING*** In order for the object to move with constant velocity, the net force on the object must be zero. Therefore, the north/south component of the third force must be equal in magnitude and opposite in direction to the 80.0 N force, while the east/west component of the third force must be equal in magnitude and opposite in direction to the 60.0 N force. Therefore, the third force has components: 80.0 N due south and 60.0 N due east. We can use the Pythagorean theorem and trigonometry to find the magnitude and direction of this third force.

SOLUTION The magnitude of the third force is

$$F_3 = \sqrt{(80.0 \text{ N})^2 + (60.0 \text{ N})^2} = \boxed{1.00 \times 10^2 \text{ N}}$$

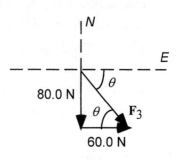

The direction of F_3 is specified by the angle θ where

$$\theta = \tan^{-1}\left(\frac{80.0 \text{ N}}{60.0 \text{ N}}\right) = \boxed{53.1°, \text{ south of east}}$$

---

57. *REASONING AND SOLUTION* The free body diagram for the plane is shown below to the left. The figure at the right shows the forces resolved into components parallel to and perpendicular to the line of motion of the plane.

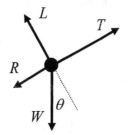

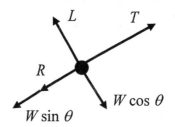

If the plane is to continue at constant velocity, the resultant force must still be zero after the fuel is jettisoned. Therefore (using the directions of T and L to define the positive directions),

$$T - R - W(\sin\theta) = 0 \qquad\qquad (1)$$
$$L - W(\cos\theta) = 0 \qquad\qquad (2)$$

From Example 13, before the fuel is jettisoned, the weight of the plane is 86 500 N, the thrust is 103 000 N, and the lift is 74 900 N. The force of air resistance is the same before and after the fuel is jettisoned and is given in Example 13 as $R = 59$ 800 N.

After the fuel is jettisoned, $W = 86$ 500 N $- 2800$ N $= 83$ 700 N

From Equation (1) above, the thrust after the fuel is jettisoned is

$$T = R + W(\sin\theta) = [(59 \text{ 800 N}) + (83 \text{ 700 N})(\sin 30.0°)] = 101 \text{ 600 N}$$

From Equation (2), the lift after the fuel is jettisoned is

$$L = W(\cos\theta) = (83 \text{ 700 N})(\cos 30.0°) = 72 \text{ 500 N}$$

a. The pilot must, therefore, reduce the thrust by

$$103\ 000\ \text{N} - 101\ 600\ \text{N} = \boxed{1400\ \text{N}}$$

b. The pilot must reduce the lift by

$$74\ 900\ \text{N} - 72\ 500\ \text{N} = \boxed{2400\ \text{N}}$$

___

63. **REASONING** There are four forces that act on the chandelier; they are the forces of tension T in each of the three wires, and the downward force of gravity mg. Under the influence of these forces, the chandelier is at rest and, therefore, in equilibrium. Consequently, the sum of the x components as well as the sum of the y components of the forces must each be zero. The figure below shows a quasi-free-body diagram for the chandelier and the force components for a suitable system of x, y axes. Note that the diagram only shows one of the forces of tension; the second and third tension forces are not shown in the interest of clarity. The triangle at the right shows the geometry of one of the cords, where ℓ is the length of the cord, and d is the distance from the ceiling.

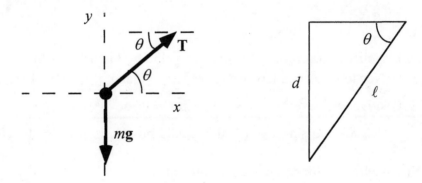

We can use the forces in the y direction to find the magnitude T of the tension in any one wire.

SOLUTION Remembering that there are three tension forces, we see from the diagram that

$$3T\sin\theta = mg \qquad \text{or} \qquad T = \frac{mg}{3\sin\theta} = \frac{mg}{3(d/\ell)} = \frac{mg\ell}{3d}$$

Therefore, the magnitude of the tension in any one of the cords is

$$T = \frac{(44\ \text{kg})(9.80\ \text{m/s}^2)(2.0\ \text{m})}{3(1.5\ \text{m})} = \boxed{1.9\times10^2\,\text{N}}$$

___

67. ***REASONING***   When the bicycle is coasting straight down the hill, the forces that act on it are the normal force F_N exerted by the surface of the hill, the force of gravity mg, and the force of air resistance R. When the bicycle climbs the hill, there is one additional force; it is the applied force that is required for the bicyclist to climb the hill at constant speed. We can use our knowledge of the motion of the bicycle down the hill to find R. Once R is known, we can analyze the motion of the bicycle as it climbs the hill.

SOLUTION   The figure to the left below shows the free-body diagram for the forces during the downhill motion. The hill is inclined at an angle θ above the horizontal. The figure to the right shows these forces resolved into components parallel to and perpendicular to the line of motion.

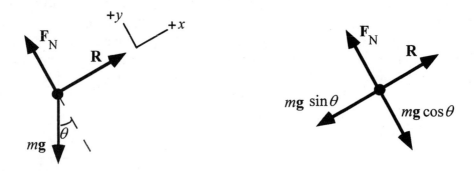

Since the bicyclist is traveling at a constant velocity, his acceleration is zero. Therefore, according to Newton's second law, we have $\sum F_x = 0$ and $\sum F_y = 0$. Taking the direction up the hill as positive, we have $\sum F_x = R - mg \sin \theta = 0$, or

$$R = mg \sin \theta = (80.0 \text{ kg})(9.80 \text{ m/s}^2) \sin 15.0° = 203 \text{ N}$$

When the bicyclist climbs the same hill at constant speed, an applied force P must push the system up the hill. Since the speed is the same, the magnitude of the force of air resistance will remain 203 N. However, the air resistance will oppose the motion by pointing down the hill. The figure at the right shows the resolved forces that act on the system during the uphill motion.

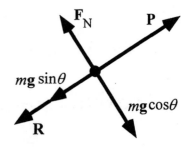

Using the same sign convention as above, we have $\sum F_x = P - mg \sin \theta - R = 0$, or

$$P = R + mg \sin \theta = 203 \text{ N} + 203 \text{ N} = \boxed{406 \text{ N}}$$

73. ***REASONING*** If we assume that the acceleration is constant, we can use Equation 2.4 ($v = v_0 + at$) to find the acceleration of the car. Once the acceleration is known, Newton's second law ($\sum \mathbf{F} = m\mathbf{a}$) can be used to find the magnitude and direction of the net force that produces the deceleration of the car.

SOLUTION The average acceleration of the car is, according to Equation 2.4,

$$a = \frac{v - v_0}{t} = \frac{17.0 \text{ m/s} - 27.0 \text{ m/s}}{8.00 \text{ s}} = -1.25 \text{ m/s}^2$$

where the minus sign indicates that the direction of the acceleration is opposite to the direction of motion; therefore, the acceleration points due west.

According to Newton's Second law, the net force on the car is

$$\sum F = ma = (1380 \text{ kg})(-1.25 \text{ m/s}^2) = -1730 \text{ N}$$

The magnitude of the net force is $\boxed{1730 \text{ N}}$. From Newton's second law, we know that the direction of the force is the same as the direction of the acceleration, so the force also points $\boxed{\text{due west}}$.

---

75. ***REASONING AND SOLUTION***
a. Each cart has the same mass and acceleration; therefore, the net force acting on any one of the carts is, according to Newton's second law

$$\sum F = ma = (26 \text{ kg})(0.050 \text{ m/s}^2) = \boxed{1.3 \text{ N}}$$

b. The fifth cart must essentially push the sixth, seventh, eight, ninth and tenth cart. In other words, it must exert on the sixth cart a total force of

$$\sum F = ma = 5(26 \text{ kg})(0.050 \text{ m/s}^2) = \boxed{6.5 \text{ N}}$$

---

77. ***REASONING*** The speed of the skateboarder at the bottom of the ramp can be found by solving Equation 2.9 ($v^2 = v_0^2 + 2ax$, where x is the distance that the skater moves down the ramp) for v. The figure at the right shows the free-body diagram for the skateboarder. The net force ΣF, which accelerates the skateboarder down the ramp, is the component of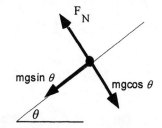

the weight that is parallel to the incline: $\Sigma F = mg \sin\theta$. Therefore, we know from Newton's second law that the acceleration of the skateboarder down the ramp is

$$a = \frac{\Sigma F}{m} = \frac{mg \sin \theta}{m} = g \sin \theta$$

SOLUTION   Thus, the speed of the skateboarder at the bottom of the ramp is

$$v = \sqrt{v_0^2 + 2ax} = \sqrt{v_0^2 + 2gx \, \sin \theta} = \sqrt{(2.6 \text{ m/s})^2 + 2(9.80 \text{ m/s}^2)(6.0 \text{ m}) \sin 18°} = \boxed{6.6 \text{ m/s}}$$

---

83. **REASONING**   The free-body diagrams for Robin (mass $= m$) and for the chandelier (mass $= M$) are given at the right. The tension T in the rope applies an upward force to both.   Robin accelerates upward, while the chandelier accelerates downward, each acceleration having the same magnitude.   Our solution is based on separate applications of Newton's second law to Robin and the chandelier.

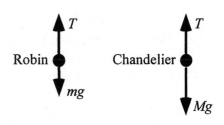

SOLUTION Applying Newton's second law, we find

$$\underbrace{T - mg = ma}_{\text{Robin Hood}} \qquad \text{and} \qquad \underbrace{T - Mg = -Ma}_{\text{Chandelier}}$$

In these applications we have taken upward as the positive direction, so that Robin's acceleration is a, while the chandelier's acceleration is $-a$.   Solving the Robin-Hood equation for T gives

$$T = mg + ma$$

Substituting this expression for T into the Chandelier equation gives

$$mg + ma - Mg = -Ma \qquad \text{or} \qquad a = \left(\frac{M - m}{M + m} \right) g$$

a. Robin's acceleration is

$$a = \left(\frac{M - m}{M + m} \right) g = \left[\frac{(195 \text{ kg}) - (77.0 \text{ kg})}{(195 \text{ kg}) + (77.0 \text{ kg})} \right] (9.80 \text{ m/s}^2) = \boxed{4.25 \text{ m/s}^2}$$

b. Substituting the value of a into the expression for T gives

$$T = mg + ma = (77.0 \text{ kg}) \left(9.80 \text{ m/s}^2 + 4.25 \text{ m/s}^2 \right) = \boxed{1080 \text{ N}}$$

89. **REASONING** The tension in each coupling bar is responsible for accelerating the objects behind it. The masses of the cars are m_1, m_2, and m_3. We can use Newton's second law to express the tension in each coupling bar, since friction is negligible:

$$\underbrace{T_A = (m_1 + m_2 + m_3)a}_{\text{Coupling bar A}} \qquad \underbrace{T_B = (m_2 + m_3)a}_{\text{Coupling bar B}} \qquad \underbrace{T_C = m_3 a}_{\text{Coupling bar C}}$$

In these expressions $a = 0.12 \text{ m/s}^2$ remains constant. Consequently, the tension in a given bar will change only if the total mass of the objects accelerated by that bar changes as a result of the luggage transfer. Using Δ (Greek capital delta) to denote a change in the usual fashion, we can express the changes in the above tensions as follows:

$$\underbrace{\Delta T_A = \left[\Delta(m_1 + m_2 + m_3)\right]a}_{\text{Coupling bar A}} \qquad \underbrace{\Delta T_B = \left[\Delta(m_2 + m_3)\right]a}_{\text{Coupling bar B}} \qquad \underbrace{\Delta T_C = (\Delta m_3)a}_{\text{Coupling bar C}}$$

SOLUTION
a. Moving luggage from car 2 to car 1 does not change the total mass $m_1 + m_2 + m_3$, so $\Delta(m_1 + m_2 + m_3) = 0$ kg and $\boxed{\Delta T_A = 0 \text{ N}}$.

The transfer from car 2 to car 1 causes the total mass $m_2 + m_3$ to decrease by 39 kg, so $\Delta(m_2 + m_3) = -39$ kg and

$$\Delta T_B = \left[\Delta(m_2 + m_3)\right]a = (-39 \text{ kg})(0.12 \text{ m/s}^2) = \boxed{-4.7 \text{ N}}$$

The transfer from car 2 to car 1 does not change the mass m_3, so $\Delta m_3 = 0$ kg and $\boxed{\Delta T_C = 0 \text{ N}}$.

b. Moving luggage from car 2 to car 3 does not change the total mass $m_1 + m_2 + m_3$, so $\Delta(m_1 + m_2 + m_3) = 0$ kg and $\boxed{\Delta T_A = 0 \text{ N}}$.

The transfer from car 2 to car 3 does not change the total mass $m_2 + m_3$, so $\Delta(m_2 + m_3) = 0$ kg and $\boxed{\Delta T_B = 0 \text{ N}}$.

The transfer from car 2 to car 3 causes the mass m_3 to increase by 39 kg, so $\Delta m_3 = +39$ kg and

$$\Delta T_C = (\Delta m_3)a = (+39 \text{ kg})(0.12 \text{ m/s}^2) = \boxed{+4.7 \text{ N}}$$

93. ***REASONING*** *AND* ***SOLUTION***

a. The left mass (mass 1) has a tension T_1 pulling it up. Newton's second law gives

$$T_1 - m_1 g = m_1 a \qquad (1)$$

The right mass (mass 3) has a different tension, T_3, trying to pull it up. Newton's second for it is

$$T_3 - m_3 g = -m_3 a \qquad (2)$$

The middle mass (mass 2) has both tensions acting on it along with friction. Newton's second law for its horizontal motion is

$$T_3 - T_1 - \mu_k m_2 g = m_2 a \qquad (3)$$

Solving Equation (1) and Equation (2) for T_1 and T_3, respectively, and substituting into Equation (3) gives

$$a = \frac{\left(m_3 - m_1 - \mu_k m_2\right) g}{m_1 + m_2 + m_3}$$

Hence,

$$a = \frac{\left[25.0 \text{ kg} - 10.0 \text{ kg} - (0.100)(80.0 \text{ kg})\right]\left(9.80 \text{ m/s}^2\right)}{10.0 \text{ kg} + 80.0 \text{ kg} + 25.0 \text{ kg}} = \boxed{0.60 \text{ m/s}^2}$$

b. From part a:

$$T_1 = m_1(g + a) = (10.0 \text{ kg})\left(9.80 \text{ m/s}^2 + 0.60 \text{ m/s}^2\right) = \boxed{104 \text{ N}}$$

$$T_3 = m_3(g - a) = (25.0 \text{ kg})\left(9.80 \text{ m/s}^2 - 0.60 \text{ m/s}^2\right) = \boxed{230 \text{ N}}$$

---

95. ***REASONING*** The magnitude of the gravitational force that each part exerts on the other is given by Newton's law of gravitation as $F = G m_1 m_2 / r^2$. To use this expression, we need the masses m_1 and m_2 of the parts, whereas the problem statement gives the weights W_1 and W_2. However, the weight is related to the mass by $W = mg$, so that for each part we know that $m = W/g$.

SOLUTION The gravitational force that each part exerts on the other is

$$F = \frac{Gm_1m_2}{r^2} = \frac{G(W_1/g)(W_2/g)}{r^2}$$

$$= \frac{(6.67\times10^{-11}\ \text{N}\cdot\text{m}^2/\text{kg}^2)(11\,000\ \text{N})(3400\ \text{N})}{(9.80\ \text{m/s}^2)^2(12\ \text{m})^2} = \boxed{1.8\times10^{-7}\ \text{N}}$$

97. **REASONING AND SOLUTION** According to Equation 3.3b, the acceleration of the astronaut is $a_y = (v_y - v_{0y})/t = v_y/t$. The apparent weight and the true weight of the astronaut are related according to Equation 4.6. Direct substitution gives

$$\underbrace{F_N}_{\substack{\text{Apparent} \\ \text{weight}}} = \underbrace{mg}_{\substack{\text{True} \\ \text{weight}}} + ma_y = m\,(g + a_y) = m\left(g + \frac{v_y}{t}\right)$$

$$= (57\ \text{kg})\left(9.80\ \text{m/s}^2 + \frac{45\ \text{m/s}}{15\ \text{s}}\right) = \boxed{7.3\times10^2\ \text{N}}$$

99. **REASONING** In order to start the crate moving, an external agent must supply a force that is at least as large as the maximum value $f_s^{\text{MAX}} = \mu_s F_N$, where μ_s is the coefficient of static friction (see Equation 4.7). Once the crate is moving, the magnitude of the frictional force is very nearly constant at the value $f_k = \mu_k F_N$, where μ_k is the coefficient of kinetic friction (see Equation 4.8). In both cases described in the problem statement, there are only two vertical forces that act on the crate; they are the upward normal force F_N, and the downward pull of gravity (the weight) mg. Furthermore, the crate has no vertical acceleration in either case. Therefore, if we take upward as the positive direction, Newton's second law in the vertical direction gives $F_N - mg = 0$, and we see that, in both cases, the magnitude of the normal force is $F_N = mg$.

SOLUTION
a. Therefore, the applied force needed to start the crate moving is

$$f_s^{\text{MAX}} = \mu_s mg = (0.760)(60.0\ \text{kg})(9.80\ \text{m/s}^2) = \boxed{447\ \text{N}}$$

b. When the crate moves in a straight line at constant speed, its velocity does not change, and it has zero acceleration. Thus, Newton's second law in the horizontal direction becomes $P - f_k = 0$, where P is the required pushing force. Thus, the applied force required to keep the crate sliding across the dock at a constant speed is

$$P = f_k = \mu_k mg = (0.410)(60.0\ \text{kg})(9.80\ \text{m/s}^2) = \boxed{241\ \text{N}}$$

103. **REASONING** We can use the appropriate equation of kinematics to find the acceleration of the bullet. Then Newton's second law can be used to find the average net force on the bullet.

SOLUTION According to Equation 2.4, the acceleration of the bullet is

$$a = \frac{v - v_0}{t} = \frac{715 \text{ m/s} - 0 \text{ m/s}}{2.50 \times 10^{-3} \text{ s}} = 2.86 \times 10^5 \text{ m/s}^2$$

Therefore, the net average force on the bullet is

$$\Sigma F = ma = (15 \times 10^{-3} \text{ kg})(2.86 \times 10^5 \text{ m/s}^2) = \boxed{4290 \text{ N}}$$

107. **REASONING AND SOLUTION**   The system is shown in the drawing. We will let $m_1 = 21.0 \text{ kg}$, and $m_2 = 45.0 \text{ kg}$. Then, m_1 will move upward, and m_2 will move downward. There are two forces that act on each object; they are the tension T in the cord and the weight mg of the object. The forces are shown in the free-body diagrams at the far right.

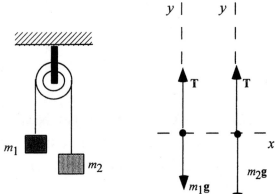

We will take up as the positive direction. If the acceleration of m_1 is a, then the acceleration of m_2 must be $-a$.

From Newton's second law, we have for m_1

$$\Sigma F_y = T - m_1 g = m_1 a \qquad (1)$$

and for m_2

$$\Sigma F_y = T - m_2 g = -m_2 a \qquad (2)$$

a. Eliminating T between these two equations, we obtain

$$a = \frac{m_2 - m_1}{m_2 + m_1} g = \left(\frac{45.0 \text{ kg} - 21.0 \text{ kg}}{45.0 \text{ kg} + 21.0 \text{ kg}} \right) (9.80 \text{ m/s}^2) = \boxed{3.56 \text{ m/s}^2}$$

b. Eliminating a between Equations (1) and (2), we find

$$T = \frac{2m_1 m_2}{m_1 + m_2}g = \left[\frac{2(21.0 \text{ kg})(45.0 \text{ kg})}{21.0 \text{ kg} + 45.0 \text{ kg}}\right](9.80 \text{ m/s}^2) = \boxed{281 \text{ N}}$$

---

111. **REASONING**   The shortest time to pull the person from the cave corresponds to the maximum acceleration, a_y, that the rope can withstand.   We first determine this acceleration and then use kinematic Equation 3.5b ($y = v_{0y}t + \frac{1}{2}a_y t^2$) to find the time t.

SOLUTION   As the person is being pulled from the cave, there are two forces that act on him; they are the tension T in the rope that points vertically upward, and the weight of the person mg that points vertically downward.   Thus, if we take upward as the positive direction, Newton's second law gives $\sum F_y = T - mg = ma_y$. Solving for a_y, we have

$$a_y = \frac{T}{m} - g = \frac{T}{W/g} - g = \frac{569 \text{ N}}{(5.20 \times 10^2 \text{ N})/(9.80 \text{ m/s}^2)} - 9.80 \text{ m/s}^2 = 0.92 \text{ m/s}^2$$

Therefore, from Equation 3.5b with $v_{0y} = 0$ m/s, we have $y = \frac{1}{2}a_y t^2$. Solving for t, we find

$$t = \sqrt{\frac{2y}{a_y}} = \sqrt{\frac{2(35.1 \text{ m})}{0.92 \text{ m/s}^2}} = \boxed{8.7 \text{ s}}$$

---

115. **REASONING AND SOLUTION**   The free-body diagram is shown at the right. The forces that act on the picture are the pressing force P, the normal force $\mathbf{F}_N$ exerted on the picture by the wall, the weight mg of the picture, and the force of static friction $\mathbf{f}_s^{MAX}$.   The maximum magnitude for the frictional force is given by Equation 4.7: $f_s^{MAX} = \mu_s F_N$.   The picture is in equilibrium, and, if we take the directions to the right and up as positive, we have in the x direction

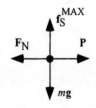

$$\sum F_x = P - F_N = 0 \qquad \text{or} \qquad P = F_N$$

and in the y direction

$$\sum F_y = f_s^{MAX} - mg = 0 \qquad \text{or} \qquad f_s^{MAX} = mg$$

Therefore,

$$f_s^{MAX} = \mu_s F_N = mg$$

But since $F_N = P$, we have

$$\mu_s P = mg$$

Solving for P, we have

$$P = \frac{mg}{\mu_s} = \frac{(1.10 \text{ kg})(9.80 \text{ m/s}^2)}{0.660} = \boxed{16.3 \text{ N}}$$

---

117. ***REASONING AND SOLUTION*** The penguin comes to a halt on the horizontal surface because the kinetic frictional force opposes the motion and causes it to slow down. The time required for the penguin to slide to a halt ($v = 0$ m/s) after entering the horizontal patch of ice is, according to Equation 2.4,

$$t = \frac{v - v_0}{a_x} = \frac{-v_0}{a_x}$$

We must, therefore, determine the acceleration of the penguin as it slides along the horizontal patch (see the following drawing).

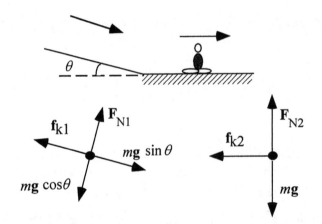

Free-body diagram A   **Free-body diagram B**

For the penguin sliding on the horizontal patch of ice, we find from free-body diagram B and Newton's second law in the x direction (motion to the right is taken as positive) that

$$\sum F_x = -f_{k2} = ma_x \qquad \text{or} \qquad a_x = \frac{-f_{k2}}{m} = \frac{-\mu_k F_{N2}}{m}$$

In the y direction in free-body diagram B, we have $\sum F_y = F_{N2} - mg = 0$, or $F_{N2} = mg$. Therefore, the acceleration of the penguin is

$$a_x = \frac{-\mu_k mg}{m} = -\mu_k g \qquad\qquad (1)$$

Equation (1) indicates that, in order to find the acceleration a_x, we must find the coefficient of kinetic friction.

We are told in the problem statement that the coefficient of kinetic friction between the penguin and the ice is the same for the incline as for the horizontal patch. Therefore, we can use the motion of the penguin on the incline to determine the coefficient of friction and use it in Equation (1).

For the penguin sliding down the incline, we find from free-body diagram A (see the previous drawing) and Newton's second law (taking the direction of motion as positive) that

$$\Sigma F_x = mg \sin \theta - f_{k1} = ma_x = 0 \qquad \text{or} \qquad f_{k1} = mg \sin \theta \qquad (2)$$

Here, we have used the fact that the penguin slides down the incline with a constant velocity, so that it has zero acceleration. From Equation 4.8, we know that $f_{k1} = \mu_k F_{N1}$. Applying Newton's second law in the direction perpendicular to the incline, we have

$$\Sigma F_y = F_{N1} - mg \cos \theta = 0 \qquad \text{or} \qquad F_{N1} = mg \cos \theta$$

Therefore, $f_{k1} = \mu_k mg \cos \theta$, so that according to Equation (2), we find

$$f_{k1} = \mu_k mg \cos \theta = mg \sin \theta$$

Solving for the coefficient of kinetic friction, we have

$$\mu_k = \frac{\sin \theta}{\cos \theta} = \tan \theta$$

Finally, the time required for the penguin to slide to a halt after entering the horizontal patch of ice is

$$t = \frac{-v_0}{a_x} = \frac{-v_0}{-\mu_k g} = \frac{v_0}{g \tan \theta} = \frac{1.4 \text{ m/s}}{(9.80 \text{ m/s}^2) \tan 6.9°} = \boxed{1.2 \text{ s}}$$

CHAPTER 5 | DYNAMICS OF UNIFORM CIRCULAR MOTION

1. **REASONING** The speed of the plane is given by Equation 5.1: $v = 2\pi r / T$, where T is the period or the time required for the plane to complete one revolution.

   SOLUTION Solving Equation 5.1 for T we have

   $$T = \frac{2\pi r}{v} = \frac{2\pi (2850 \text{ m})}{110 \text{ m/s}} = \boxed{160 \text{ s}}$$

5. **REASONING** The magnitude a_c of the car's centripetal acceleration is given by Equation 5.2 as $a_c = v^2 / r$, where v is the speed of the car and r is the radius of the track. The radius is $r = 2.6 \times 10^3$ m. The speed can be obtained from Equation 5.1 as the circumference $(2\pi r)$ of the track divided by the period T of the motion. The period is the time for the car to go once around the track ($T = 360$ s).

   SOLUTION Since $a_c = v^2 / r$ and $v = (2\pi r) / T$, the magnitude of the car's centripetal acceleration is

   $$a_c = \frac{v^2}{r} = \frac{\left(\dfrac{2\pi r}{T}\right)^2}{r} = \frac{4\pi^2 r}{T^2} = \frac{4\pi^2 \left(2.6 \times 10^3 \text{ m}\right)}{(360 \text{ s})^2} = \boxed{0.79 \text{ m/s}^2}$$

9. **REASONING AND SOLUTION** Since the magnitude of the centripetal acceleration is given by Equation 5.2, $a_C = v^2 / r$, we can solve for r and find that

   $$r = \frac{v^2}{a_C} = \frac{(98.8 \text{ m/s})^2}{3.00(9.80 \text{ m/s}^2)} = \boxed{332 \text{ m}}$$

15. **REASONING AND SOLUTION** The magnitude of the centripetal force on the ball is given by Equation 5.3: $F_C = mv^2 / r$. Solving for v, we have

    $$v = \sqrt{\frac{F_c r}{m}} = \sqrt{\frac{(0.028 \text{ N})(0.25 \text{ m})}{0.015 \text{ kg}}} = \boxed{0.68 \text{ m/s}}$$

33. **REASONING** Equation 5.5 gives the orbital speed for a satellite in a circular orbit around the earth. It can be modified to determine the orbital speed around any planet **P** by replacing the mass of the earth M_E by the mass of the planet M_P: $v = \sqrt{GM_P / r}$.

SOLUTION The ratio of the orbital speeds is, therefore,

$$\frac{v_2}{v_1} = \frac{\sqrt{GM_P / r_2}}{\sqrt{GM_P / r_1}} = \sqrt{\frac{r_1}{r_2}}$$

Solving for v_2 gives

$$v_2 = v_1 \sqrt{\frac{r_1}{r_2}} = (1.70 \times 10^4 \text{ m/s}) \sqrt{\frac{5.25 \times 10^6 \text{ m}}{8.60 \times 10^6 \text{ m}}} = \boxed{1.33 \times 10^4 \text{ m/s}}$$

---

37. **REASONING** Equation 5.2 for the centripetal acceleration applies to both the plane and the satellite, and the centripetal acceleration is the same for each. Thus, we have

$$a_c = \frac{v_{plane}^2}{r_{plane}} = \frac{v_{satellite}^2}{r_{satellite}} \quad \text{or} \quad v_{plane} = \left(\sqrt{\frac{r_{plane}}{r_{satellite}}} \right) v_{satellite}$$

The speed of the satellite can be obtained directly from Equation 5.5.

SOLUTION Using Equation 5.5, we can express the speed of the satellite as

$$v_{satellite} = \sqrt{\frac{Gm_E}{r_{satellite}}}$$

Substituting this expression into the expression obtained in the reasoning for the speed of the plane gives

$$v_{plane} = \left(\sqrt{\frac{r_{plane}}{r_{satellite}}} \right) v_{satellite} = \left(\sqrt{\frac{r_{plane}}{r_{satellite}}} \right) \sqrt{\frac{Gm_E}{r_{satellite}}} = \frac{\sqrt{r_{plane}} \sqrt{Gm_E}}{r_{satellite}}$$

$$v_{plane} = \frac{\sqrt{(15 \text{ m})(6.67 \times 10^{-11} \text{ N} \cdot \text{m}^2 / \text{kg}^2)(5.98 \times 10^{24} \text{ kg})}}{6.7 \times 10^6 \text{ m}} = \boxed{12 \text{ m/s}}$$

41. ***REASONING*** According to Equation 5.3, the magnitude F_c of the centripetal force that acts on each passenger is $F_c = mv^2/r$, where m and v are the mass and speed of a passenger and r is the radius of the turn. From this relation we see that the speed is given by $v = \sqrt{F_c r/m}$. The centripetal force is the net force required to keep each passenger moving on the circular path and points toward the center of the circle. With the aid of a free-body diagram, we will evaluate the net force and, hence, determine the speed.

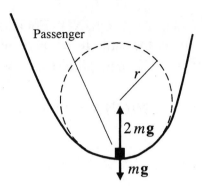

SOLUTION The free-body diagram shows a passenger at the bottom of the circular dip. There are two forces acting: her downward-acting weight $m\mathbf{g}$ and the upward-acting force $2m\mathbf{g}$ that the seat exerts on her. The net force is $+2mg - mg = +mg$, where we have taken "up" as the positive direction. Thus, $F_c = mg$. The speed of the passenger can be found by using this result in the equation above.

Substituting $F_c = mg$ into the relation $v = \sqrt{F_c r/m}$ yields

$$v = \sqrt{\frac{F_c r}{m}} = \sqrt{\frac{(mg)r}{m}} = \sqrt{gr} = \sqrt{(9.80 \text{ m/s}^2)(20.0 \text{ m})} = \boxed{14.0 \text{ m/s}}$$

---

43. ***REASONING*** The centripetal force is the name given to the net force pointing toward the center of the circular path. At point 3 at the top the net force pointing toward the center of the circle consists of the normal force and the weight, both pointing toward the center. At point 1 at the bottom the net force consists of the normal force pointing upward toward the center and the weight pointing downward or away from the center. In either case the centripetal force is given by Equation 5.3 as $F_c = mv^2/r$.

SOLUTION At point 3 we have

$$F_c = F_N + mg = \frac{mv_3^2}{r}$$

At point 1 we have

$$F_c = F_N - mg = \frac{mv_1^2}{r}$$

Subtracting the second equation from the first gives

$$2mg = \frac{mv_3^2}{r} - \frac{mv_1^2}{r}$$

Rearranging gives

$$v_3^2 = 2gr + v_1^2$$

Thus, we find that

$$v_3 = \sqrt{2(9.80 \text{ m/s}^2)(3.0 \text{ m}) + (15 \text{ m/s})^2} = \boxed{17 \text{ m/s}}$$

---

49. **REASONING**   In Example 3, it was shown that the magnitudes of the centripetal acceleration for the two cases are

$$[\text{Radius} = 33 \text{ m}] \qquad\qquad a_C = 35 \text{ m/s}^2$$
$$[\text{Radius} = 24 \text{ m}] \qquad\qquad a_C = 48 \text{ m/s}^2$$

According to Newton's second law, the centripetal force is $F_C = ma_C$ (see Equation 5.3).

SOLUTION a. Therefore, when the sled undergoes the turn of radius 33 m,

$$F_C = ma_C = (350 \text{ kg})(35 \text{ m/s}^2) = \boxed{1.2 \times 10^4 \text{ N}}$$

b. Similarly, when the radius of the turn is 24 m,

$$F_C = ma_C = (350 \text{ kg})(48 \text{ m/s}^2) = \boxed{1.7 \times 10^4 \text{ N}}$$

---

55. **REASONING**   As the motorcycle passes over the top of the hill, it will experience a centripetal force, the magnitude of which is given by Equation 5.3: $F_C = mv^2/r$. The centripetal force is provided by the net force on the cycle + driver system. At that instant, the net force on the system is composed of the normal force, which points upward, and the weight, which points downward. Taking the direction toward the center of the circle (downward) as the positive direction, we have $F_C = mg - F_N$. This expression can be solved for F_N, the normal force.

SOLUTION
a. The magnitude of the centripetal force is

$$F_C = \frac{mv^2}{r} = \frac{(342 \text{ kg})(25.0 \text{ m/s})^2}{126 \text{ m}} = \boxed{1.70 \times 10^3 \text{ N}}$$

b. The magnitude of the normal force is

$$F_N = mg - F_C = (342 \text{ kg})(9.80 \text{ m/s}^2) - 1.70 \times 10^3 \text{ N} = \boxed{1.66 \times 10^3 \text{ N}}$$

57. **REASONING AND SOLUTION** The centripetal acceleration for any point on the blade a distance r from center of the circle, according to Equation 5.2, is $a_c = v^2/r$. From Equation 5.1, we know that $v = 2\pi r/T$ where T is the period of the motion. Combining these two equations, we obtain

$$a_c = \frac{(2\pi r/T)^2}{r} = \frac{4\pi^2 r}{T^2}$$

a. Since the turbine blades rotate at 617 rev/s, all points on the blades rotate with a period of $T = (1/617)\,\text{s} = 1.62 \times 10^{-3}$ s. Therefore, for a point with $r = 0.020$ m, the magnitude of the centripetal acceleration is

$$a_c = \frac{4\pi^2 (0.020 \text{ m})}{(1.62 \times 10^{-3} \text{ s})^2} = \boxed{3.0 \times 10^5 \text{ m/s}^2}$$

b. Expressed as a multiple of g, this centripetal acceleration is

$$a_c = \left(3.0 \times 10^5 \text{ m/s}^2\right) \left(\frac{1.00 \text{ g}}{9.80 \text{ m/s}^2}\right) = \boxed{3.1 \times 10^4 \text{ g}}$$

---

59. **REASONING** Let v_0 be the initial speed of the ball as it begins its projectile motion. Then, the centripetal force is given by Equation 5.3: $F_C = mv_0^2/r$. We are given the values for m and r; however, we must determine the value of v_0 from the details of the projectile motion after the ball is released.

In the absence of air resistance, the x component of the projectile motion has zero acceleration, while the y component of the motion is subject to the acceleration due to gravity. The horizontal distance traveled by the ball is given by Equation 3.5a (with $a_x = 0 \text{ m/s}^2$):

$$x = v_{0x}t = (v_0 \cos\theta)t$$

with t equal to the flight time of the ball while it exhibits projectile motion. The time t can be found by considering the vertical motion. From Equation 3.3b,

$$v_y = v_{0y} + a_y t$$

After a time t, $v_y = -v_{0y}$. Assuming that up and to the right are the positive directions, we have

$$t = \frac{-2v_{0y}}{a_y} = \frac{-2v_0 \sin\theta}{a_y}$$

and

$$x = (v_0 \cos\theta)\left(\frac{-2v_0 \sin\theta}{a_y}\right)$$

Using the fact that $2\sin\theta\cos\theta = \sin 2\theta$, we have

$$x = -\frac{2v_0^2 \cos\theta \sin\theta}{a_y} = -\frac{v_0^2 \sin 2\theta}{a_y} \qquad (1)$$

Equation (1) (with upward and to the right chosen as the positive directions) can be used to determine the speed v_0 with which the ball begins its projectile motion. Then Equation 5.3 can be used to find the centripetal force.

SOLUTION   Solving equation (1) for v_0, we have

$$v_0 = \sqrt{\frac{-x\,a_y}{\sin 2\theta}} = \sqrt{\frac{-(86.75 \text{ m})(-9.80 \text{ m/s}^2)}{\sin 2(41°)}} = 29.3 \text{ m/s}$$

Then, from Equation 5.3,

$$F_C = \frac{mv_0^2}{r} = \frac{(7.3 \text{ kg})(29.3 \text{ m/s})^2}{1.8 \text{ m}} = \boxed{3500 \text{ N}}$$

---

61. **REASONING**   If the effects of gravity are not ignored in Example 5, the plane will make an angle θ with the vertical as shown in figure **A** below. The figure **B** shows the forces that act on the plane, and figure **C** shows the horizontal and vertical components of these forces.

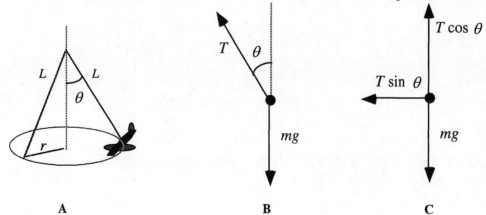

A   **B**   **C**

From figure **C** we see that the resultant force in the horizontal direction is the horizontal component of the tension in the guideline and provides the centripetal force. Therefore,

$$T\sin\theta = \frac{mv^2}{r}$$

From figure **A**, the radius r is related to the length L of the guideline by $r = L \sin\theta$; therefore,

$$T \sin \theta = \frac{mv^2}{L \sin \theta} \tag{1}$$

The resultant force in the vertical direction is zero: $T\cos\theta - mg = 0$, so that

$$T\cos\theta = mg \tag{2}$$

From equation (2) we have

$$T = \frac{mg}{\cos \theta} \tag{3}$$

Equation (3) contains two unknown, T and θ. First we will solve equations (1) and (3) simultaneously to determine the value(s) of the angle θ. Once θ is known, we can calculate the tension using equation (3).

SOLUTION Substituting equation (3) into equation (1):

$$\left(\frac{mg}{\cos \theta}\right) \sin \theta = \frac{mv^2}{L \sin \theta}$$

Thus,

$$\frac{\sin^2 \theta}{\cos \theta} = \frac{v^2}{gL} \tag{4}$$

Using the fact that $\cos^2 \theta + \sin^2 \theta = 1$, equation (4) can be written

$$\frac{1 - \cos^2 \theta}{\cos \theta} = \frac{v^2}{gL}$$

or

$$\frac{1}{\cos \theta} - \cos \theta = \frac{v^2}{gL}$$

This can be put in the form of an equation that is quadratic in $\cos \theta$. Multiplying both sides by $\cos \theta$ and rearranging yields:

$$\cos^2 \theta + \frac{v^2}{gL}\cos \theta - 1 = 0 \tag{5}$$

Equation (5) is of the form

$$ax^2 + bx + c = 0 \tag{6}$$

with $x = \cos \theta$, $a = 1$, $b = v^2/(gL)$, and $c = -1$. The solution to equation (6) is found from the quadratic formula:

$$x = \frac{-b \pm \sqrt{b^2 - 4ac}}{2a}$$

When $v = 19.0$ m/s, $b = 2.17$. The positive root from the quadratic formula gives $x = \cos \theta = 0.391$. Substitution into equation (3) yields

$$T = \frac{mg}{\cos \theta} = \frac{(0.900 \text{ kg})(9.80 \text{ m}/\text{s}^2)}{0.391} = \boxed{23 \text{ N}}$$

When $v = 38.0$ m/s, $b = 8.67$. The positive root from the quadratic formula gives $x = \cos \theta = 0.114$. Substitution into equation (3) yields

$$T = \frac{mg}{\cos \theta} = \frac{(0.900 \text{ kg})(9.80 \text{ m}/\text{s}^2)}{0.114} = \boxed{77 \text{ N}}$$

CHAPTER 6 | *WORK AND ENERGY*

1. **REASONING** The work W done by the tension in the tow rope is given by Equation 6.1 as $W = (F\cos\theta)s$, where F is the magnitude of the tension, s is the magnitude of the skier's displacement, and θ is the angle between the tension and the displacement vectors. The magnitude of the displacement (or the distance) is the speed v of the skier multiplied by the time t (see Equation 2.1), or $s = vt$.

   SOLUTION Substituting $s = vt$ into the expression for the work, $W = (F\cos\theta)s$, we have $W = (F\cos\theta)vt$. Since the skier moves parallel to the boat and since the tow rope is parallel to the water, the angle between the tension and the skier's displacement is $\theta = 37.0°$. Thus, the work done by the tension is

$$W = (F\cos\theta)vt = [(135 \text{ N})\cos 37.0°](9.30 \text{ m/s})(12.0 \text{ s}) = \boxed{1.20\times10^4\text{ J}}$$

---

5. **REASONING AND SOLUTION** Solving Equation 6.1 for the angle θ, we obtain

$$\theta = \cos^{-1}\left(\frac{W}{Fs}\right) = \cos^{-1}\left[\frac{1.10\times10^3 \text{ J}}{(30.0 \text{ N})(50.0 \text{ m})}\right] = \boxed{42.8°}$$

---

9. **REASONING AND SOLUTION**
   a. According to Equation 6.1, the work done by the applied force is

$$W = Fs\cos\theta = (2.40 \times 10^2 \text{ N})(8.00 \text{ m})\cos 20.0° = \boxed{1.80 \times 10^3 \text{ J}}$$

   b. According to Equation 6.1, the work done by the frictional force is $W_f = f_k s \cos\theta$, where

$$f_k = \mu_s(mg - F\sin\theta)$$

$$= (0.200)\left[(85.0 \text{ kg})(9.80 \text{ m/s}^2) - (2.40\times10^2 \text{ N})(\sin 20.0°)\right] = 1.50\times10^2 \text{ N}$$

   Therefore,

$$W_f = (1.50 \times 10^2 \text{ N})(8.00 \text{ m})\cos 180° = \boxed{-1.20 \times 10^3 \text{ J}}$$

---

15. ***REASONING AND SOLUTION*** The work done on the arrow by the bow is given by

$$W = Fs \cos 0° = Fs$$

This work is converted into kinetic energy according to the work energy theorem.

$$W = \tfrac{1}{2} m v_{\rm f}^2 - \tfrac{1}{2} m v_0^2$$

Solving for $v_{\rm f}$, we find that

$$v_{\rm f} = \sqrt{\frac{2W}{m} + v_0^2} = \sqrt{\frac{2Fs}{m} + v_0^2} = \sqrt{\frac{2(65\ {\rm N})(0.90\ {\rm m})}{75 \times 10^{-3}\ {\rm kg}} + (0\ {\rm m/s})^2} = \boxed{39\ {\rm m/s}}$$

---

19. ***REASONING*** The work done to launch either object can be found from Equation 6.3, the work-energy theorem, $W = {\rm KE}_{\rm f} - {\rm KE}_0 = \tfrac{1}{2} m v_{\rm f}^2 - \tfrac{1}{2} m v_0^2$.

SOLUTION
a. The work required to launch the hammer is

$$W = \tfrac{1}{2} m v_{\rm f}^2 - \tfrac{1}{2} m v_0^2 = \tfrac{1}{2} m \left(v_{\rm f}^2 - v_0^2 \right) = \tfrac{1}{2}(7.3\ {\rm kg})\left[(29\ {\rm m/s})^2 - (0\ {\rm m/s})^2\right] = \boxed{3.1 \times 10^3\ {\rm J}}$$

b. Similarly, the work required to launch the bullet is

$$W = \tfrac{1}{2} m \left(v_{\rm f}^2 - v_0^2 \right) = \tfrac{1}{2}(0.0026\ {\rm kg})\left[(410\ {\rm m/s})^2 - (0\ {\rm m/s})^2\right] = \boxed{2.2 \times 10^2\ {\rm J}}$$

---

23. ***REASONING*** According to the work-energy theorem, the kinetic energy of the sled increases in each case because work is done on the sled. The work-energy theorem is given by Equation 6.3: $W = {\rm KE}_{\rm f} - {\rm KE}_0 = \tfrac{1}{2} m v_{\rm f}^2 - \tfrac{1}{2} m v_0^2$. The work done on the sled is given by Equation 6.1: $W = (F \cos \theta)s$. The work done in each case can, therefore, be expressed as

$$W_1 = (F \cos 0°)s = \tfrac{1}{2} m v_{\rm f}^2 - \tfrac{1}{2} m v_0^2 = \Delta {\rm KE}_1$$

and

$$W_2 = (F \cos 62°)s = \tfrac{1}{2} m v_{\rm f}^2 - \tfrac{1}{2} m v_0^2 = \Delta {\rm KE}_2$$

The fractional increase in the kinetic energy of the sled when $\theta = 0°$ is

$$\frac{\Delta {\rm KE}_1}{{\rm KE}_0} = \frac{(F \cos 0°)s}{{\rm KE}_0} = 0.38$$

Therefore,

$$Fs = (0.38)\,\mathrm{KE}_0 \qquad (1)$$

The fractional increase in the kinetic energy of the sled when $\theta = 62°$ is

$$\frac{\Delta\mathrm{KE}_2}{\mathrm{KE}_0} = \frac{(F\cos 62°)s}{\mathrm{KE}_0} = \frac{Fs}{\mathrm{KE}_0}(\cos 62°) \qquad (2)$$

Equation (1) can be used to substitute for Fs in Equation (2).

SOLUTION Combining Equations (1) and (2), we have

$$\frac{\Delta\mathrm{KE}_2}{\mathrm{KE}_0} = \frac{Fs}{\mathrm{KE}_0}(\cos 62°) = \frac{(0.38)\,\mathrm{KE}_0}{\mathrm{KE}_0}(\cos 62°) = (0.38)(\cos 62°) = 0.18$$

Thus, the sled's kinetic energy would increase by $\boxed{18\ \%}$.

---

25. **REASONING** When the satellite goes from the first to the second orbit, its kinetic energy changes. The net work that the external force must do to change the orbit can be found from the work-energy theorem: $W = \mathrm{KE}_f - \mathrm{KE}_0 = \frac{1}{2}mv_f^2 - \frac{1}{2}mv_0^2$. The speeds v_f and v_0 can be obtained from Equation 5.5 for the speed of a satellite in a circular orbit of radius r. Given the speeds, the work energy theorem can be used to obtain the work.

SOLUTION According to Equation 5.5, $v = \sqrt{GM_E/r}$. Substituting into the work-energy theorem, we have

$$W = \frac{1}{2}mv_f^2 - \frac{1}{2}mv_0^2 = \frac{1}{2}m\left(v_f^2 - v_0^2\right) = \frac{1}{2}m\left[\left(\sqrt{\frac{GM_E}{r_f}}\right)^2 - \left(\sqrt{\frac{GM_E}{r_0}}\right)^2\right] = \frac{GM_E m}{2}\left(\frac{1}{r_f} - \frac{1}{r_0}\right)$$

Therefore,

$$W = \frac{(6.67\times 10^{-11}\ \mathrm{N\cdot m^2/kg^2})(5.98\times 10^{24}\ \mathrm{kg})(6200\ \mathrm{kg})}{2}$$

$$\times\left(\frac{1}{7.0\times 10^6\ \mathrm{m}} - \frac{1}{3.3\times 10^7\ \mathrm{m}}\right) = \boxed{1.4\times 10^{11}\ \mathrm{J}}$$

31. ***REASONING*** During each portion of the trip, the work done by the resistive force is given by Equation 6.1, $W = (F \cos \theta)s$. Since the resistive force always points opposite to the displacement of the bicyclist, $\theta = 180°$; hence, on each part of the trip, $W = (F \cos 180°)s = -Fs$. The work done by the resistive force during the round trip is the algebraic sum of the work done during each portion of the trip.

SOLUTION
a. The work done during the round trip is, therefore,

$$W_{total} = W_1 + W_2 = -F_1 s_1 - F_2 s_2$$

$$= -(3.0 \text{ N})(5.0 \times 10^3 \text{ m}) - (3.0 \text{ N})(5.0 \times 10^3 \text{ m}) = \boxed{-3.0 \times 10^4 \text{ J}}$$

b. Since the work done by the resistive force over the closed path is *not* zero, we can conclude that $\boxed{\text{the resistive force is } not \text{ a conservative force}}$.

---

35. ***REASONING AND SOLUTION***
a. The work done by non-conservative forces is given by Equation 6.7b as

$$W_{nc} = \Delta KE + \Delta PE \qquad \text{so} \qquad \Delta PE = W_{nc} - \Delta KE$$

Now

$$\Delta KE = \tfrac{1}{2}mv_f^2 - \tfrac{1}{2}mv_0^2 = \tfrac{1}{2}(55.0 \text{ kg})[(6.00 \text{ m/s})^2 - (1.80 \text{ m/s})^2] = 901 \text{ J}$$

and

$$\Delta PE = 80.0 \text{ J} - 265 \text{ J} - 901 \text{ J} = \boxed{-1086 \text{ J}}$$

b. $\Delta PE = mg(h - h_0)$ so

$$h - h_0 = \frac{\Delta PE}{mg} = \frac{-1086 \text{ J}}{(55.0 \text{ kg})(9.80 \text{ m/s}^2)} = -2.01 \text{ m}$$

This answer is negative, so that the skater's vertical position has changed by $\boxed{2.01 \text{ m}}$ and the skater is $\boxed{\text{below the starting point}}$.

---

37. ***REASONING*** No forces other than gravity act on the rock since air resistance is being ignored. Thus, the net work done by nonconservative forces is zero, $W_{nc} = 0$ J. Consequently, the principle of conservation of mechanical energy holds, so the total mechanical energy remains constant as the rock falls.

If we take $h = 0$ m at ground level, the gravitational potential energy at any height h is, according to Equation 6.5, $PE = mgh$. The kinetic energy of the rock is given by Equation 6.2: $KE = \frac{1}{2}mv^2$. In order to use Equation 6.2, we must have a value for v^2 at each desired height h. The quantity v^2 can be found from Equation 2.9 with $v_0 = 0$ m/s, since the rock is released from rest. At a height h, the rock has fallen through a distance $(20.0 \text{ m}) - h$, and according to Equation 2.9, $v^2 = 2ay = 2a[(20.0 \text{ m}) - h]$. Therefore, the kinetic energy at any height h is given by $KE = ma[(20.0 \text{ m}) - h]$. The total energy at any height h is the sum of the kinetic energy and potential energy at the particular height.

SOLUTION The calculations are performed below for $h = 10.0$ m. The table that follows also shows the results for $h = 20.0$ m and $h = 0$ m.

$$PE = mgh = (2.00 \text{ kg})(9.80 \text{ m/s}^2)(10.0 \text{ m}) = \boxed{196 \text{ J}}$$

$$KE = ma[(20.0 \text{ m}) - h] = (2.00 \text{ kg})(9.80 \text{ m/s}^2)(20.0 \text{ m} - 10.0 \text{ m}) = \boxed{196 \text{ J}}$$

$$E = KE + PE = 196 \text{ J} + 196 \text{ J} = \boxed{392 \text{ J}}$$

h (m)	KE (J)	PE (J)	E (J)
20.0	0	392	392
10.0	196	196	392
0	392	0	392

Note that in each case, the value of E is the same, because mechanical energy is conserved.

---

45. **REASONING** Friction and air resistance are being ignored. The normal force from the slide is perpendicular to the motion, so it does no work. Thus, no net work is done by nonconservative forces, and the principle of conservation of mechanical energy applies.

SOLUTION Applying the principle of conservation of mechanical energy to the swimmer at the top and the bottom of the slide, we have

$$\underbrace{\frac{1}{2}mv_f^2 + mgh_f}_{E_f} = \underbrace{\frac{1}{2}mv_0^2 + mgh_0}_{E_0}$$

If we let h be the height of the bottom of the slide above the water, $h_f = h$, and $h_0 = H$. Since the swimmer starts from rest, $v_0 = 0$ m/s, and the above expression becomes

$$\frac{1}{2}v_f^2 + gh = gH$$

Solving for H, we obtain

$$H = h + \frac{v_f^2}{2g}$$

Before we can calculate H, we must find v_f and h. Since the velocity in the horizontal direction is constant,

$$v_f = \frac{\Delta x}{\Delta t} = \frac{5.00 \text{ m}}{0.500 \text{ s}} = 10.0 \text{ m/s}$$

The vertical displacement of the swimmer after leaving the slide is, from Equation 3.5b (with down being negative),

$$y = \frac{1}{2} a_y t^2 = \frac{1}{2}(-9.80 \text{ m/s}^2)(0.500 \text{ s})^2 = -1.23 \text{ m}$$

Therefore, $h = 1.23$ m. Using these values of v_f and h in the above expression for H, we find

$$H = h + \frac{v_f^2}{2g} = 1.23 \text{ m} + \frac{(10.0 \text{ m/s})^2}{2(9.80 \text{ m/s}^2)} = \boxed{6.33 \text{ m}}$$

---

53. ***REASONING AND SOLUTION*** The work-energy theorem can be used to determine the change in kinetic energy of the car according to Equation 6.8:

$$W_{nc} = \underbrace{\left(\tfrac{1}{2} mv_f^2 + mgh_f\right)}_{E_f} - \underbrace{\left(\tfrac{1}{2} mv_0^2 + mgh_0\right)}_{E_0}$$

The nonconservative forces are the forces of friction and the force due to the chain mechanism. Thus, we can rewrite Equation 6.8 as

$$W_{friction} + W_{chain} = \left(KE_f + mgh_f\right) - \left(KE_0 + mgh_0\right)$$

We will measure the heights from ground level. Solving for the change in kinetic energy of the car, we have

$$KE_f - KE_0 = W_{friction} + W_{chain} - mgh_f + mgh_0$$

$$= -2.00 \times 10^4 \text{ J} + 3.00 \times 10^4 \text{ J} - (375 \text{ kg})(9.80 \text{ m/s}^2)(20.0 \text{ m})$$

$$+ (375 \text{ kg})(9.80 \text{ m/s}^2)(5.00 \text{ m}) = \boxed{-4.51 \times 10^4 \text{ J}}$$

55. **REASONING** The work-energy theorem can be used to determine the net work done on the car by the nonconservative forces of friction and air resistance. The work-energy theorem is, according to Equation 6.8,

$$W_{nc} = \left(\tfrac{1}{2} m v_f^2 + m g h_f \right) - \left(\tfrac{1}{2} m v_0^2 + m g h_0 \right)$$

The nonconservative forces are the force of friction, the force of air resistance, and the force provided by the engine. Thus, we can rewrite Equation 6.8 as

$$W_{friction} + W_{air} + W_{engine} = \left(\tfrac{1}{2} m v_f^2 + m g h_f \right) - \left(\tfrac{1}{2} m v_0^2 + m g h_0 \right)$$

This expression can be solved for $W_{friction} + W_{air}$.

SOLUTION We will measure the heights from sea level, where $h_0 = 0$ m. Since the car starts from rest, $v_0 = 0$ m/s. Thus, we have

$$W_{friction} + W_{air} = m \left(\tfrac{1}{2} v_f^2 + g h_f \right) - W_{engine}$$

$$= (1.50 \times 10^3 \text{ kg}) \left[\tfrac{1}{2} (27.0 \text{ m/s})^2 + (9.80 \text{ m/s}^2)(2.00 \times 10^2 \text{ m}) \right] - (4.70 \times 10^6 \text{ J})$$

$$= \boxed{-1.21 \times 10^6 \text{ J}}$$

---

61. **REASONING** After the wheels lock, the only nonconservative force acting on the truck is friction. The work done by this conservative force can be determined from the work-energy theorem. According to Equation 6.8,

$$W_{nc} = E_f - E_0 = \left(\tfrac{1}{2} m v_f^2 + m g h_f \right) - \left(\tfrac{1}{2} m v_0^2 + m g h_0 \right) \qquad (1)$$

where W_{nc} is the work done by friction. According to Equation 6.1, $W = (F \cos\theta)s$; since the force of kinetic friction points opposite to the displacement, $\theta = 180°$. According to Equation 4.8, the kinetic frictional force has a magnitude of $f_k = \mu_k F_N$, where μ_k is the coefficient of kinetic friction and F_N is the magnitude of the normal force. Thus,

$$W_{nc} = (f_k \cos\theta)s = \mu_k F_N (\cos 180°)s = -\mu_k F_N s \qquad (2)$$

Since the truck is sliding down an incline, we refer to the free-body diagram in order to determine the magnitude of the normal force F_N. The free-body diagram at the right shows the three forces that act on the truck. They are the normal force $\mathbf{F_N}$, the force of kinetic friction $\mathbf{f_k}$, and the weight $m\mathbf{g}$. The weight has been resolved into its components, and these vectors are shown as dashed arrows. From the free body diagram and the fact that the truck does not accelerate perpendicular to the incline, we can see that the magnitude of the normal force is given by

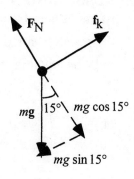

$$F_N = mg\cos 15.0°$$

Therefore, Equation (2) becomes

$$W_{nc} = -\mu_k\left(mg\cos 15.0°\right)s \qquad (3)$$

Combining Equations (1), (2), and (3), we have

$$-\mu_k\left(mg\cos 15.0°\right)s = \left(\tfrac{1}{2}mv_f^2 + mgh_f\right) - \left(\tfrac{1}{2}mv_0^2 + mgh_0\right) \qquad (4)$$

This expression may be solved for s.

SOLUTION When the car comes to a stop, $v_f = 0$ m/s. If we take $h_f = 0$ m at ground level, then, from the drawing at the right, we see that $h_0 = s\sin 15.0°$. Equation (4) then gives

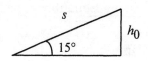

$$s = \frac{v_0^2}{2g\left(\mu_k\cos 15.0° - \sin 15.0°\right)} = \frac{\left(11.1\text{ m/s}\right)^2}{2\left(9.80\text{ m/s}^2\right)\left[\left(0.750\right)\cos 15.0° - \sin 15.0°\right]} = \boxed{13.5\text{ m}}$$

---

63. **REASONING AND SOLUTION** One kilowatt·hour is the amount of work or energy generated when one kilowatt of power is supplied for a time of one hour. From Equation 6.10a, we know that $W = \overline{P}t$. Using the fact that $1\text{ kW} = 1.0 \times 10^3$ J/s and that $1\text{h} = 3600$ s, we have

$$1.0\text{ kWh} = \left(1.0 \times 10^3\text{ J/s}\right)\left(1\text{ h}\right) = \left(1.0 \times 10^3\text{ J/s}\right)\left(3600\text{ s}\right) = \boxed{3.6 \times 10^6\text{ J}}$$

---

69. ***REASONING AND SOLUTION*** In the drawings below, the positive direction is to the right. When the boat is not pulling a skier, the engine must provide a force F_1 to overcome the resistive force of the water F_R. Since the boat travels with constant speed, these forces must be equal in magnitude and opposite in direction.

$$-F_R + F_1 = ma = 0$$

or

$$F_R = F_1 \qquad (1)$$

When the boat is pulling a skier, the engine must provide a force F_2 to balance the resistive force of the water F_R and the tension in the tow rope T.

$$-F_R + F_2 - T = ma = 0$$

or

$$F_2 = F_R + T \qquad (2)$$

The average power is given by $\overline{P} = F\overline{v}$, according to Equation 6.11 in the text. Since the boat moves with the same speed in both cases, $v_1 = v_2$, and we have

$$\frac{\overline{P}_1}{F_1} = \frac{\overline{P}_2}{F_2} \qquad \text{or} \qquad F_2 = F_1 \frac{\overline{P}_2}{\overline{P}_1}$$

Using Equations (1) and (2), this becomes

$$F_1 + T = F_1 \frac{\overline{P}_2}{\overline{P}_1}$$

Solving for T gives

$$T = F_1 \left(\frac{\overline{P}_2}{\overline{P}_1} - 1 \right)$$

The force F_1 can be determined from $\overline{P}_1 = F_1 v_1$, thereby giving

$$T = \frac{\overline{P}_1}{v_1} \left(\frac{\overline{P}_2}{\overline{P}_1} - 1 \right) = \frac{7.50 \times 10^4 \text{ W}}{12 \text{ m/s}} \left(\frac{8.30 \times 10^4 \text{ W}}{7.50 \times 10^4 \text{ W}} - 1 \right) = \boxed{6.7 \times 10^2 \text{ N}}$$

71. **REASONING** The area under the force-versus-displacement graph over any displacement interval is equal to the work done over that displacement interval.

SOLUTION
a. Since the area under the force-versus-displacement graph from $s = 0$ to 0.50 m is greater for bow 1, it follows that $\boxed{\text{bow 1 requires more work to draw the bow fully}}$.

b. The work required to draw bow 1 is equal to the area of the triangular region under the force-versus-displacement curve for bow 1. Since the area of a triangle is equal one-half times the base of the triangle times the height of the triangle, we have

$$W_1 = \frac{1}{2}(0.50\text{m})(350 \text{ N}) = 88 \text{ J}$$

For bow 2, we note that each small square under the force-versus-displacement graph has an area of

$$(0.050 \text{ m})(40.0 \text{ N}) = 2.0 \text{ J}$$

We estimate that there are approximately 31.3 squares under the force-versus-displacement graph for bow 2; therefore, the total work done is

$$(31.3 \text{ squares})\left(\frac{2.0 \text{ J}}{\text{square}}\right) = 63 \text{ J}$$

Therefore, the additional work required to stretch bow 1 as compared to bow 2 is

$$88 \text{ J} - 63 \text{ J} = \boxed{25 \text{ J}}$$

---

77. **REASONING** The only two forces that act on the gymnast are his weight and the force exerted on his hands by the high bar. The latter is the (non-conservative) reaction force to the force exerted on the bar by the gymnast, as predicted by Newton's third law. This force, however, does no work because it points perpendicular to the circular path of motion. Thus, $W_{\text{nc}} = 0$ J, and we can apply the principle of conservation of mechanical energy.

SOLUTION The conservation principle gives

$$\underbrace{\frac{1}{2}mv_{\text{f}}^2 + mgh_{\text{f}}}_{E_{\text{f}}} = \underbrace{\frac{1}{2}mv_0^2 + mgh_0}_{E_0}$$

Since the gymnast's speed is momentarily zero at the top of the swing, $v_0 = 0$ m/s. If we take $h_{\text{f}} = 0$ m at the bottom of the swing, then $h_0 = 2r$, where r is the radius of the circular path followed by the gymnast's waist. Making these substitutions in the above expression and solving for v_{f}, we obtain

$$v_{\text{f}} = \sqrt{2gh_0} = \sqrt{2g(2r)} = \sqrt{2(9.80 \text{ m/s}^2)(2 \times 1.1 \text{ m})} = \boxed{6.6 \text{ m/s}}$$

79. ***REASONING AND SOLUTION*** We will assume that the tug-of-war rope remains parallel to the ground, so that the force that moves team B is in the same direction as the displacement. According to Equation 6.1, the work done by team A is

$$W = (F\cos\theta)s = (1100\text{ N})(\cos 0°)(2.0\text{ m}) = \boxed{2.2 \times 10^3\text{ J}}$$

---

83. ***REASONING AND SOLUTION***
a. The lost mechanical energy is

$$E_{\text{lost}} = E_0 - E_f$$

The ball is dropped from rest, so its initial energy is purely potential. The ball is momentarily at rest at the highest point in its rebound, so its final energy is also purely potential. Then

$$E_{\text{lost}} = mgh_0 - mgh_f = (0.60\text{ kg})(9.80\text{ m/s}^2)\big[(1.05\text{ m}) - (0.57\text{ m})\big] = \boxed{2.8\text{ J}}$$

b. The work done by the player must compensate for this loss of energy.

$$E_{\text{lost}} = \underbrace{(F\cos\theta)s}_{\substack{\text{work done by}\\\text{player}}} \quad\Rightarrow\quad F = \frac{E_{\text{lost}}}{(\cos\theta)s} = \frac{2.8\text{ J}}{(\cos 0°)(0.080\text{ m})} = \boxed{35\text{ N}}$$

---

85. ***REASONING*** Gravity is the only force acting on the vaulters, since friction and air resistance are being ignored. Therefore, the net work done by the nonconservative forces is zero, and the principle of conservation of mechanical energy holds.

SOLUTION Let E_{2f} represent the total mechanical energy of the second vaulter at the ground, and E_{20} represent the total mechanical energy of the second vaulter at the bar. Then, the principle of mechanical energy is written for the second vaulter as

$$\underbrace{\tfrac{1}{2}mv_{2f}^2 + mgh_{2f}}_{E_{2f}} = \underbrace{\tfrac{1}{2}mv_{20}^2 + mgh_{20}}_{E_{20}}$$

Since the mass m of the vaulter appears in every term of the equation, m can be eliminated algebraically. The quantity h_{20} is $h_{20} = h$, where h is the height of the bar. Furthermore, when the vaulter is at ground level, $h_{2f} = 0$ m. Solving for v_{20} we have

$$v_{20} = \sqrt{v_{2f}^2 - 2gh} \tag{1}$$

In order to use Equation (1), we must first determine the height h of the bar. The height h can be determined by applying the principle of conservation of mechanical energy to the first vaulter on the ground and at the bar. Using notation similar to that above, we have

$$\underbrace{\frac{1}{2}mv_{1f}^2 + mgh_{1f}}_{E_{1f}} = \underbrace{\frac{1}{2}mv_{10}^2 + mgh_{10}}_{E_{10}}$$

Where E_{10} corresponds to the total mechanical energy of the first vaulter at the bar. The height of the bar is, therefore,

$$h = h_{10} = \frac{v_{1f}^2 - v_{10}^2}{2g} = \frac{(8.90 \text{ m/s})^2 - (1.00 \text{ m/s})^2}{2(9.80 \text{ m/s}^2)} = 3.99 \text{ m}$$

The speed at which the second vaulter clears the bar is, from Equation (1),

$$v_{20} = \sqrt{v_{2f}^2 - 2gh} = \sqrt{(9.00 \text{ m/s})^2 - 2(9.80 \text{ m/s}^2)(3.99 \text{ m})} = \boxed{1.7 \text{ m/s}}$$

CHAPTER 7 | IMPULSE AND MOMENTUM

1. **REASONING** According to Equation 7.1, the impulse $\mathbf{J}$ produced by an average force $\overline{\mathbf{F}}$ is $\mathbf{J} = \overline{\mathbf{F}}\Delta t$, where Δt is the time interval during which the force acts. We will apply this definition for each of the forces and then set the two impulses equal to one another. The fact that one average force has a magnitude that is three times as large as that of the other average force will then be used to obtain the desired time interval.

   SOLUTION Applying Equation 7.1, we write the impulse of each average force as follows:

   $$\mathbf{J}_1 = \overline{\mathbf{F}}_1\Delta t_1 \quad \text{and} \quad \mathbf{J}_2 = \overline{\mathbf{F}}_2\Delta t_2$$

   But the impulses $\mathbf{J}_1$ and $\mathbf{J}_2$ are the same, so we have that $\overline{\mathbf{F}}_1\Delta t_1 = \overline{\mathbf{F}}_2\Delta t_2$. Writing this result in terms of the magnitudes of the forces gives

   $$\overline{F}_1\Delta t_1 = \overline{F}_2\Delta t_2 \quad \text{or} \quad \Delta t_2 = \left(\frac{\overline{F}_1}{\overline{F}_2}\right)\Delta t_1$$

   The ratio of the force magnitudes is given as $\overline{F}_1 / \overline{F}_2 = 3$, so we find that

   $$\Delta t_2 = \left(\frac{\overline{F}_1}{\overline{F}_2}\right)\Delta t_1 = 3(3.2 \text{ ms}) = \boxed{9.6 \text{ ms}}$$

5. **REASONING** The impulse that the volleyball player applies to the ball can be found from the impulse-momentum theorem, Equation 7.4. Two forces act on the volleyball while it's being spiked: an average force $\overline{\mathbf{F}}$ exerted by the player, and the weight of the ball. As in Example 1, we will assume that $\overline{\mathbf{F}}$ is much greater than the weight of the ball, so the weight can be neglected. Thus, the net average force $(\Sigma\overline{\mathbf{F}})$ is equal to $\overline{\mathbf{F}}$.

   SOLUTION From Equation 7.4, the impulse that the player applies to the volleyball is

   $$\underbrace{\overline{\mathbf{F}}\,\Delta t}_{\text{Impulse}} = \underbrace{m\mathbf{v}_f}_{\substack{\text{Final} \\ \text{momentum}}} - \underbrace{m\mathbf{v}_0}_{\substack{\text{Initial} \\ \text{momentum}}}$$

   $$= m(\mathbf{v}_f - \mathbf{v}_0) = (0.35 \text{ kg})\big[(-21 \text{ m/s}) - (+4.0 \text{ m/s})\big] = \boxed{-8.7 \text{ kg}\cdot\text{m/s}}$$

   The minus sign indicates that the direction of the impulse is the same as that of the final velocity of the ball.

13. **REASONING** The impulse applied to the golf ball by the floor can be found from Equation 7.4, the impulse-momentum theorem: $(\Sigma\overline{\mathbf{F}})\Delta t = m\mathbf{v}_f - m\mathbf{v}_0$. Two forces act on the golf ball, the average force $\overline{\mathbf{F}}$ exerted by the floor, and the weight of the golf ball. Since $\overline{\mathbf{F}}$ is much greater than the weight of the golf ball, the net average force $(\Sigma\overline{\mathbf{F}})$ is equal to $\overline{\mathbf{F}}$.

Only the vertical component of the ball's momentum changes during impact with the floor. In order to use Equation 7.4 directly, we must first find the vertical components of the initial and final velocities. We begin, then, by finding these velocity components.

SOLUTION The figures below show the initial and final velocities of the golf ball.

If we take up as the positive direction, then the vertical components of the initial and final velocities are, respectively, $\mathbf{v}_{0y} = -v_0 \cos 30.0°$ and $\mathbf{v}_{fy} = +v_f \cos 30.0°$. Then, from Equation 7.4 the impulse is

$$\overline{\mathbf{F}}\Delta t = m(\mathbf{v}_{fy} - \mathbf{v}_{0y}) = m\left[(+v_f \cos 30.0°) - (-v_0 \cos 30.0°)\right]$$

Since $v_0 = v_f = 45$ m/s, the impulse applied to the golf ball by the floor is

$$\overline{\mathbf{F}}\Delta t = 2mv_0 \cos 30.0° = 2(0.047 \text{ kg})(45 \text{ m/s})(\cos 30.0°) = \boxed{3.7 \text{ N} \cdot \text{s}}$$

---

19. **REASONING** Let m be Al's mass, which means that Jo's mass is 168 kg − m. Since friction is negligible and since the downward-acting weight of each person is balanced by the upward-acting normal force from the ice, the net external force acting on the two-person system is zero. Therefore, the system is isolated, and the conservation of linear momentum applies. The initial total momentum must be equal to the final total momentum.

SOLUTION Applying the principle of conservation of linear momentum and assuming that the direction in which Al moves is the positive direction, we find

$$\underbrace{m(0 \text{ m/s}) + (168 \text{ kg} - m)(0 \text{ m/s})}_{\text{Initial total momentum}} = \underbrace{m(0.90 \text{ m/s}) + (168 \text{ kg} - m)(-1.2 \text{ m/s})}_{\text{Final total momentum}}$$

Solving this equation for m, we find that

$$0 = m(0.90 \text{ m/s}) - (168 \text{ kg})(1.2 \text{ m/s}) + m(1.2 \text{ m/s})$$

$$m = \frac{(168 \text{ kg})(1.2 \text{ m/s})}{0.90 \text{ m/s} + 1.2 \text{ m/s}} = \boxed{96 \text{ kg}}$$

---

21. **REASONING**   The two-stage rocket constitutes the system. The forces that act to cause the separation during the explosion are, therefore, forces that are internal to the system. Since no external forces act on this system, it is isolated and the principle of conservation of linear momentum applies:

$$\underbrace{m_1 v_{f1} + m_2 v_{f2}}_{\substack{\text{Total momentum} \\ \text{after separation}}} = \underbrace{(m_1 + m_2)v_0}_{\substack{\text{Total momentum} \\ \text{before separation}}}$$

where the subscripts "1" and "2" refer to the lower and upper stages, respectively. This expression can be solved for v_{f1}.

SOLUTION   Solving for v_{f1} gives

$$v_{f1} = \frac{(m_1 + m_2)v_0 - m_2 v_{f2}}{m_1}$$

$$= \frac{[2400 \text{ kg} + 1200 \text{ kg}](4900 \text{ m/s}) - (1200 \text{ kg})(5700 \text{ m/s})}{2400 \text{ kg}} = \boxed{+4500 \text{ m/s}}$$

Since v_{f1} is positive, $\boxed{\text{its direction is the same as the rocket before the explosion}}$.

---

23. **REASONING**   No net external force acts on the plate parallel to the floor; therefore, the component of the momentum of the plate that is parallel to the floor is conserved as the plate breaks and flies apart. Initially, the total momentum parallel to the floor is zero. After the collision with the floor, the component of the total momentum parallel to the floor must remain zero. The drawing in the text shows the pieces in the plane parallel to the floor just after the collision. Clearly, the linear momentum in the plane parallel to the floor has two components; therefore the linear momentum of the plate must be conserved in each of these two mutually perpendicular directions. Using the drawing in the text, with the positive directions taken to be up and to the right, we have

$$[x \text{ direction}] \qquad -m_1 v_1 (\sin 25.0°) + m_2 v_2 (\cos 45.0°) = 0 \qquad (1)$$

$[y \text{ direction}]$ $\qquad m_1 v_1 (\cos 25.0°) + m_2 v_2 (\sin 45.0°) - m_3 v_3 = 0$ $\qquad$ (2)

These equations can be solved simultaneously for the masses m_1 and m_2.

SOLUTION Using the values given in the drawing for the velocities after the plate breaks, we have,

$$-m_1 (3.00 \text{ m/s}) \sin 25.0° + m_2 (1.79 \text{ m/s}) \cos 45.0° = 0 \qquad (1)$$

$$m_1 (3.00 \text{ m/s}) \cos 25.0° + m_2 (1.79 \text{ m/s}) \sin 45.0° - (1.30 \text{ kg})(3.07 \text{ m/s}) = 0 \qquad (2)$$

Subtracting (2) from (1), and noting that $\cos 45.0° = \sin 45.0°$, gives $\boxed{m_1 = 1.00 \text{ kg}}$.

Substituting this value into either (1) or (2) then yields $\boxed{m_2 = 1.00 \text{ kg}}$.

---

27. **REASONING** The cannon and the shell constitute the system. Since no external force hinders the motion of the system after the cannon is unbolted, conservation of linear momentum applies in that case. If we assume that the burning gun powder imparts the same kinetic energy to the system in each case, we have sufficient information to develop a mathematical description of this situation, and solve it for the velocity of the shell fired by the loose cannon.

SOLUTION For the case where the cannon is unbolted, momentum conservation gives

$$\underbrace{m_1 v_{f1} + m_2 v_{f2}}_{\substack{\text{Total momentum} \\ \text{after shell is fired}}} = \underbrace{0}_{\substack{\text{Initial momentum} \\ \text{of system}}} \qquad (1)$$

where the subscripts "1" and "2" refer to the cannon and shell, respectively. In both cases, the burning gun power imparts the same kinetic energy to the system. When the cannon is bolted to the ground, only the shell moves and the kinetic energy imparted to the system is

$$KE = \tfrac{1}{2} m_{\text{shell}} v_{\text{shell}}^2 = \tfrac{1}{2}(85.0 \text{ kg})(551 \text{ m/s})^2 = 1.29 \times 10^7 \text{ J}$$

The kinetic energy imparted to the system when the cannon is unbolted has the same value and can be written using the same notation as in equation (1):

$$KE = \tfrac{1}{2} m_1 v_{f1}^2 + \tfrac{1}{2} m_2 v_{f2}^2 \qquad (2)$$

Solving equation (1) for v_{f1}, the velocity of the cannon after the shell is fired, and substituting the resulting expression into Equation (2) gives

$$KE = \frac{m_2^2 v_{f2}^2}{2m_1} + \frac{1}{2}m_2 v_{f2}^2 \qquad (3)$$

Solving equation (3) for v_{f2} gives

$$v_{f2} = \sqrt{\frac{2KE}{m_2\left(\frac{m_2}{m_1}+1\right)}} = \sqrt{\frac{2(1.29 \times 10^7 \text{ J})}{(85.0 \text{ kg})\left(\frac{85.0 \text{ kg}}{5.80 \times 10^3 \text{ kg}}+1\right)}} = \boxed{+547 \text{ m/s}}$$

---

31. ***REASONING*** We obtain the desired percentage in the usual way, as the kinetic energy of the target (with the projectile in it) divided by the projectile's incident kinetic energy, multiplied by a factor of 100. Each kinetic energy is given by Equation 6.2 as $\frac{1}{2}mv^2$, where m and v are mass and speed, respectively. Data for the masses are given, but the speeds are not provided. However, information about the speeds can be obtained by using the principle of conservation of linear momentum.

SOLUTION We define the following quantities:

KE_{TP} = kinetic energy of the target with the projectile in it

KE_{0P} = kinetic energy of the incident projectile

m_P = mass of incident projectile = 0.20 kg

m_T = mass of target = 2.50 kg

v_f = speed at which the target with the projectile in it flies off after being struck

v_{0P} = speed of incident projectile

The desired percentage is

$$\text{Percentage} = \frac{KE_{TP}}{KE_{0P}} \times 100 = \frac{\frac{1}{2}\left(m_T + m_P\right)v_f^2}{\frac{1}{2}m_P v_{0P}^2} \times 100\% \qquad (1)$$

According to the momentum-conservation principle, we have

$$\underbrace{\left(m_T + m_P\right)v_f}_{\substack{\text{Total momentum of target and} \\ \text{projectile after target is struck}}} = \underbrace{0 + m_P v_{0P}}_{\substack{\text{Total momentum of target and} \\ \text{projectile before target is struck}}}$$

Note that the target is stationary before being struck and, hence, has zero initial momentum. Solving for the ratio v_f / v_{0P}, we find that

$$\frac{v_f}{v_{0P}} = \frac{m_P}{m_T + m_P}$$

Substituting this result into Equation (1) gives

$$\text{Percentage} = \frac{\frac{1}{2}\left(m_T + m_P\right)v_f^2}{\frac{1}{2}m_P v_{0P}^2} \times 100\% = \frac{\frac{1}{2}\left(m_T + m_P\right)}{\frac{1}{2}m_P}\left(\frac{m_P}{m_T + m_P}\right)^2 \times 100\%$$

$$= \frac{m_P}{m_T + m_P} \times 100\% = \frac{0.20 \text{ kg}}{2.50 \text{ kg} + 0.20 \text{ kg}} \times 100\% = \boxed{7.4\%}$$

---

33. **REASONING** The system consists of the two balls. The total linear momentum of the two-ball system is conserved because the net external force acting on it is zero. The principle of conservation of linear momentum applies whether or not the collision is elastic.

$$\underbrace{m_1 v_{f1} + m_2 v_{f2}}_{\substack{\text{Total momentum} \\ \text{after collision}}} = \underbrace{m_1 v_{01} + 0}_{\substack{\text{Total momentum} \\ \text{before collision}}}$$

When the collision is elastic, the kinetic energy is also conserved during the collision

$$\underbrace{\tfrac{1}{2}m_1 v_{f1}^2 + \tfrac{1}{2}m_2 v_{f2}^2}_{\substack{\text{Total kinetic energy} \\ \text{after collision}}} = \underbrace{\tfrac{1}{2}m_1 v_{01}^2 + 0}_{\substack{\text{Total kinetic energy} \\ \text{before collision}}}$$

SOLUTION

a. The final velocities for an elastic collision are determined by simultaneously solving the above equations for the final velocities. The procedure is discussed in Example 7 in the text, and leads to Equations 7.8a and 7.8b. According to Equation 7.8:

$$v_{f1} = \left(\frac{m_1 - m_2}{m_1 + m_2}\right)v_{01} \qquad \text{and} \qquad v_{f2} = \left(\frac{2m_1}{m_1 + m_2}\right)v_{01}$$

Let the initial direction of motion of the 5.00-kg ball define the positive direction. Substituting the values given in the text, these equations give

$$\left[5.00\text{-kg ball}\right] \qquad v_{f1} = \left(\frac{5.00 \text{ kg} - 7.50 \text{ kg}}{5.00 \text{ kg} + 7.50 \text{ kg}}\right)(2.00 \text{ m/s}) = \boxed{-0.400 \text{ m/s}}$$

$$\left[7.50\text{-kg ball}\right] \qquad v_{f2} = \left(\frac{2(5.00 \text{ kg})}{5.00 \text{ kg} + 7.50 \text{ kg}}\right)(2.00 \text{ m/s}) = \boxed{+1.60 \text{ m/s}}$$

The signs indicate that, after the collision, the 5.00-kg ball reverses its direction of motion, while the 7.50-kg ball moves in the direction in which the 5.00-kg ball was initially moving.

b. When the collision is completely inelastic, the balls stick together, giving a composite body of mass $m_1 + m_2$ that moves with a velocity v_f. The statement of conservation of linear momentum then becomes

$$\underbrace{(m_1 + m_2)v_f}_{\substack{\text{Total momentum} \\ \text{after collision}}} = \underbrace{m_1 v_{01} + 0}_{\substack{\text{Total momentum} \\ \text{before collision}}}$$

The final velocity of the two balls after the collision is, therefore,

$$v_f = \frac{m_1 v_{01}}{m_1 + m_2} = \frac{(5.00 \text{ kg})(2.00 \text{ m/s})}{5.00 \text{ kg} + 7.50 \text{ kg}} = \boxed{+0.800 \text{ m/s}}$$

---

35. **REASONING** Batman and the boat with the criminal constitute the system. Gravity acts on this system as an external force; however, gravity acts vertically, and we are concerned only with the horizontal motion of the system. If we neglect air resistance and friction, there are no external forces that act horizontally; therefore, the total linear momentum in the horizontal direction is conserved. When Batman collides with the boat, the horizontal component of his velocity is zero, so the statement of conservation of linear momentum in the horizontal direction can be written as

$$\underbrace{(m_1 + m_2)v_f}_{\substack{\text{Total horizontal momentum} \\ \text{after collision}}} = \underbrace{m_1 v_{01} + 0}_{\substack{\text{Total horizontal momentum} \\ \text{before collision}}}$$

Here, m_1 is the mass of the boat, and m_2 is the mass of Batman. This expression can be solved for v_f, the velocity of the boat after Batman lands in it.

SOLUTION Solving for v_f gives

$$v_f = \frac{m_1 v_{01}}{m_1 + m_2} = \frac{(510 \text{ kg})(+11 \text{ m/s})}{510 \text{ kg} + 91 \text{ kg}} = \boxed{+9.3 \text{ m/s}}$$

The plus sign indicates that the boat continues to move in its initial direction of motion.

---

41. **REASONING** The two skaters constitute the system. Since the net external force acting on the system is zero, the total linear momentum of the system is conserved. In the x direction (the east/west direction), conservation of linear momentum gives $P_{fx} = P_{0x}$, or

$$(m_1 + m_2)v_f \cos\theta = m_1 v_{01}$$

Note that since the skaters hold onto each other, they move away with a common velocity v_f. In the y direction, $P_{fy} = P_{0y}$, or

$$(m_1 + m_2)v_f \sin\theta = m_2 v_{02}$$

These equations can be solved simultaneously to obtain both the angle θ and the velocity v_f.

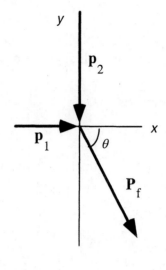

SOLUTION

a. Division of the equations above gives

$$\theta = \tan^{-1}\left(\frac{m_2 v_{02}}{m_1 v_{01}}\right) = \tan^{-1}\left[\frac{(70.0 \text{ kg})(7.00 \text{ m/s})}{(50.0 \text{ kg})(3.00 \text{ m/s})}\right] = \boxed{73.0°}$$

b. Solution of the first of the momentum equations gives

$$v_f = \frac{m_1 v_{01}}{(m_1 + m_2)\cos\theta} = \frac{(50.0 \text{ kg})(3.00 \text{ m/s})}{(50.0 \text{ kg} + 70.0 \text{ kg})(\cos 73.0°)} = \boxed{4.28 \text{ m/s}}$$

---

45. **REASONING** The two balls constitute the system. The tension in the wire is the only nonconservative force that acts on the ball. The tension does no work since it is perpendicular to the displacement of the ball. Since $W_{nc} = 0$ J, the principle of conservation of mechanical energy holds and can be used to find the speed of the 1.50-kg ball just before the collision. Momentum is conserved during the collision, so the principle of conservation of momentum can be used to find the velocities of both balls just after the collision. Once the collision has occurred, energy conservation can be used to determine how high each ball rises.

SOLUTION

a. Applying the principle of energy conservation to the 1.50-kg ball, we have

$$\underbrace{\tfrac{1}{2}mv_f^2 + mgh_f}_{E_f} = \underbrace{\tfrac{1}{2}mv_0^2 + mgh_0}_{E_0}$$

If we measure the heights from the lowest point in the swing, $h_f = 0$ m, and the expression above simplifies to

$$\tfrac{1}{2}mv_f^2 = \tfrac{1}{2}mv_0^2 + mgh_0$$

Solving for v_f, we have

$$v_f = \sqrt{v_0^2 + 2gh_0} = \sqrt{(5.00 \text{ m/s})^2 + 2(9.80 \text{ m/s}^2)(0.300 \text{ m})} = \boxed{5.56 \text{ m/s}}$$

b. If we assume that the collision is elastic, then the velocities of both balls just after the collision can be obtained from Equations 7.8a and 7.8b:

$$v_{f1} = \left(\frac{m_1 - m_2}{m_1 + m_2} \right) v_{01} \qquad \text{and} \qquad v_{f2} = \left(\frac{2m_1}{m_1 + m_2} \right) v_{01}$$

Since v_{01} corresponds to the speed of the 1.50-kg ball just before the collision, it is equal to the quantity v_f calculated in part (a). With the given values of $m_1 = 1.50$ kg and $m_2 = 4.60$ kg, and the value of $v_{01} = 5.56$ m/s obtained in part (a), Equations 7.8a and 7.8b yield the following values:

$$\boxed{v_{f1} = -2.83 \text{ m/s}} \qquad \text{and} \qquad \boxed{v_{f2} = +2.73 \text{ m/s}}$$

The minus sign in v_{f1} indicates that the first ball reverses its direction as a result of the collision.

c. If we apply the conservation of mechanical energy to either ball after the collision we have

$$\underbrace{\tfrac{1}{2}mv_f^2 + mgh_f}_{E_f} = \underbrace{\tfrac{1}{2}mv_0^2 + mgh_0}_{E_0}$$

where v_0 is the speed of the ball just after the collision, and h_f is the final height to which the ball rises. For either ball, $h_0 = 0$ m, and when either ball has reached its maximum height, $v_f = 0$ m/s. Therefore, the expression of energy conservation reduces to

$$gh_f = \tfrac{1}{2}v_0^2 \qquad \text{or} \qquad h_f = \frac{v_0^2}{2g}$$

Thus, the heights to which each ball rises after the collision are

$$\left[1.50\text{-kg ball}\right] \qquad h_f = \frac{v_0^2}{2g} = \frac{(2.83 \text{ m/s})^2}{2(9.80 \text{ m/s}^2)} = \boxed{0.409 \text{ m}}$$

$$\left[4.60\text{-kg ball}\right] \qquad h_f = \frac{v_0^2}{2g} = \frac{(2.73 \text{ m/s})^2}{2(9.80 \text{ m/s}^2)} = \boxed{0.380 \text{ m}}$$

49. ***REASONING AND SOLUTION*** The velocity of the center of mass of a system is given by Equation 7.11. Using the data and the results obtained in Example 5, we obtain the following:

a. The velocity of the center of mass of the two-car system before the collision is

$$\left(v_{cm}\right)_{before} = \frac{m_1 v_{01} + m_2 v_{02}}{m_1 + m_2}$$

$$= \frac{(65 \times 10^3 \text{ kg})(+0.80 \text{ m/s}) + (92 \times 10^3 \text{ kg})(+1.2 \text{ m/s})}{65 \times 10^3 \text{ kg} + 92 \times 10^3 \text{ kg}} = \boxed{+1.0 \text{ m/s}}$$

b. The velocity of the center of mass of the two-car system after the collision is

$$\left(v_{cm}\right)_{after} = \frac{m_1 v_f + m_2 v_f}{m_1 + m_2} = v_f = \boxed{+1.0 \text{ m/s}}$$

c. The answer in part (b) $\boxed{\text{should be the same}}$ as the common velocity v_f. Since the cars are coupled together, every point of the two-car system, including the center of mass, must move with the same velocity.

---

53. ***REASONING*** The system consists of the lumberjack and the log. For this system, the sum of the external forces is zero. This is because the weight of the system is balanced by the corresponding normal force (provided by the buoyant force of the water) and the water is assumed to be frictionless. The lumberjack and the log, then, constitute an isolated system, and the principle of conservation of linear momentum holds.

SOLUTION
a. The total linear momentum of the system before the lumberjack begins to move is zero, since all parts of the system are at rest. Momentum conservation requires that the total momentum remains zero during the motion of the lumberjack.

$$\underbrace{m_1 v_{f1} + m_2 v_{f2}}_{\substack{\text{Total momentum} \\ \text{just before the jump}}} = \underbrace{0}_{\text{Initial momentum}}$$

Here the subscripts "1" and "2" refer to the first log and lumberjack, respectively. Let the direction of motion of the lumberjack be the positive direction. Then, solving for v_{1f} gives

$$v_{f1} = -\frac{m_2 v_{f2}}{m_1} = -\frac{(98 \text{ kg})(+3.6 \text{ m/s})}{230 \text{ kg}} = \boxed{-1.5 \text{ m/s}}$$

The minus sign indicates that the first log recoils as the lumberjack jumps off.

b. Now the system is composed of the lumberjack, just before he lands on the second log, and the second log. Gravity acts on the system, but for the short time under consideration while the lumberjack lands, the effects of gravity in changing the linear momentum of the system are negligible. Therefore, to a very good approximation, we can say that the linear momentum of the system is very nearly conserved. In this case, the initial momentum is not zero as it was in part (a); rather the initial momentum of the system is the momentum of the lumberjack just before he lands on the second log. Therefore,

$$\underbrace{m_1 v_{f1} + m_2 v_{f2}}_{\substack{\text{Total momentum} \\ \text{just after lumberjack lands}}} = \underbrace{m_1 v_{01} + m_2 v_{02}}_{\text{Initial momentum}}$$

In this expression, the subscripts "1" and "2" now represent the second log and lumberjack, respectively. Since the second log is initially at rest, $v_{01} = 0$. Furthermore, since the lumberjack and the second log move with a common velocity, $v_{f1} = v_{f2} = v_f$. The statement of momentum conservation then becomes

$$m_1 v_f + m_2 v_f = m_2 v_{02}$$

Solving for v_f, we have

$$v_f = \frac{m_2 v_{02}}{m_1 + m_2} = \frac{(98 \text{ kg})(+3.6 \text{ m/s})}{230 \text{ kg} + 98 \text{ kg}} = \boxed{+1.1 \text{ m/s}}$$

The positive sign indicates that the system moves in the same direction as the original direction of the lumberjack's motion.

---

61. ***REASONING AND SOLUTION*** The comet piece and Jupiter constitute an isolated system, since no external forces act on them. Therefore, the head-on collision obeys the conservation of linear momentum:

$$\underbrace{(m_{\text{comet}} + m_{\text{Jupiter}})v_f}_{\substack{\text{Total momentum} \\ \text{after collision}}} = \underbrace{m_{\text{comet}} v_{\text{comet}} + m_{\text{Jupiter}} v_{\text{Jupiter}}}_{\substack{\text{Total momentum} \\ \text{before collision}}}$$

where v_f is the common velocity of the comet piece and Jupiter after the collision. We assume initially that Jupiter is moving in the $+x$ direction so $v_{\text{Jupiter}} = +1.3 \times 10^4$ m/s. The comet piece must be moving in the opposite direction so $v_{\text{comet}} = -6.0 \times 10^4$ m/s. The final velocity v_f of Jupiter and the comet piece can be written as $v_f = v_{\text{Jupiter}} + \Delta v$ where Δv is the change in velocity of Jupiter due to the collision. Substituting this expression into the conservation of momentum equation gives

$$(m_{comet} + m_{Jupiter})(v_{Jupiter} + \Delta v) = m_{comet}v_{comet} + m_{Jupiter}v_{Jupiter}$$

Multiplying out the left side of this equation, algebraically canceling the term $m_{Jupiter}v_{Jupiter}$ from both sides of the equation, and solving for Δv yields

$$\Delta v = \frac{m_{comet}(v_{comet} - v_{Jupiter})}{m_{comet} + m_{Jupiter}}$$

$$= \frac{(4.0 \times 10^{12} \text{ kg})(-6.0 \times 10^{4} \text{ m/s} - 1.3 \times 10^{4} \text{ m/s})}{4.0 \times 10^{12} \text{ kg} + 1.9 \times 10^{27} \text{ kg}} = -1.5 \times 10^{-10} \text{ m/s}$$

The change in Jupiter's speed is $\boxed{1.5 \times 10^{-10} \text{ m/s}}$.

---

65. **REASONING** We will define the system to be the platform, the two people and the ball. Since the ball travels nearly horizontally, the effects of gravity are negligible. Momentum is conserved. Since the initial momentum of the system is zero, it must remain zero as the ball is thrown and caught. While the ball is in motion, the platform will recoil in such a way that the total momentum of the system remains zero. As the ball is caught, the system must come to rest so that the total momentum remains zero. The distance that the platform moves before coming to rest again can be determined by using the expressions for momentum conservation and the kinematic description for this situation.

SOLUTION While the ball is in motion, we have

$$MV + mv = 0 \qquad (1)$$

where M is the combined mass of the platform and the two people, V is the recoil velocity of the platform, m is the mass of the ball, and v is the velocity of the ball.

The distance that the platform moves is given by

$$x = Vt \qquad (2)$$

where t is the time that the ball is in the air. The time that the ball is in the air is given by

$$t = \frac{L}{v - V} \qquad (3)$$

where L is the length of the platform, and the quantity $(v - V)$ is the velocity of the ball relative to the platform. Remember, both the ball and the platform are moving while the ball is in the air. Combining equations (2) and (3) gives

$$x = \left(\frac{V}{v-V}\right)L \qquad (4)$$

From equation (1) the ratio of the velocities is $V/v = -m/M$. Equation (4) then gives

$$x = \frac{(V/v)L}{1-(V/v)} = \frac{(-m/M)L}{1+(m/M)} = -\frac{mL}{M+m} = -\frac{(6.0 \text{ kg})(2.0 \text{ m})}{118 \text{ kg} + 6.0 \text{ kg}} = -0.097 \text{ m}$$

The minus sign indicates that displacement of the platform is in the opposite direction to the displacement of the ball. The distance moved by the platform is the magnitude of this displacement, or $\boxed{0.097 \text{ m}}$.

CHAPTER 8 | ROTATIONAL KINEMATICS

1. **REASONING AND SOLUTION** Since there are 2π radians per revolution and it is stated in the problem that there are 100 grads in one-quarter of a circle, we find that the number of grads in one radian is

$$\left(1.00 \ \text{rad}\right)\left(\frac{1 \ \text{rev}}{2\pi \ \text{rad}}\right)\left(\frac{100 \ \text{grad}}{0.250 \ \text{rev}}\right) = \boxed{63.7 \ \text{grad}}$$

5. **REASONING** The average angular velocity is equal to the angular displacement divided by the elapsed time (Equation 8.2). Thus, the angular displacement of the baseball is equal to the product of the average angular velocity and the elapsed time. However, the problem gives the travel time in seconds and asks for the displacement in radians, while the angular velocity is given in revolutions per minute. Thus, we will begin by converting the angular velocity into radians per second.

   SOLUTION Since 2π rad = 1 rev and 1 min = 60 s, the average angular velocity $\bar{\omega}$ (in rad/s) of the baseball is

$$\bar{\omega} = \left(\frac{330 \ \text{rev}}{\text{min}}\right)\left(\frac{2\pi \ \text{rad}}{1 \ \text{rev}}\right)\left(\frac{1 \ \text{min}}{60 \ \text{s}}\right) = 35 \ \text{rad/s}$$

   Since the average angular velocity of the baseball is equal to the angular displacement $\Delta\theta$ divided by the elapsed time Δt, the angular displacement is

$$\Delta\theta = \bar{\omega} \ \Delta t = \left(35 \ \text{rad/s}\right)\left(0.60 \ \text{s}\right) = \boxed{21 \ \text{rad}} \tag{8.2}$$

9. **REASONING** Equation 8.4 $\left[\bar{\alpha} = (\omega - \omega_0)/t\right]$ indicates that the average angular acceleration is equal to the change in the angular velocity divided by the elapsed time. Since the wheel starts from rest, its initial angular velocity is $\omega_0 = 0$ rad/s. Its final angular velocity is given as $\omega = 0.24$ rad/s. Since the average angular acceleration is given as $\bar{\alpha} = 0.030$ rad/s^2, Equation 8.4 can be solved to determine the elapsed time t.

   SOLUTION Solving Equation 8.4 for the elapsed time gives

$$t = \frac{\omega - \omega_0}{\bar{\alpha}} = \frac{0.24 \ \text{rad/s} - 0 \ \text{rad/s}}{0.030 \ \text{rad/s}^2} = \boxed{8.0 \ \text{s}}$$

17. **REASONING AND SOLUTION**

a. If the propeller is to appear stationary, each blade must move through an angle of 120° or $2\pi/3$ rad between flashes. The time required is

$$t = \frac{\theta}{\omega} = \frac{(2\pi/3)\,\text{rad}}{(16.7\,\text{rev/s})\left(\dfrac{2\pi\,\text{rad}}{1\,\text{rev}}\right)} = \boxed{2.00 \times 10^{-2}\,\text{s}}$$

b. The next shortest time occurs when each blade moves through an angle of 240°, or $4\pi/3$ rad, between successive flashes. This time is twice the value that we found in part a, or $\boxed{4.00 \times 10^{-2}\,\text{s}}$.

---

21. **REASONING AND SOLUTION**

a. From Equation 8.7 we obtain

$$\theta = \omega_0 t + \tfrac{1}{2}\alpha t^2 = (5.00\,\text{rad/s})(4.00\,\text{s}) + \tfrac{1}{2}(2.50\,\text{rad/s}^2)(4.00\,\text{s})^2 = \boxed{4.00 \times 10^1\ \text{rad}}$$

b. From Equation 8.4, we obtain

$$\omega = \omega_0 + \alpha t = 5.00\,\text{rad/s} + (2.50\,\text{rad/s}^2)(4.00\,\text{s}) = \boxed{15.0\,\text{rad/s}}$$

---

23. **REASONING AND SOLUTION**

a. $$\omega = \omega_0 + \alpha t = 0\,\text{rad/s} + (3.00\,\text{rad/s}^2)(18.0\,\text{s}) = \boxed{54.0\,\text{rad/s}}$$

b. $$\theta = \tfrac{1}{2}(\omega_0 + \omega)t = \tfrac{1}{2}(0\,\text{rad/s} + 54.0\,\text{rad/s})(18.0\,\text{s}) = \boxed{486\,\text{rad}}$$

---

27. **REASONING** The equations of kinematics for rotational motion cannot be used directly to find the angular displacement, because the final angular velocity (not the initial angular velocity), the acceleration, and the time are known. We will combine two of the equations, Equations 8.4 and 8.6 to obtain an expression for the angular displacement that contains the three known variables.

SOLUTION The angular displacement of each wheel is equal to the average angular velocity multiplied by the time

$$\theta = \underbrace{\tfrac{1}{2}\left(\omega_0 + \omega\right)}_{\bar{\omega}} t \qquad\qquad (8.6)$$

The initial angular velocity ω_0 is not known, but it can be found in terms of the angular acceleration and time, which are known. The angular acceleration is defined as (with $t_0 = 0$ s)

$$\bar{\alpha} = \frac{\omega - \omega_0}{t} \quad \text{or} \quad \omega_0 = \omega - \alpha t \tag{8.4}$$

Substituting this expression for ω_0 into Equation 8.6 gives

$$\theta = \tfrac{1}{2}\left[\underbrace{\left(\omega - \alpha t\right)}_{\omega_0} + \omega \right] t = \omega t - \tfrac{1}{2}\alpha t^2$$

$$= \left(+74.5 \text{ rad/s}\right)\left(4.50 \text{ s}\right) - \tfrac{1}{2}\left(+6.70 \text{ rad/s}^2\right)\left(4.50 \text{ s}\right)^2 = \boxed{+267 \text{ rad}}$$

---

33. **REASONING** The angular displacement of the child when he catches the horse is, from Equation 8.2, $\theta_c = \omega_c t$. In the same time, the angular displacement of the horse is, from Equation 8.7 with $\omega_0 = 0$ rad/s, $\theta_h = \tfrac{1}{2}\alpha t^2$. If the child is to catch the horse $\theta_c = \theta_h + (\pi/2)$.

SOLUTION Using the above conditions yields

$$\tfrac{1}{2}\alpha t^2 - \omega_c t + \tfrac{1}{2}\pi = 0$$

or

$$\tfrac{1}{2}(0.0100 \text{ rad/s}^2)t^2 - \left(0.250 \text{ rad/s}\right)t + \tfrac{1}{2}\left(\pi \text{ rad}\right) = 0$$

The quadratic formula yields $t = 7.37$ s and 42.6 s; therefore, the shortest time needed to catch the horse is $\boxed{t = 7.37 \text{ s}}$.

---

37. **REASONING** The angular speed ω and tangential speed v_T are related by Equation 8.9 ($v_T = r\omega$), and this equation can be used to determine the radius r. However, we must remember that this relationship is only valid if we use radian measure. Therefore, it will be necessary to convert the given angular speed in rev/s into rad/s.

SOLUTION Solving Equation 8.9 for the radius gives

$$r = \frac{v_T}{\omega} = \frac{54 \text{ m/s}}{\underbrace{(47 \text{ rev/s})\left(\dfrac{2\pi \text{ rad}}{1 \text{ rev}}\right)}_{\text{Conversion from rev/s into rad/s}}} = \boxed{0.18 \text{ m}}$$

where we have used the fact that 1 rev corresponds to 2π rad to convert the given angular speed from rev/s into rad/s.

---

41. **REASONING AND SOLUTION**

a. From Equation 8.9, and the fact that 1 revolution $= 2\pi$ radians, we obtain

$$v_T = r\omega = (0.0568 \text{ m})\left(3.50 \frac{\text{rev}}{\text{s}}\right)\left(\frac{2\pi \text{ rad}}{1 \text{ rev}}\right) = \boxed{1.25 \text{ m/s}}$$

b. Since the disk rotates at *constant* tangential speed,

$$v_{T1} = v_{T2} \qquad \text{or} \qquad \omega_1 r_1 = \omega_2 r_2$$

Solving for ω_2, we obtain

$$\omega_2 = \frac{\omega_1 r_1}{r_2} = \frac{(3.50 \text{ rev/s})(0.0568 \text{ m})}{0.0249 \text{ m}} = \boxed{7.98 \text{ rev/s}}$$

---

45. **REASONING** The magnitude ω of each car's angular speed can be evaluated from $a_c = r\omega^2$ (Equation 8.11), where r is the radius of the turn and a_c is the magnitude of the centripetal acceleration. We are given that the centripetal acceleration of each car is the same. In addition, the radius of each car's turn is known. These facts will enable us to determine the ratio of the angular speeds.

SOLUTION Solving Equation 8.11 for the angular speed gives $\omega = \sqrt{a_c / r}$. Applying this relation to each car yields:

Car A:
$$\omega_A = \sqrt{a_{c, A} / r_A}$$

Car B:
$$\omega_B = \sqrt{a_{c, B} / r_B}$$

Taking the ratio of these two angular speeds, and noting that $a_{c,\,A} = a_{c,\,B}$, gives

$$\frac{\omega_A}{\omega_B} = \frac{\sqrt{\dfrac{a_{c,\,A}}{r_A}}}{\sqrt{\dfrac{a_{c,\,B}}{r_B}}} = \sqrt{\frac{\cancel{a_{c,\,A}}\,r_B}{\cancel{a_{c,\,B}}\,r_A}} = \sqrt{\frac{36\ \text{m}}{48\ \text{m}}} = \boxed{0.87}$$

---

47. *REASONING AND SOLUTION*

a. The tangential acceleration of the train is given by Equation 8.10 as

$$a_T = r\alpha = (2.00\times10^2\ \text{m})(1.50\times10^{-3}\ \text{rad/s}^2) = 0.300\ \text{m/s}^2$$

The centripetal acceleration of the train is given by Equation 8.11 as

$$a_c = r\omega^2 = (2.00\times10^2\ \text{m})(0.0500\ \text{rad/s})^2 = 0.500\ \text{m/s}^2$$

The magnitude of the total acceleration is found from the Pythagorean theorem to be

$$a = \sqrt{a_T^2 + a_c^2} = \boxed{0.583\ \text{m/s}^2}$$

b. The total acceleration vector makes an angle relative to the radial acceleration of

$$\theta = \tan^{-1}\!\left(\frac{a_T}{a_c}\right) = \tan^{-1}\!\left(\frac{0.300\ \text{m/s}^2}{0.500\ \text{m/s}^2}\right) = \boxed{31.0^\circ}$$

---

51. *REASONING*

a. The tangential speed v_T of the sun as it orbits about the center of the Milky Way is related to the orbital radius r and angular speed ω by Equation 8.9, $v_T = r\omega$. Before we use this relation, however, we must first convert r to meters from light-years.

b. The centripetal force is the net force required to keep an object, such as the sun, moving on a circular path. According to Newton's second law of motion, the magnitude F_c of the centripetal force is equal to the product of the object's mass m and the magnitude a_c of its centripetal acceleration (see Section 5.3): $F_c = ma_c$. The magnitude of the centripetal acceleration is expressed by Equation 8.11 as $a_c = r\omega^2$, where r is the radius of the circular path and ω is the angular speed of the object.

SOLUTION

a. The radius of the sun's orbit about the center of the Milky Way is

$$r = \left(2.3 \times 10^4 \text{ light-years}\right)\left(\frac{9.5 \times 10^{15} \text{ m}}{1 \text{ light-year}}\right) = 2.2 \times 10^{20} \text{ m}$$

The tangential speed of the sun is

$$v_T = r\omega = \left(2.2 \times 10^{20} \text{ m}\right)\left(1.1 \times 10^{-15} \text{ rad/s}\right) = \boxed{2.4 \times 10^5 \text{ m/s}} \qquad (8.9)$$

b. The magnitude of the centripetal force that acts on the sun is

$$\underbrace{F_c}_{\substack{\text{Centripetal} \\ \text{force}}} = ma_c = mr\,\omega^2$$

$$= \left(1.99 \times 10^{30} \text{ kg}\right)\left(2.2 \times 10^{20} \text{ m}\right)\left(1.1 \times 10^{-15} \text{ rad/s}\right)^2 = \boxed{5.3 \times 10^{20} \text{ N}}$$

---

53. ***REASONING AND SOLUTION*** From Equation 2.4, the linear acceleration of the motorcycle is

$$a = \frac{v - v_0}{t} = \frac{22.0 \text{ m/s} - 0 \text{ m/s}}{9.00 \text{ s}} = 2.44 \text{ m/s}^2$$

Since the tire rolls without slipping, the linear acceleration equals the tangential acceleration of a point on the outer edge of the tire: $a = a_T$. Solving Equation 8.13 for α gives

$$\alpha = \frac{a_T}{r} = \frac{2.44 \text{ m/s}^2}{0.280 \text{ m}} = \boxed{8.71 \text{ rad/s}^2}$$

---

63. ***REASONING AND SOLUTION*** By inspection, the distance traveled by the "axle" or the center of the moving quarter is

$$d = 2\pi(2r) = 4\pi r$$

where r is the radius of the quarter. The distance d traveled by the "axle" of the moving quarter must be equal to the circular arc length s along the outer edge of the quarter. This arc length is $s = r\theta$, where θ is the angle through which the quarter rotates. Thus,

$$4\pi r = r\theta$$

so that $\theta = 4\pi \text{ rad}$. This is equivalent to

$$\left(4\pi \text{ rad}\right)\left(\frac{1 \text{ rev}}{2\pi \text{ rad}}\right) = \boxed{2 \text{ revolutions}}$$

65. **REASONING** The tangential acceleration $\mathbf{a}_T$ of the speedboat can be found by using Newton's second law, $\mathbf{F}_T = m\mathbf{a}_T$, where $\mathbf{F}_T$ is the net tangential force. Once the tangential acceleration of the boat is known, Equation 2.4 can be used to find the tangential speed of the boat 2.0 s into the turn. With the tangential speed and the radius of the turn known, Equation 5.2 can then be used to find the centripetal acceleration of the boat.

SOLUTION
a. From Newton's second law, we obtain

$$a_T = \frac{F_T}{m} = \frac{550 \text{ N}}{220 \text{ kg}} = \boxed{2.5 \text{ m/s}^2}$$

b. The tangential speed of the boat 2.0 s into the turn is, according to Equation 2.4,

$$v_T = v_{0T} + a_T t = 5.0 \text{ m/s} + (2.5 \text{ m/s}^2)(2.0 \text{ s}) = 1.0 \times 10^1 \text{ m/s}$$

The centripetal acceleration of the boat is then

$$a_c = \frac{v_T^2}{r} = \frac{(1.0 \times 10^1 \text{ m/s})^2}{32 \text{ m}} = \boxed{3.1 \text{ m/s}^2}$$

67. **REASONING AND SOLUTION** Since the angular speed of the fan decreases, the sign of the angular acceleration must be opposite to the sign for the angular velocity. Taking the angular velocity to be positive, the angular acceleration, therefore, must be a negative quantity. Using Equation 8.4 we obtain

$$\omega_0 = \omega - \alpha t = 83.8 \text{ rad/s} - (-42.0 \text{ rad/s}^2)(1.75 \text{ s}) = \boxed{157.3 \text{ rad/s}}$$

71. **REASONING** The tangential speed v_T of a point on the "equator" of the baseball is given by Equation 8.9 as $v_T = r\omega$, where r is the radius of the baseball and ω is its angular speed. The radius is given in the statement of the problem. The (constant) angular speed is related to that angle θ through which the ball rotates by Equation 8.2 as $\omega = \theta / t$, where we have assumed for convenience that $\theta_0 = 0$ rad when $t_0 = 0$ s. Thus, the tangential speed of the ball is

$$v_T = r\omega = r\left(\frac{\theta}{t}\right)$$

The time t that the ball is in the air is equal to the distance x it travels divided by its linear speed v, $t = x/v$, so the tangential speed can be written as

$$v_T = r\left(\frac{\theta}{t}\right) = r\left(\frac{\theta}{\dfrac{x}{v}}\right) = \frac{r\theta v}{x}$$

SOLUTION The tangential speed of a point on the equator of the baseball is

$$v_T = \frac{r\theta v}{x} = \frac{(3.67 \times 10^{-2}\ \text{m})(49.0\ \text{rad})(42.5\ \text{m/s})}{16.5\ \text{m}} = \boxed{4.63\ \text{m/s}}$$

CHAPTER 9 | *ROTATIONAL DYNAMICS*

3. ***REASONING*** According to Equation 9.1, we have

$$\text{Magnitude of torque} = F\ell$$

where F is the magnitude of the applied force and ℓ is the lever arm. From the figure in the text, the lever arm is given by $\ell = (0.28 \text{ m}) \sin 50.0°$. Since both the magnitude of the torque and ℓ are known, Equation 9.1 can be solved for F.

SOLUTION Solving Equation 9.1 for F, we have

$$F = \frac{\text{Magnitude of torque}}{\ell} = \frac{45 \text{ N} \cdot \text{m}}{(0.28 \text{ m}) \sin 50.0°} = \boxed{2.1 \times 10^2 \text{ N}}$$

5. ***REASONING*** In both parts of the problem, the magnitude of the torque is given by Equation 9.1 as the magnitude F of the force times the lever arm ℓ. In part (a), the lever arm is just the distance of 0.55 m given in the drawing. However, in part (b), the lever arm is less than the given distance and must be expressed using trigonometry as $\ell = (0.55 \text{ m}) \sin \theta$. See the drawing at the right.

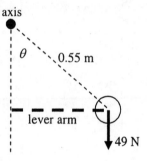

SOLUTION
a. Using Equation 9.1, we find that

$$\text{Magnitude of torque} = F\ell = (49 \text{ N})(0.55 \text{ m}) = \boxed{27 \text{ N} \cdot \text{m}}$$

b. Again using Equation 9.1, this time with a lever arm of $\ell = (0.55 \text{ m}) \sin \theta$, we obtain

$$\text{Magnitude of torque} = 15 \text{ N} \cdot \text{m} = F\ell = (49 \text{ N})(0.55 \text{ m}) \sin \theta$$

$$\sin \theta = \frac{15 \text{ N} \cdot \text{m}}{(49 \text{ N})(0.55 \text{ m})} \quad \text{or} \quad \theta = \sin^{-1} \left[\frac{15 \text{ N} \cdot \text{m}}{(49 \text{ N})(0.55 \text{ m})} \right] = \boxed{34°}$$

7. ***REASONING*** Each of the two forces produces a torque about the axis of rotation, one clockwise and the other counterclockwise. By setting the sum of the torques equal to zero, we will be able to determine the angle θ in the drawing.

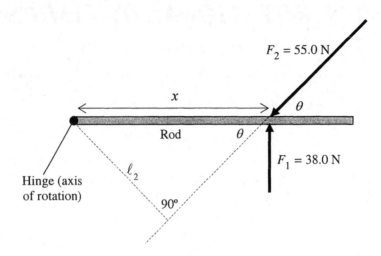

Table (Top view)

SOLUTION The two forces act on the rod at a distance x from the hinge. The torque τ_1 produced by the force $\mathbf{F}_1$ is given by $\tau_1 = +F_1 \ell_1$ (see Equation 9.1), where F_1 is the magnitude of the force and ℓ_1 is the lever arm. It is a positive torque, since it tends to produce a counterclockwise rotation. Since $\mathbf{F}_1$ is applied perpendicular to the rod, $\ell_1 = x$. The torque τ_2 produced by $\mathbf{F}_2$ is $\tau_2 = -F_2 \ell_2$, where $\ell_2 = x \sin \theta$ (see the drawing). It is a negative torque, since it tends to produce a clockwise rotation. Setting the sum of the torques equal to zero, we have

$$+F_1\ell_1 + \left(-F_2\ell_2\right) = 0 \qquad \text{or} \qquad +F_1 x - F_2 \underbrace{\left(x \sin \theta\right)}_{\ell_2} = 0$$

The distance x in this relation can be eliminated algebraically. Solving for the angle θ gives

$$\sin \theta = \frac{F_1}{F_2} \qquad \text{or} \qquad \theta = \sin^{-1}\left(\frac{F_1}{F_2}\right) = \sin^{-1}\left(\frac{38.0 \text{ N}}{55.0 \text{ N}}\right) = \boxed{43.7°}$$

---

13. **REASONING** The drawing shows the bridge and the four forces that act on it: the upward force $\mathbf{F}_1$ exerted on the left end by the support, the force due to the weight $\mathbf{W}_h$ of the hiker, the weight $\mathbf{W}_b$ of the bridge, and the upward force $\mathbf{F}_2$ exerted on the right side by the support. Since the bridge is in equilibrium, the sum of the torques about any axis of rotation must be zero $\left(\Sigma \tau = 0\right)$, and the sum of the forces in the vertical direction must be zero $\left(\Sigma F_y = 0\right)$. These two conditions will allow us to determine the magnitudes of $\mathbf{F}_1$ and $\mathbf{F}_2$.

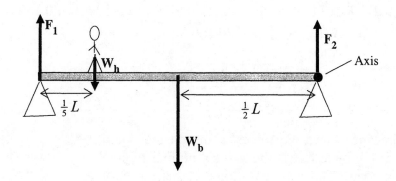

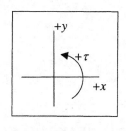

SOLUTION

a. We will begin by taking the axis of rotation about the right end of the bridge. The torque produced by $\mathbf{F}_2$ is zero, since its lever arm is zero. When we set the sum of the torques equal to zero, the resulting equation will have only one unknown, $\mathbf{F}_1$, in it. Setting the sum of the torques produced by the three forces equal to zero gives

$$\Sigma \tau = -F_1 L + W_h \left(\tfrac{4}{5} L\right) + W_b \left(\tfrac{1}{2} L\right) = 0$$

Algebraically eliminating the length L of the bridge from this equation and solving for F_1 gives

$$F_1 = \tfrac{4}{5} W_h + \tfrac{1}{2} W_b = \tfrac{4}{5}\left(985 \text{ N}\right) + \tfrac{1}{2}\left(3610 \text{ N}\right) = \boxed{2590 \text{ N}}$$

b. Since the bridge is in equilibrium, the sum of the forces in the vertical direction must be zero:

$$\Sigma F_y = F_1 - W_h - W_b + F_2 = 0$$

Solving for F_2 gives

$$F_2 = -F_1 + W_h + W_b = -2590 \text{ N} + 985 \text{ N} + 3610 \text{ N} = \boxed{2010 \text{ N}}$$

---

19. **REASONING AND SOLUTION** The net torque about an axis through the contact point between the tray and the thumb is

$$\Sigma \tau = F(0.0400 \text{ m}) - (0.250 \text{ kg})(9.80 \text{ m/s}^2)(0.320 \text{ m}) - (1.00 \text{ kg})(9.80 \text{ m/s}^2)(0.180 \text{ m})$$
$$- (0.200 \text{ kg})(9.80 \text{ m/s}^2)(0.140 \text{ m}) = 0$$

$$F = \boxed{70.6 \text{ N, up}}$$

Similarly, the net torque about an axis through the point of contact between the tray and the finger is

$$\Sigma\tau = T(0.0400 \text{ m}) - (0.250 \text{ kg})(9.80 \text{ m/s}^2)(0.280 \text{ m}) - (1.00 \text{ kg})(9.80 \text{ m/s}^2)(0.140 \text{ m})$$
$$- (0.200 \text{ kg})(9.80 \text{ m/s}^2)(0.100 \text{ m}) = 0$$

$$T = \boxed{56.4 \text{ N, down}}$$

---

23. ***REASONING*** The drawing shows the forces acting on the board, which has a length L. The ground exerts the vertical normal force **V** on the lower end of the board. The maximum force of static friction has a magnitude of $\mu_s V$ and acts horizontally on the lower end of the board. The weight **W** acts downward at the board's center. The vertical wall applies a force **P** to the upper end of the board, this force being perpendicular to the wall since the wall is smooth (i.e., there is no friction along the wall). We take upward and to the right as our positive directions. Then, since the horizontal forces balance to zero, we have

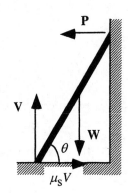

$$\mu_s V - P = 0 \tag{1}$$

The vertical forces also balance to zero giving

$$V - W = 0 \tag{2}$$

Using an axis through the lower end of the board, we express the fact that the torques balance to zero as

$$PL \sin \theta - W\left(\frac{L}{2}\right)\cos \theta = 0 \tag{3}$$

Equations (1), (2), and (3) may then be combined to yield an expression for θ.

SOLUTION Rearranging Equation (3) gives

$$\tan \theta = \frac{W}{2P} \tag{4}$$

But, $P = \mu_s V$ according to Equation (1), and $W = V$ according to Equation (2). Substituting these results into Equation (4) gives

$$\tan \theta = \frac{V}{2\mu_s V} = \frac{1}{2\mu_s}$$

Therefore,

$$\theta = \tan^{-1}\left(\frac{1}{2\mu_s}\right) = \tan^{-1}\left[\frac{1}{2(0.650)}\right] = \boxed{37.6°}$$

---

27. **REASONING** Since the man holds the ball motionless, the ball and the arm are in equilibrium. Therefore, the net force, as well as the net torque about any axis, must be zero.

    SOLUTION Using Equation 9.1, the net torque about an axis through the elbow joint is

    $$\Sigma\tau = M(0.0510 \text{ m}) - (22.0 \text{ N})(0.140 \text{ m}) - (178 \text{ N})(0.330 \text{ m}) = 0$$

    Solving this expression for M gives $M = \boxed{1.21 \times 10^3 \text{ N}}$.

    The net torque about an axis through the center of gravity is

    $$\Sigma\tau = -(1210 \text{ N})(0.0890 \text{ m}) + F(0.140 \text{ m}) - (178 \text{ N})(0.190 \text{ m}) = 0$$

    Solving this expression for F gives $F = \boxed{1.01 \times 10^3 \text{ N}}$. Since the forces must add to give a net force of zero, we know that the direction of **F** is $\boxed{\text{downward}}$.

---

29. **REASONING AND SOLUTION** Consider the left board, which has a length L and a weight of $(356 \text{ N})/2 = 178 \text{ N}$. Let $\mathbf{F_V}$ be the upward normal force exerted by the ground on the board. This force balances the weight, so $F_V = 178 \text{ N}$. Let $\mathbf{f_s}$ be the force of static friction, which acts horizontally on the end of the board in contact with the ground. $\mathbf{f_s}$ points to the right. Since the board is in equilibrium, the net torque acting on the board through any axis must be zero. Measuring the torques with respect to an axis through the apex of the triangle formed by the boards, we have

    $$+ (178 \text{ N})(\sin 30.0°)\left(\frac{L}{2}\right) + f_s(L \cos 30.0°) - F_V (L \sin 30.0°) = 0$$

    or

    $$44.5 \text{ N} + f_s \cos 30.0° - F_V \sin 30.0° = 0$$

    so that

    $$f_s = \frac{(178 \text{ N})(\sin 30.0°) - 44.5 \text{ N}}{\cos 30.0°} = \boxed{51.4 \text{ N}}$$

---

37. **REASONING** The rotational analog of Newton's second law is given by Equation 9.7, $\Sigma\tau = I\alpha$. Since the person pushes on the outer edge of one section of the door with a force **F** that is directed perpendicular to the section, the torque exerted on the door has a magnitude of FL, where the lever arm L is equal to the width of one section. Once the moment of inertia is known, Equation 9.7 can be solved for the angular acceleration α.

The moment of inertia of the door relative to the rotation axis is $I = 4I_P$, where I_P is the moment of inertia for one section. According to Table 9.1, we find $I_P = \frac{1}{3}ML^2$, so that the rotational inertia of the door is $I = \frac{4}{3}ML^2$.

SOLUTION Solving Equation 9.7 for α, and using the expression for I determined above, we have

$$\alpha = \frac{FL}{\frac{4}{3}ML^2} = \frac{F}{\frac{4}{3}ML} = \frac{68\ \text{N}}{\frac{4}{3}(85\ \text{kg})(1.2\ \text{m})} = \boxed{0.50\ \text{rad/s}^2}$$

---

39. ***REASONING*** The figure below shows eight particles, each one located at a different corner of an imaginary cube. As shown, if we consider an axis that lies along one edge of the cube, two of the particles lie on the axis, and for these particles $r = 0$. The next four particles closest to the axis have $r = \ell$, where ℓ is the length of one edge of the cube. The remaining two particles have $r = d$, where d is the length of the diagonal along any one of the faces. From the Pythagorean theorem, $d = \sqrt{\ell^2 + \ell^2} = \ell\sqrt{2}$.

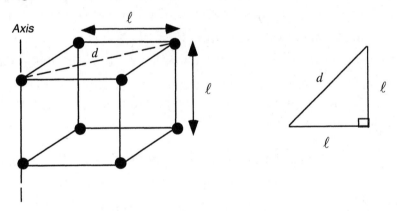

According to Equation 9.6, the moment of inertia of a system of particles is given by $I = \sum mr^2$.

SOLUTION Direct application of Equation 9.6 gives

$$I = \sum mr^2 = 4(m\ell^2) + 2(md^2) = 4(m\ell^2) + 2(2m\ell^2) = 8m\ell^2$$

or

$$I = 8(0.12\ \text{kg})(0.25\ \text{m})^2 = \boxed{0.060\ \text{kg}\cdot\text{m}^2}$$

41. **REASONING** The drawing shows the two identical sheets and the axis of rotation for each.

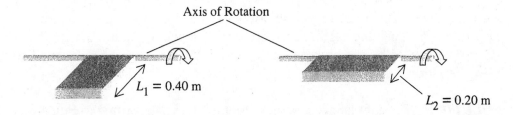

Axis of Rotation

$L_1 = 0.40$ m

$L_2 = 0.20$ m

The time t it takes for each sheet to reach its final angular velocity depends on the angular acceleration α of the sheet. This relation is given by Equation 8.4 as $t = (\omega - \omega_0)/\alpha$, where ω and ω_0 are the final and initial angular velocities, respectively. We know that $\omega_0 = 0$ rad/s in each case and that the final angular velocities are the same. The angular acceleration can be determined by using Newton's second law for rotational motion, Equation 9.7, as $\alpha = \tau/I$, where τ is the torque applied to a sheet and I is its moment of inertia.

SOLUTION Substituting the relation $\alpha = \tau/I$ into $t = (\omega - \omega_0)/\alpha$ gives

$$t = \frac{(\omega - \omega_0)}{\alpha} = \frac{(\omega - \omega_0)}{\left(\dfrac{\tau}{I}\right)} = \frac{I(\omega - \omega_0)}{\tau}$$

The time it takes for each sheet to reach its final angular velocity is:

$$t_{\text{Left}} = \frac{I_{\text{Left}}(\omega - \omega_0)}{\tau} \qquad \text{and} \qquad t_{\text{Right}} = \frac{I_{\text{Right}}(\omega - \omega_0)}{\tau}$$

The moments of inertia I of the left and right sheets about the axes of rotation are given by the following relations, where M is the mass of each sheet (see Table 9.1 and the drawings above): $I_{\text{Left}} = \frac{1}{3}ML_1^2$ and $I_{\text{Right}} = \frac{1}{3}ML_2^2$. Note that the variables M, ω, ω_0, and τ are the same for both sheets. Dividing the time-expression for the right sheet by that for the left sheet gives

$$\frac{t_{\text{Right}}}{t_{\text{Left}}} = \frac{\dfrac{I_{\text{Right}}(\omega - \omega_0)}{\tau}}{\dfrac{I_{\text{Left}}(\omega - \omega_0)}{\tau}} = \frac{I_{\text{Right}}}{I_{\text{Left}}} = \frac{\frac{1}{3}ML_2^2}{\frac{1}{3}ML_1^2} = \frac{L_2^2}{L_1^2}$$

Solving this expression for t_{Right} yields

$$t_{\text{Right}} = t_{\text{Left}}\left(\frac{L_2^2}{L_1^2}\right) = (8.0 \text{ s})\frac{(0.20 \text{ m})^2}{(0.40 \text{ m})^2} = \boxed{2.0 \text{ s}}$$

45. **REASONING** The angular acceleration of the bicycle wheel can be calculated from Equation 8.4. Once the angular acceleration is known, Equation 9.7 can be used to find the net torque caused by the brake pads. The normal force can be calculated from the torque using Equation 9.1.

SOLUTION The angular acceleration of the wheel is, according to Equation 8.4,

$$\alpha = \frac{\omega - \omega_0}{t} = \frac{3.7 \text{ rad/s} - 13.1 \text{ rad/s}}{3.0 \text{ s}} = -3.1 \text{ rad/s}^2$$

If we assume that all the mass of the wheel is concentrated in the rim, we may treat the wheel as a hollow cylinder. From Table 9.1, we know that the moment of inertia of a hollow cylinder of mass m and radius r about an axis through its center is $I = mr^2$. The net torque that acts on the wheel due to the brake pads is, therefore,

$$\sum \tau = I\alpha = (mr^2)\alpha \tag{1}$$

From Equation 9.1, the net torque that acts on the wheel due to the action of the two brake pads is

$$\sum \tau = -2 f_k \ell \tag{2}$$

where f_k is the kinetic frictional force applied to the wheel by each brake pad, and $\ell = 0.33$ m is the lever arm between the axle of the wheel and the brake pad (see the drawing in the text). The factor of 2 accounts for the fact that there are two brake pads. The minus sign arises because the net torque must have the same sign as the angular acceleration. The kinetic frictional force can be written as (see Equation 4.8)

$$f_k = \mu_k F_N \tag{3}$$

where μ_k is the coefficient of kinetic friction and F_N is the magnitude of the normal force applied to the wheel by each brake pad.

Combining Equations (1), (2), and (3) gives

$$-2(\mu_k F_N)\ell = (mr^2)\alpha$$

$$F_N = \frac{-mr^2\alpha}{2\mu_k \ell} = \frac{-(1.3 \text{ kg})(0.33 \text{ m})^2(-3.1 \text{ rad/s}^2)}{2(0.85)(0.33 \text{ m})} = \boxed{0.78 \text{ N}}$$

49. *REASONING* The kinetic energy of the flywheel is given by Equation 9.9. The moment of inertia of the flywheel is the same as that of a solid disk, and, according to Table 9.1 in the text, is given by $I = \frac{1}{2}MR^2$. Once the moment of inertia of the flywheel is known, Equation 9.9 can be solved for the angular speed ω in rad/s. This quantity can then be converted to rev/min.

SOLUTION Solving Equation 9.9 for ω, we obtain,

$$\omega = \sqrt{\frac{2(\mathrm{KE}_R)}{I}} = \sqrt{\frac{2(\mathrm{KE}_R)}{\frac{1}{2}MR^2}} = \sqrt{\frac{4(1.2 \times 10^9 \text{ J})}{(13 \text{ kg})(0.30 \text{ m})^2}} = 6.4 \times 10^4 \text{ rad/s}$$

Converting this answer into rev/min, we find that

$$\omega = \left(6.4 \times 10^4 \text{ rad/s}\right)\left(\frac{1 \text{ rev}}{2\pi \text{ rad}}\right)\left(\frac{60 \text{ s}}{1 \text{ min}}\right) = \boxed{6.1 \times 10^5 \text{ rev/min}}$$

---

51. *REASONING AND SOLUTION*

a. The tangential speed of each object is given by Equation 8.9, $v_T = r\omega$. Therefore,

For object 1: $\qquad v_{T1} = (2.00 \text{ m})(6.00 \text{ rad/s}) = \boxed{12.0 \text{ m/s}}$

For object 2: $\qquad v_{T2} = (1.50 \text{ m})(6.00 \text{ rad/s}) = \boxed{9.00 \text{ m/s}}$

For object 3: $\qquad v_{T3} = (3.00 \text{ m})(6.00 \text{ rad/s}) = \boxed{18.0 \text{ m/s}}$

b. The total kinetic energy of this system can be calculated by computing the sum of the kinetic energies of each object in the system. Therefore,

$$\mathrm{KE} = \frac{1}{2}m_1 v_1^2 + \frac{1}{2}m_2 v_2^2 + \frac{1}{2}m_3 v_3^2$$

$$\mathrm{KE} = \frac{1}{2}\left[(6.00 \text{ kg})(12.0 \text{ m/s})^2 + (4.00 \text{ kg})(9.00 \text{ m/s})^2 + (3.00 \text{ kg})(18.0 \text{ m/s})^2\right] = \boxed{1.08 \times 10^3 \text{ J}}$$

c. The total moment of inertia of this system can be calculated by computing the sum of the moments of inertia of each object in the system. Therefore,

$$I = \sum mr^2 = m_1 r_1^2 + m_2 r_2^2 + m_3 r_3^2$$

$$I = (6.00 \text{ kg})(2.00 \text{ m})^2 + (4.00 \text{ kg})(1.50 \text{ m})^2 + (3.00 \text{ kg})(3.00 \text{ m})^2 = \boxed{60.0 \text{ kg} \cdot \text{m}^2}$$

d. The rotational kinetic energy of the system is, according to Equation 9.9,

$$KE_R = \tfrac{1}{2}I\omega^2 = \tfrac{1}{2}(60.0 \text{ kg}\cdot\text{m}^2)(6.00 \text{ rad}/\text{s})^2 = \boxed{1.08\times10^3 \text{ J}}$$

This agrees, as it should, with the result for part (b).

---

55. ***REASONING AND SOLUTION*** The only force that does work on the cylinders as they move up the incline is the conservative force of gravity; hence, the total mechanical energy is conserved as the cylinders ascend the incline. We will let $h = 0$ m on the horizontal plane at the bottom of the incline. Applying the principle of conservation of mechanical energy to the solid cylinder, we have

$$mgh_s = \tfrac{1}{2}mv_0^2 + \tfrac{1}{2}I_s\omega_0^2 \tag{1}$$

where, from Table 9.1, $I_s = \tfrac{1}{2}mr^2$. In this expression, v_0 and ω_0 are the initial translational and rotational speeds, and h_s is the final height attained by the solid cylinder. Since the cylinder rolls without slipping, the rotational speed ω_0 and the translational speed v_0 are related according to Equation 8.12, $\omega_0 = v_0/r$. Then, solving Equation (1) for h_s, we obtain

$$h_s = \frac{3v_0^2}{4g}$$

Repeating the above for the hollow cylinder and using $I_h = mr^2$ we have

$$h_h = \frac{v_0^2}{g}$$

The height h attained by each cylinder is related to the distance s traveled along the incline and the angle θ of the incline by

$$s_s = \frac{h_s}{\sin\theta} \qquad \text{and} \qquad s_h = \frac{h_h}{\sin\theta}$$

Dividing these gives

$$\frac{s_s}{s_h} = \boxed{3/4}$$

---

59. **REASONING** Let the two disks constitute the system. Since there are no external torques acting on the system, the principle of conservation of angular momentum applies. Therefore we have $L_{\text{initial}} = L_{\text{final}}$, or

$$I_A \omega_A + I_B \omega_B = (I_A + I_B)\omega_{\text{final}}$$

This expression can be solved for the moment of inertia of disk B.

SOLUTION Solving the above expression for I_B, we obtain

$$I_B = I_A \left(\frac{\omega_{\text{final}} - \omega_A}{\omega_B - \omega_{\text{final}}} \right) = (3.4 \text{ kg} \cdot \text{m}^2) \left[\frac{-2.4 \text{ rad/s} - 7.2 \text{ rad/s}}{-9.8 \text{ rad/s} - (-2.4 \text{ rad/s})} \right] = \boxed{4.4 \text{ kg} \cdot \text{m}^2}$$

75. **REASONING AND SOLUTION** The figure below shows the massless board and the forces that act on the board due to the person and the scales.

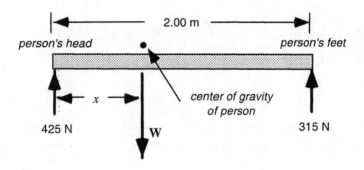

a. Applying Newton's second law to the vertical direction gives

$$315 \text{ N} + 425 \text{ N} - W = 0 \qquad \text{or} \qquad W = \boxed{7.40 \times 10^2 \text{ N, downward}}$$

b. Let x be the position of the center of gravity relative to the scale at the person's head. Taking torques about an axis through the contact point between the scale at the person's head and the board gives

$$(315 \text{ N})(2.00 \text{ m}) - (7.40 \times 10^2 \text{ N})x = 0 \qquad \text{or} \qquad x = \boxed{0.851 \text{ m}}$$

77. **REASONING** When the modules pull together, they do so by means of forces that are internal. These pulling forces, therefore, do not create a net external torque, and the angular momentum of the system is conserved. In other words, it remains constant. We will use the conservation of angular momentum to obtain a relationship between the initial and final

angular speeds. Then, we will use Equation 8.9 ($v = r\omega$) to relate the angular speeds ω_0 and ω_f to the tangential speeds v_0 and v_f.

SOLUTION Let L be the initial length of the cable between the modules and ω_0 be the initial angular speed. Relative to the center-of-mass axis, the initial momentum of inertia of the two-module system is $I_0 = 2M(L/2)^2$, according to Equation 9.6. After the modules pull together, the length of the cable is $L/2$, the final angular speed is ω_f, and the momentum of inertia is $I_f = 2M(L/4)^2$. The conservation of angular momentum indicates that

$$\underbrace{I_f \omega_f}_{\substack{\text{Final angular} \\ \text{momentum}}} = \underbrace{I_0 \omega_0}_{\substack{\text{Initial angular} \\ \text{momentum}}}$$

$$\left[2M\left(\frac{L}{4}\right)^2 \right]\omega_f = \left[2M\left(\frac{L}{2}\right)^2 \right]\omega_0$$

$$\omega_f = 4\omega_0$$

According to Equation 8.9, $\omega_f = v_f/(L/4)$ and $\omega_0 = v_0/(L/2)$. With these substitutions, the result that $\omega_f = 4\omega_0$ becomes

$$\frac{v_f}{L/4} = 4\left(\frac{v_0}{L/2}\right) \quad \text{or} \quad v_f = 2v_0 = 2(17 \text{ m/s}) = \boxed{34 \text{ m/s}}$$

CHAPTER 10 | SIMPLE HARMONIC MOTION AND ELASTICITY

1. **REASONING AND SOLUTION** Using Equation 10.1, we first determine the spring constant:

$$k = \frac{F_x^{\text{Applied}}}{x} = \frac{89.0 \text{ N}}{0.0191 \text{ m}} = 4660 \text{ N/m}$$

Again using Equation 10.1, we find that the force needed to compress the spring by 0.0508 m is

$$F_x^{\text{Applied}} = kx = (4660 \text{ N/m})(0.0508 \text{ m}) = \boxed{237 \text{ N}}$$

5. **REASONING AND SOLUTION** According to Newton's second law, the force required to accelerate the trailer is $F_x = ma_x$ (Equation 4.2a). The displacement x of the spring is given by $F_x = -kx$ (Equation 10.2). Solving Equation 10.2 for x and using $F_x = ma_x$, we obtain

$$x = -\frac{F_x}{k} = -\frac{ma_x}{k} = -\frac{(92 \text{ kg})(0.30 \text{ m/s}^2)}{2300 \text{ N/m}} = -0.012 \text{ m}$$

The amount that the spring stretches is $\boxed{0.012 \text{ m}}$.

9. **REASONING AND SOLUTION** The force that acts on the block is given by Newton's Second law, $F_x = ma_x$ (Equation 4.2a). Since the block has a constant acceleration, the acceleration is given by Equation 2.8 with $v_0 = 0 \text{ m/s}$; that is, $a_x = 2d/t^2$, where d is the distance through which the block is pulled. Therefore, the force that acts on the block is given by

$$F_x = ma_x = \frac{2md}{t^2}$$

The force acting on the block is the restoring force of the spring. Thus, according to Equation 10.2, $F_x = -kx$, where k is the spring constant and x is the displacement. Solving Equation 10.2 for x and using the expression above for F_x, we obtain

$$x = -\frac{F_x}{k} = -\frac{2md}{kt^2} = -\frac{2(7.00 \text{ kg})(4.00 \text{ m})}{(415 \text{ N/m})(0.750 \text{ s})^2} = -0.240 \text{ m}$$

The amount that the spring stretches is $\boxed{0.240 \text{ m}}$.

11. **REASONING**   When the ball is whirled in a horizontal circle of radius r at speed v, the centripetal force is provided by the restoring force of the spring. From Equation 5.3, the magnitude of the centripetal force is mv^2/r, while the magnitude of the restoring force is kx (see Equation 10.2). Thus,

$$\frac{mv^2}{r} = kx \qquad (1)$$

The radius of the circle is equal to $(L_0 + \Delta L)$, where L_0 is the unstretched length of the spring and ΔL is the amount that the spring stretches. Equation (1) becomes

$$\frac{mv^2}{L_0 + \Delta L} = k\Delta L \qquad (1')$$

If the spring were attached to the ceiling and the ball were allowed to hang straight down, motionless, the net force must be zero: $mg - kx = 0$, where $-kx$ is the restoring force of the spring. If we let Δy be the displacement of the spring in the vertical direction, then

$$mg = k\Delta y$$

Solving for Δy, we obtain

$$\Delta y = \frac{mg}{k} \qquad (2)$$

SOLUTION   According to equation (1') above, the spring constant k is given by

$$k = \frac{mv^2}{\Delta L(L_0 + \Delta L)}$$

Substituting this expression for k into equation (2) gives

$$\Delta y = \frac{mg\Delta L(L_0 + \Delta L)}{mv^2} = \frac{g\Delta L(L_0 + \Delta L)}{v^2}$$

or

$$\Delta y = \frac{(9.80 \text{ m/s}^2)(0.010 \text{ m})(0.200 \text{ m} + 0.010 \text{ m})}{(3.00 \text{ m/s})^2} = \boxed{2.29 \times 10^{-3} \text{ m}}$$

15. **REASONING AND SOLUTION** According to Equations 10.6 and 10.11, $2\pi f = \sqrt{k/m}$. According to the data in the problem, the frequency of vibration of either spring is

$$f = \frac{5.0 \text{ cycles}}{3.0 \text{ s}} = \frac{5.0}{3.0} \text{ Hz}$$

Squaring both sides of the equation $2\pi f = \sqrt{k/m}$ and solving for k, we obtain

$$k = 4\pi^2 f^2 m = 4\pi^2 \left(\frac{5.0}{3.0} \text{ Hz}\right)^2 (320 \text{ kg}) = \boxed{3.5\times10^4 \text{ N/m}}$$

---

21. **REASONING** The frequency of vibration of the spring is related to the added mass m by Equations 10.6 and 10.11:

$$f = \frac{1}{2\pi}\sqrt{\frac{k}{m}} \tag{1}$$

The spring constant can be determined from Equation 10.1.

SOLUTION Since the spring stretches by 0.018 m when a 2.8-kg object is suspended from its end, the spring constant is, according to Equation 10.1,

$$k = \frac{F_x^{\text{Applied}}}{x} = \frac{mg}{x} = \frac{(2.8 \text{ kg})(9.80 \text{ m/s}^2)}{0.018 \text{ m}} = 1.52\times10^3 \text{ N/m}$$

Solving Equation (1) for m, we find that the mass required to make the spring vibrate at 3.0 Hz is

$$m = \frac{k}{4\pi^2 f^2} = \frac{1.52\times10^3 \text{ N/m}}{4\pi^2(3.0 \text{ Hz})^2} = \boxed{4.3 \text{ kg}}$$

---

29. **REASONING AND SOLUTION** If we neglect air resistance, only the conservative forces of the spring and gravity act on the object. Therefore, the principle of conservation of mechanical energy applies.

When the 2.00 kg object is hung on the end of the vertical spring, it stretches the spring by an amount y, where

$$y = \frac{F}{k} = \frac{mg}{k} = \frac{(2.00 \text{ kg})(9.80 \text{ m/s}^2)}{50.0 \text{ N/m}} = 0.392 \text{ m} \tag{10.1}$$

This position represents the equilibrium position of the system with the 2.00-kg object suspended from the spring. The object is then pulled down another 0.200 m and released from rest ($v_0 = 0$ m/s). At this point the spring is stretched by an amount of $0.392 \text{ m} + 0.200 \text{ m} = 0.592 \text{ m}$. This point represents the zero reference level ($h = 0$ m) for the gravitational potential energy.

h = 0 m: The kinetic energy, the gravitational potential energy, and the elastic potential energy at the point of release are:

$$\text{KE} = \tfrac{1}{2} m v_0^2 = \tfrac{1}{2} m (0 \text{ m/s})^2 = \boxed{0 \text{ J}}$$

$$\text{PE}_{\text{gravity}} = mgh = mg(0 \text{ m}) = \boxed{0 \text{ J}}$$

$$\text{PE}_{\text{elastic}} = \tfrac{1}{2} k y_0^2 = \tfrac{1}{2}(50.0 \text{ N/m})(0.592 \text{ m})^2 = \boxed{8.76 \text{ J}}$$

The total mechanical energy E_0 at the point of release is the sum of the three energies above: $E_0 = \boxed{8.76 \text{ J}}$.

h = 0.200 m: When the object has risen a distance of $h = 0.200$ m above the release point, the spring is stretched by an amount of $0.592 \text{ m} - 0.200 \text{ m} = 0.392 \text{ m}$. Since the total mechanical energy is conserved, its value at this point is still $E = \boxed{8.76 \text{ J}}$. The gravitational and elastic potential energies are:

$$\text{PE}_{\text{gravity}} = mgh = (2.00 \text{ kg})(9.80 \text{ m/s}^2)(0.200 \text{ m}) = \boxed{3.92 \text{ J}}$$

$$\text{PE}_{\text{elastic}} = \tfrac{1}{2} k y^2 = \tfrac{1}{2}(50.0 \text{ N/m})(0.392 \text{ m})^2 = \boxed{3.84 \text{ J}}$$

Since $\text{KE} + \text{PE}_{\text{gravity}} + \text{PE}_{\text{elastic}} = E$,

$$\text{KE} = E - \text{PE}_{\text{gravity}} - \text{PE}_{\text{elastic}} = 8.76 \text{ J} - 3.92 \text{ J} - 3.84 \text{ J} = \boxed{1.00 \text{ J}}$$

h = 0.400 m: When the object has risen a distance of $h = 0.400$ m above the release point, the spring is stretched by an amount of $0.592 \text{ m} - 0.400 \text{ m} = 0.192 \text{ m}$. At this point, the total mechanical energy is still $E = \boxed{8.76 \text{ J}}$. The gravitational and elastic potential energies are:

$$\text{PE}_{\text{gravity}} = mgh = (2.00 \text{ kg})(9.80 \text{ m/s}^2)(0.400 \text{ m}) = \boxed{7.84 \text{ J}}$$

$$PE_{elastic} = \tfrac{1}{2}ky^2 = \tfrac{1}{2}(50.0 \text{ N/m})(0.192 \text{ m})^2 = \boxed{0.92 \text{ J}}$$

The kinetic energy is

$$KE = E - PE_{gravity} - PE_{elastic} = 8.76 \text{ J} - 7.84 \text{ J} - 0.92 \text{ J} = \boxed{0 \text{ J}}$$

The results are summarized in the table below:

h	KE	PE_{grav}	$PE_{elastic}$	E
0 m	0 J	0 J	8.76 J	8.76 J
0.200 m	1.00 J	3.92 J	3.84 J	8.76 J
0.400 m	0.00 J	7.84 J	0.92 J	8.76 J

33. **REASONING** The only force that acts on the block along the line of motion is the force due to the spring. Since the force due to the spring is a conservative force, the principle of conservation of mechanical energy applies. Initially, when the spring is unstrained, all of the mechanical energy is kinetic energy, $(1/2)mv_0^2$. When the spring is fully compressed, all of the mechanical energy is in the form of elastic potential energy, $(1/2)kx_{max}^2$, where x_{max}, the maximum compression of the spring, is the amplitude A. Therefore, the statement of energy conservation can be written as

$$\tfrac{1}{2}mv_0^2 = \tfrac{1}{2}kA^2$$

This expression may be solved for the amplitude A.

SOLUTION Solving for the amplitude A, we obtain

$$A = \sqrt{\frac{mv_0^2}{k}} = \sqrt{\frac{(1.00 \times 10^{-2} \text{ kg})(8.00 \text{ m/s})^2}{124 \text{ N/m}}} = \boxed{7.18 \times 10^{-2} \text{ m}}$$

37. **REASONING** The angular frequency ω (in rad/s) is given by $\omega = \sqrt{\dfrac{k}{m}}$ (Equation 10.11), where k is the spring constant and m is the mass of the object. However, we are given neither k nor m. Instead, we are given information about how much the spring is compressed and the launch speed of the object. Once launched, the object has kinetic energy, which is related to its speed. Before launching, the spring/object system has elastic potential energy, which is related to the amount by which the spring is compressed. This suggests that we apply the principle of conservation of mechanical energy in order to use the

given information. This principle indicates that the total mechanical energy of the system is the same after the object is launched as it is before the launch. The resulting equation will provide us with the value of k/m that we need in order to determine the angular frequency from $\omega = \sqrt{\dfrac{k}{m}}$.

SOLUTION The conservation of mechanical energy states that the final total mechanical energy E_f is equal to the initial total mechanical energy E_0. The expression for the total mechanical energy for a spring/mass system is given by Equation 10.14, so that we have

$$\underbrace{\tfrac{1}{2}mv_f^2 + \tfrac{1}{2}I\omega_f^2 + mgh_f + \tfrac{1}{2}kx_f^2}_{E_f} = \underbrace{\tfrac{1}{2}mv_0^2 + \tfrac{1}{2}I\omega_0^2 + mgh_0 + \tfrac{1}{2}kx_0^2}_{E_0}$$

Since the object does not rotate, the angular speeds ω_f and ω_0 are zero. Since the object is initially at rest, the initial translational speed v_0 is also zero. Moreover, the motion takes place horizontally, so that the final height h_f is the same as the initial height h_0. Lastly, the spring is unstrained after the launch, so that x_f is zero. Thus, the above expression can be simplified as follows:

$$\tfrac{1}{2}mv_f^2 = \tfrac{1}{2}kx_0^2 \qquad \text{or} \qquad \frac{k}{m} = \frac{v_f^2}{x_0^2}$$

Substituting this result into Equation 10.11 shows that

$$\omega = \sqrt{\frac{k}{m}} = \sqrt{\frac{v_f^2}{x_0^2}} = \frac{v_f}{x_0} = \frac{1.50 \text{ m/s}}{0.0620 \text{ m}} = \boxed{24.2 \text{ rad/s}}$$

---

41. ***REASONING*** Using the principle of conservation of mechanical energy, the initial elastic potential energy stored in the elastic bands must be equal to the sum of the kinetic energy and the gravitational potential energy of the performer at the point of ejection:

$$\frac{1}{2}kx^2 = \frac{1}{2}mv_0^2 + mgh$$

where v_0 is the speed of the performer at the point of ejection and, from the figure at the right, $h = x \sin \theta$.

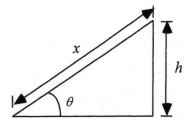

Thus,

$$\frac{1}{2}kx^2 = \frac{1}{2}mv_0^2 + mgx\sin\theta \qquad (1)$$

From the horizontal motion of the performer

$$v_{0x} = v_0 \cos\theta \qquad (2)$$

where

$$v_{0x} = \frac{s_x}{t} \qquad (3)$$

and $s_x = 26.8$ m. Combining equations (2) and (3) gives

$$v_0 = \frac{s_x}{t(\cos\theta)}$$

Equation (1) becomes:

$$\frac{1}{2}kx^2 = \frac{1}{2}m\frac{s_x^2}{t^2\cos^2\theta} + mgx\sin\theta$$

This expression can be solved for k, the spring constant of the firing mechanism.

SOLUTION Solving for k yields:

$$k = m\left(\frac{s_x}{xt\cos\theta}\right)^2 + \frac{2mg(\sin\theta)}{x}$$

$$k = (70.0 \text{ kg})\left[\frac{26.8 \text{ m}}{(3.00 \text{ m})(2.14 \text{ s})(\cos 40.0°)}\right]^2$$

$$+ \frac{2(70.0 \text{ kg})(9.80 \text{ m/s}^2)(\sin 40.0°)}{3.00 \text{ m}} = \boxed{2.37\times10^3 \text{ N/m}}$$

---

51. **REASONING** When the tow truck pulls the car out of the ditch, the cable stretches and a tension exists in it. This tension is the force that acts on the car. The amount ΔL that the cable stretches depends on the tension F, the length L_0 and cross-sectional area A of the cable, as well as Young's modulus Y for steel. All of these quantities are given in the statement of the problem, except for Young's modulus, which can be found by consulting Table 10.1.

SOLUTION Solving Equation 10.17, $F = Y\left(\dfrac{\Delta L}{L_0}\right)A$, for the change in length, we have

$$\Delta L = \frac{F L_0}{A Y} = \frac{(890\ \text{N})(9.1\ \text{m})}{\pi\left(0.50 \times 10^{-2}\ \text{m}\right)^2 \left(2.0 \times 10^{11}\ \text{N/m}^2\right)} = \boxed{5.2 \times 10^{-4}\ \text{m}}$$

---

53. **REASONING AND SOLUTION** According to Equation 10.20, it follows that

$$\Delta P = -B\frac{\Delta V}{V_0} = -\left(2.6 \times 10^{10}\ \text{N/m}^2\right)\frac{-1.0 \times 10^{-10}\ \text{m}^3}{1.0 \times 10^{-6}\ \text{m}^3} = 2.6 \times 10^6\ \text{N/m}^2 \qquad (10.20)$$

where we have expressed the volume V_0 of the cube at the ocean's surface as $V_0 = \left(1.0 \times 10^{-2}\ \text{m}\right)^3 = 1.0 \times 10^{-6}\ \text{m}^3$.

Since the pressure increases by $1.0 \times 10^4\ \text{N/m}^2$ per meter of depth, the depth is

$$\frac{2.6 \times 10^6\,\text{N/m}^2}{1.0 \times 10^4\,\dfrac{\text{N/m}^2}{\text{m}}} = \boxed{260\ \text{m}}$$

---

59. **REASONING AND SOLUTION** The shearing stress is equal to the force per unit area applied to the rivet. Thus, when a shearing stress of 5.0×10^8 Pa is applied to each rivet, the force experienced by each rivet is

$$F = (\text{Stress})A = (\text{Stress})(\pi r^2) = (5.0 \times 10^8\ \text{Pa})\left[\pi(5.0 \times 10^{-3}\ \text{m})^2\right] = 3.9 \times 10^4\ \text{N}$$

Therefore, the maximum tension T that can be applied to each beam, assuming that each rivet carries one-fourth of the total load, is $4F = \boxed{1.6 \times 10^5\ \text{N}}$.

---

65. **REASONING** Our approach is straightforward. We will begin by writing Equation 10.17 $\left[F = Y\left(\dfrac{\Delta L}{L_0}\right)A\right]$ as it applies to the composite rod. In so doing, we will use subscripts for only those variables that have different values for the composite rod and the aluminum and tungsten sections. Thus, we note that the force applied to the end of the composite rod (see Figure 10.28) is also applied to each section of the rod, with the result that the magnitude F of the force has no subscript. Similarly, the cross-sectional area A is the same for the composite rod and for the aluminum and tungsten sections. Next, we will express the

change $\Delta L_{\text{Composite}}$ in the length of the composite rod as the sum of the changes in lengths of the aluminum and tungsten sections. Lastly, we will take into account that the initial length of the composite rod is twice the initial length of either of its two sections and thereby simply our equation algebraically to the point that we can solve it for the effective value of Young's modulus that applies to the composite rod.

SOLUTION Applying Equation 10.17 to the composite rod, we obtain

$$F = Y_{\text{Composite}} \left(\frac{\Delta L_{\text{Composite}}}{L_{0,\,\text{Composite}}} \right) A \qquad (1)$$

Since the change $\Delta L_{\text{Composite}}$ in the length of the composite rod is the sum of the changes in lengths of the aluminum and tungsten sections, we have $\Delta L_{\text{Composite}} = \Delta L_{\text{Aluminum}} + \Delta L_{\text{Tungsten}}$. Furthermore, the changes in length of each section can be expressed using Equation 10.17 $\left(\Delta L = \dfrac{FL_0}{YA} \right)$, so that

$$\Delta L_{\text{Composite}} = \Delta L_{\text{Aluminum}} + \Delta L_{\text{Tungsten}} = \frac{FL_{0,\,\text{Aluminum}}}{Y_{\text{Aluminum}} A} + \frac{FL_{0,\,\text{Tungsten}}}{Y_{\text{Tungsten}} A}$$

Substituting this result into Equation (1) gives

$$F = \left(\frac{Y_{\text{Composite}} A}{L_{0,\,\text{Composite}}} \right) \Delta L_{\text{Composite}} = \left(\frac{Y_{\text{Composite}} A}{L_{0,\,\text{Composite}}} \right) \left(\frac{FL_{0,\,\text{Aluminum}}}{Y_{\text{Aluminum}} A} + \frac{FL_{0,\,\text{Tungsten}}}{Y_{\text{Tungsten}} A} \right)$$

$$1 = Y_{\text{Composite}} \left(\frac{L_{0,\,\text{Aluminum}}}{L_{0,\,\text{Composite}} Y_{\text{Aluminum}}} + \frac{L_{0,\,\text{Tungsten}}}{L_{0,\,\text{Composite}} Y_{\text{Tungsten}}} \right)$$

In this result we now use the fact that $L_{0,\,\text{Aluminum}}/L_{0,\,\text{Composite}} = L_{0,\,\text{Tungsten}}/L_{0,\,\text{Composite}} = 1/2$ and obtain

$$1 = Y_{\text{Composite}} \left(\frac{1}{2Y_{\text{Aluminum}}} + \frac{1}{2Y_{\text{Tungsten}}} \right)$$

Solving for $Y_{\text{Composite}}$ shows that

$$Y_{\text{Composite}} = \frac{2Y_{\text{Tungsten}} Y_{\text{Aluminum}}}{Y_{\text{Tungsten}} + Y_{\text{Aluminum}}} = \frac{2\left(3.6 \times 10^{11} \text{ N/m}^2\right)\left(6.9 \times 10^{10} \text{ N/m}^2\right)}{\left(3.6 \times 10^{11} \text{ N/m}^2\right) + \left(6.9 \times 10^{10} \text{ N/m}^2\right)} = \boxed{1.2 \times 10^{11} \text{ N/m}^2}$$

The values for Y_{Tungsten} and Y_{Aluminum} have been taken from Table 10.1.

69. **REASONING** The strain in the wire is given by $\Delta L / L_0$. From Equation 10.17, the strain is therefore given by

$$\frac{\Delta L}{L_0} = \frac{F}{YA} \tag{1}$$

where F must be equal to the magnitude of the centripetal force that keeps the stone moving in the circular path of radius R. Table 10.1 gives the value of Y for steel.

SOLUTION Combining Equation (1) with Equation 5.3 for the magnitude of the centripetal force, we obtain

$$\frac{\Delta L}{L_0} = \frac{F}{Y(\pi r^2)} = \frac{(mv^2 / R)}{Y(\pi r^2)} = \frac{(8.0 \text{ kg})(12 \text{ m/s})^2 / (4.0 \text{ m})}{(2.0 \times 10^{11} \text{ Pa}) \pi (1.0 \times 10^{-3} \text{ m})^2} = \boxed{4.6 \times 10^{-4}}$$

73. **REASONING** The change in length of the wire is, According to Equation 10.17, $\Delta L = FL_0 / YA$, where the force F is equal to the tension T in the wire. The tension in the wire can be found by applying Newton's second law to the two crates.

SOLUTION The drawing shows the free-body diagrams for the two crates. Taking up as the positive direction, Newton's second law for each of the two crates gives

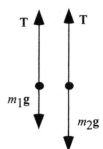

$$T - m_1 g = m_1 a \tag{1}$$

$$T - m_2 g = -m_2 a \tag{2}$$

Solving Equation (2) for a, we find $a = -\left(\dfrac{T - m_2 g}{m_2}\right)$. Substituting into Equation (1) gives

$$T - 2 m_1 g + \frac{m_1}{m_2} T = 0$$

Solving for T we find

$$T = \frac{2 m_1 m_2 g}{m_1 + m_2} = \frac{2(3.0 \text{ kg})(5.0 \text{ kg})(9.80 \text{ m/s}^2)}{3.0 \text{ kg} + 5.0 \text{ kg}} = 37 \text{ N}$$

Using the value given in Table 10.1 for Young's modulus Y of steel, we find, therefore, that the change in length of the wire is given by Equation 10.17 as

$$\Delta L = \frac{(37 \text{ N})(1.5 \text{ m})}{(2.0 \times 10^{11} \text{ N/m}^2)(1.3 \times 10^{-5} \text{ m}^2)} = \boxed{2.1 \times 10^{-5} \text{ m}}$$

77. **REASONING** Each spring supports one-quarter of the total mass m_{total} of the system (the empty car plus the four passengers), or $\frac{1}{4} m_{total}$. The mass $m_{one\ passenger}$ of one passenger is equal to $\frac{1}{4} m_{total}$ minus one-quarter of the mass $m_{empty\ car}$ of the empty car:

$$m_{one\ passenger} = \frac{1}{4} m_{total} - \frac{1}{4} m_{empty\ car} \qquad (1)$$

The mass of the empty car is known. Since the car and its passengers oscillate up and down in simple harmonic motion, the angular frequency ω of oscillation is related to the spring constant k and the mass $\frac{1}{4} m_{total}$ supported by each spring by:

$$\omega = \sqrt{\frac{k}{\frac{1}{4} m_{total}}} \qquad (10.11)$$

Solving this expression for $\frac{1}{4} m_{total}$ $\left(\frac{1}{4} m_{total} = k / \omega^2 \right)$ and substituting the result into Equation (1) gives

$$m_{one\ passenger} = \frac{k}{\omega^2} - \frac{1}{4} m_{empty\ car} \qquad (2)$$

The angular frequency ω is inversely related to the period T of oscillation by $\omega = 2\pi / T$ (see Equation 10.4). Substituting this expression for ω into Equation (2) yields

$$m_{one\ passenger} = \frac{k}{\left(\dfrac{2\pi}{T} \right)^2} - \frac{1}{4} m_{empty\ car}$$

SOLUTION The mass of one of the passengers is

$$m_{one\ passenger} = \frac{k}{\left(\dfrac{2\pi}{T} \right)^2} - \frac{1}{4} m_{empty\ car} = \frac{1.30 \times 10^5 \text{ N/m}}{\left(\dfrac{2\pi}{0.370 \text{ s}} \right)^2} - \frac{1}{4}(1560 \text{ kg}) = \boxed{61 \text{ kg}}$$

87. **REASONING AND SOLUTION** The natural frequency of the suspension system is given by Equation 10.11:

$$\omega = \sqrt{\frac{k}{m}} = \sqrt{\frac{1.50 \times 10^6 \text{ N/m}}{215 \text{ kg}}} = 83.5 \text{ rad/s}$$

Thus, the wheel will resonate when its angular speed is 83.5 rad/s. This corresponds to a linear speed of

$$v = r\omega = (0.400 \text{ m})(83.5 \text{ rad/s}) = \boxed{33.4 \text{ m/s}}$$

---

89. **REASONING**   Equation 10.20 can be used to find the fractional change in volume of the brass sphere when it is exposed to the Venusian atmosphere. Once the fractional change in volume is known, it can be used to calculate the fractional change in radius.

SOLUTION According to Equation 10.20, the fractional change in volume is

$$\frac{\Delta V}{V_0} = -\frac{\Delta P}{B} = -\frac{8.9 \times 10^6 \text{ Pa}}{6.7 \times 10^{10} \text{ Pa}} = -1.33 \times 10^{-4}$$

Here, we have used the fact that $\Delta P = 9.0 \times 10^6 \text{ Pa} - 1.0 \times 10^5 \text{ Pa} = 8.9 \times 10^6 \text{ Pa}$, and we have taken the value for the bulk modulus B of brass from Table 10.3. The initial volume of the sphere is $V_0 = \frac{4}{3}\pi r_0^3$. If we assume that the change in the radius of the sphere is very small relative to the initial radius, we can think of the sphere's change in volume as the addition or subtraction of a spherical shell of volume ΔV, whose radius is r_0 and whose thickness is Δr. Then, the change in volume of the sphere is equal to the volume of the shell and is given by $\Delta V = \left(4\pi r_0^2\right)\Delta r$. Combining the expressions for V_0 and ΔV, and solving for $\Delta r / r_0$, we have

$$\frac{\Delta r}{r_0} = \left(\frac{1}{3}\right)\frac{\Delta V}{V_0}$$

Therefore,

$$\frac{\Delta r}{r_0} = \frac{1}{3}\left(-1.33 \times 10^{-4}\right) = \boxed{-4.4 \times 10^{-5}}$$

---

91. **REASONING**   The angular frequency for simple harmonic motion is given by Equation 10.11 as $\omega = \sqrt{k/m}$. Since the frequency f is related to the angular frequency ω by $f = \omega/(2\pi)$ and f is related to the period T by $f = 1/T$, the period of the motion is given by

$$T = \frac{2\pi}{\omega} = 2\pi\sqrt{\frac{k}{m}}$$

SOLUTION

a. When $m_1 = m_2 = 3.0$ kg, we have that

$$T_1 = T_2 = 2\pi \sqrt{\frac{3.0 \text{ kg}}{120 \text{ N/m}}} = 0.99 \text{ s}$$

Both particles will pass through the position $x = 0$ m for the first time one-quarter of the way through one cycle, or

$$\Delta t = \tfrac{1}{4}T_1 = \tfrac{1}{4}T_2 = \frac{0.99 \text{ s}}{4} = \boxed{0.25 \text{ s}}$$

b. $T_1 = 0.99$ s, as in part (a) above, while

$$T_2 = 2\pi \sqrt{\frac{27.0 \text{ kg}}{120 \text{ N/m}}} = 3.0 \text{ s}$$

Each particle will pass through the position $x = 0$ m every odd-quarter of a cycle, $\tfrac{1}{4}T, \tfrac{3}{4}T, \tfrac{5}{4}T, \ldots$ Thus, the two particles will pass through $x = 0$ m when

3.0-kg particle $t = \tfrac{1}{4}T_1, \tfrac{3}{4}T_1, \tfrac{5}{4}T_1, \ldots$

27.0-kg particle $t = \tfrac{1}{4}T_2, \tfrac{3}{4}T_2, \tfrac{5}{4}T_2, \ldots$

Since $T_2 = 3T_1$, we see that both particles will be at $x = 0$ m simultaneously when $t = \tfrac{3}{4}T_1$, or $t = \tfrac{1}{4}T_2 = \tfrac{3}{4}T_1$. Thus,

$$t = \tfrac{3}{4}T_1 = \tfrac{3}{4}(0.99 \text{ s}) = \boxed{0.75 \text{ s}}$$

CHAPTER 11 | *FLUIDS*

1. **REASONING** The weight W of the water bed is equal to the mass m of water times the acceleration g due to gravity; $W = mg$ (Equation 4.5). The mass, on the other hand, is equal to the density ρ of the water times its volume V, or $m = \rho V$ (Equation 11.1).

SOLUTION Substituting $m = \rho V$ into the relation $W = mg$ gives

$$W = mg = (\rho V)g$$

$$= \left(1.00 \times 10^3 \text{ kg/m}^3\right)\left(1.83 \text{ m} \times 2.13 \text{ m} \times 0.229 \text{ m}\right)\left(9.80 \text{ m/s}^2\right) = \boxed{8750 \text{ N}}$$

We have taken the density of water from Table 11.1. Since the weight of the water bed is greater than the additional weight that the floor can tolerate, the bed should not be purchased.

3. **REASONING** Equation 11.1 can be used to find the volume occupied by 1.00 kg of silver. Once the volume is known, the area of a sheet of silver of thickness d can be found from the fact that the volume is equal to the area of the sheet times its thickness.

SOLUTION Solving Equation 11.1 for V and using a value of $\rho = 10\ 500 \text{ kg/m}^3$ for the density of silver (see Table 11.1), we find that the volume of 1.00 kg of silver is

$$V = \frac{m}{\rho} = \frac{1.00 \text{ kg}}{10\ 500 \text{ kg/m}^3} = 9.52 \times 10^{-5} \text{ m}^3$$

The area of the silver, is, therefore,

$$A = \frac{V}{d} = \frac{9.52 \times 10^{-5} \text{ m}^3}{3.00 \times 10^{-7} \text{ m}} = \boxed{317 \text{ m}^2}$$

9. **REASONING** The period T of a satellite is the time for it to make one complete revolution around the planet. The period is the circumference of the circular orbit $(2\pi R)$ divided by the speed v of the satellite, so that $T = (2\pi R)/v$ (see Equation 5.1). In Section 5.5 we saw that the centripetal force required to keep a satellite moving in a circular orbit is provided by the gravitational force. This relationship tells us that the speed of the satellite must be $v = \sqrt{GM/R}$ (Equation 5.5), where G is the universal gravitational constant and M is the mass of the planet. By combining this expression for the speed with that for the period, and using the definition of density, we can obtain the period of the satellite.

SOLUTION The period of the satellite is

$$T = \frac{2\pi R}{v} = \frac{2\pi R}{\sqrt{\dfrac{GM}{R}}} = 2\pi\sqrt{\frac{R^3}{GM}}$$

According to Equation 11.1, the mass of the planet is equal to its density ρ times its volume V. Since the planet is spherical, $V = \frac{4}{3}\pi R^3$. Thus, $M = \rho V = \rho\left(\frac{4}{3}\pi R^3\right)$. Substituting this expression for M into that for the period T gives

$$T = 2\pi\sqrt{\frac{R^3}{GM}} = 2\pi\sqrt{\frac{R^3}{G\rho\left(\frac{4}{3}\pi R^3\right)}} = \sqrt{\frac{3\pi}{G\rho}}$$

The density of iron is $\rho = 7860$ kg/m^3 (see Table 11.1), so the period of the satellite is

$$T = \sqrt{\frac{3\pi}{G\rho}} = \sqrt{\frac{3\pi}{\left(6.67\times10^{-11}\ \text{N}\cdot\text{m}^2/\text{kg}^2\right)\left(7860\ \text{kg/m}^3\right)}} = \boxed{4240\ \text{s}}$$

---

11. **REASONING** Since the inside of the box is completely evacuated; there is no air to exert an upward force on the lid from the inside. Furthermore, since the weight of the lid is negligible, there is only one force that acts on the lid; the downward force caused by the air pressure on the outside of the lid. In order to pull the lid off the box, one must supply a force that is at least equal in magnitude and opposite in direction to the force exerted on the lid by the outside air.

SOLUTION According to Equation 11.3, pressure is defined as $P = F/A$; therefore, the magnitude of the force on the lid due to the air pressure is

$$F = (0.85\times10^5\ \text{N/m}^2)(1.3\times10^{-2}\ \text{m}^2) = \boxed{1.1\times10^3\ \text{N}}$$

---

19. **REASONING AND SOLUTION** Both the cylinder and the hemisphere have circular cross sections. According to Equation 11.3, the pressure exerted on the ground by the hemisphere is

$$P = \frac{W_h}{A_h} = \frac{W_h}{\pi r_h^2}$$

where W_h and r_h are the weight and radius of the hemisphere. Similarly, the pressure exerted on the ground by the cylinder is

$$P = \frac{W_c}{A_c} = \frac{W_c}{\pi r_c^2}.$$

where W_c and r_c are the weight and radius of the cylinder. Since each object exerts the same pressure on the ground, we can equate the right-hand sides of the expressions above to obtain

$$\frac{W_h}{\pi r_h^2} = \frac{W_c}{\pi r_c^2}$$

Solving for r_h^2, we obtain

$$r_h^2 = r_c^2 \frac{W_h}{W_c} \qquad (1)$$

The weight of the hemisphere is

$$W_h = \rho g V_h = \rho g \left[\tfrac{1}{2} \left(\tfrac{4}{3} \pi r_h^3 \right) \right] = \tfrac{2}{3} \rho g \pi r_h^3$$

where ρ and V_h are the density and volume of the hemisphere, respectively. The weight of the cylinder is

$$W_c = \rho g V_c = \rho g \pi r_c^2 h$$

where ρ and V_c are the density and volume of the cylinder, respectively, and h is the height of the cylinder. Substituting these expressions for the weights into Equation (1) gives

$$r_h^2 = r_c^2 \frac{W_h}{W_c} = r_c^2 \frac{\tfrac{2}{3} \rho g \pi r_h^3}{\rho g \pi r_c^2 h}$$

Solving for r_h gives

$$r_h = \tfrac{3}{2} h = \tfrac{3}{2} (0.500 \text{ m}) = \boxed{0.750 \text{ m}}$$

---

23. **REASONING** As the depth h increases, the pressure increases according to Equation 11.4 ($P_2 = P_1 + \rho g h$). In this equation, P_1 is the pressure at the shallow end, P_2 is the pressure at the deep end, and ρ is the density of water ($1.00 \times 10^3 \text{ kg/m}^3$, see Table 11.1). We seek a value for the pressure at the deep end minus the pressure at the shallow end.

SOLUTION Using Equation 11.4, we find

$$P_{Deep} = P_{Shallow} + \rho g h \quad \text{or} \quad P_{Deep} - P_{Shallow} = \rho g h$$

The drawing at the right shows that a value for h can be obtained from the 15-m length of the pool by using the tangent of the 11° angle:

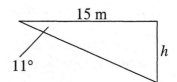

$$\tan 11° = \frac{h}{15 \text{ m}} \quad \text{or} \quad h = (15 \text{ m}) \tan 11°$$

$$P_{Deep} - P_{Shallow} = \rho g (15 \text{ m}) \tan 11°$$

$$= (1.00 \times 10^3 \text{ kg/m}^3)(9.80 \text{ m/s}^2)(15 \text{ m}) \tan 11° = \boxed{2.9 \times 10^4 \text{ Pa}}$$

---

27. ***REASONING AND SOLUTION***

a. The pressure at the level of house A is given by Equation 11.4 as $P = P_{atm} + \rho g h$. Now the height h consists of the 15.0 m and the diameter d of the tank. We first calculate the radius of the tank, from which we can infer d. Since the tank is spherical, its full mass is given by $M = \rho V = \rho[(4/3)\pi r^3]$. Therefore,

$$r^3 = \frac{3M}{4\pi \rho} \quad \text{or} \quad r = \left(\frac{3M}{4\pi \rho}\right)^{1/3} = \left[\frac{3(5.25 \times 10^5 \text{kg})}{4\pi(1.000 \times 10^3 \text{kg/m}^3)}\right]^{1/3} = 5.00 \text{ m}$$

Therefore, the diameter of the tank is 10.0 m, and the height h is given by

$$h = 10.0 \text{ m} + 15.0 \text{ m} = 25.0 \text{ m}$$

According to Equation 11.4, the gauge pressure in house A is, therefore,

$$P - P_{atm} = \rho g h = (1.000 \times 10^3 \text{ kg/m}^3)(9.80 \text{ m/s}^2)(25.0 \text{ m}) = \boxed{2.45 \times 10^5 \text{ Pa}}$$

b. The pressure at house B is $P = P_{atm} + \rho g h$, where

$$h = 15.0 \text{ m} + 10.0 \text{ m} - 7.30 \text{ m} = 17.7 \text{ m}$$

According to Equation 11.4, the gauge pressure in house B is

$$P - P_{atm} = \rho g h = (1.000 \times 10^3 \text{ kg/m}^3)(9.80 \text{ m/s}^2)(17.7 \text{ m}) = \boxed{1.73 \times 10^5 \text{ Pa}}$$

29. **REASONING** Let the length of the tube be denoted by L, and let the length of the liquid be denoted by ℓ. When the tube is whirled in a circle at a constant angular speed about an axis through one end, the liquid collects at the other end and experiences a centripetal force given by (see Equation 8.11, and use $F = ma$) $F = mr\omega^2 = mL\omega^2$.

Since there is no air in the tube, this is the only radial force experienced by the liquid, and it results in a pressure of

$$P = \frac{F}{A} = \frac{mL\omega^2}{A}$$

where A is the cross-sectional area of the tube. The mass of the liquid can be expressed in terms of its density ρ and volume V: $m = \rho V = \rho A \ell$. The pressure may then be written as

$$P = \frac{\rho A \ell L \omega^2}{A} = \rho \ell L \omega^2 \tag{1}$$

If the tube were completely filled with liquid and allowed to hang vertically, the pressure at the bottom of the tube (that is, where $h = L$) would be given by

$$P = \rho g L \tag{2}$$

SOLUTION According to the statement of the problem, the quantities calculated by Equations (1) and (2) are equal, so that $\rho \ell L \omega^2 = \rho g L$. Solving for ω gives

$$\omega = \sqrt{\frac{g}{\ell}} = \sqrt{\frac{9.80 \text{ m/s}^2}{0.0100 \text{ m}}} = \boxed{31.3 \text{ rad/s}}$$

---

33. **REASONING** According to Equation 11.4, the initial pressure at the bottom of the pool is $P_0 = \left(P_{atm}\right)_0 + \rho g h$, while the final pressure is $P_f = \left(P_{atm}\right)_f + \rho g h$. Therefore, the change in pressure at the bottom of the pool is

$$\Delta P = P_f - P_0 = \left[\left(P_{atm}\right)_f + \rho g h\right] - \left[\left(P_{atm}\right)_0 + \rho g h\right] = \left(P_{atm}\right)_f - \left(P_{atm}\right)_0$$

According to Equation 11.3, $F = PA$, the change in the force at the bottom of the pool is

$$\Delta F = (\Delta P) A = \left[\left(P_{atm}\right)_f - \left(P_{atm}\right)_0\right] A$$

SOLUTION Direct substitution of the data given in the problem into the expression above yields

$$\Delta F = \left(765 \text{ mm Hg} - 755 \text{ mm Hg}\right)(12 \text{ m})(24 \text{ m})\left(\frac{133 \text{ Pa}}{1.0 \text{ mm Hg}}\right) = \boxed{3.8 \times 10^5 \text{ N}}$$

Note that the conversion factor 133 Pa = 1.0 mm Hg is used to convert mm Hg to Pa.

---

39. **REASONING**   The pressure P' exerted on the bed of the truck by the plunger is $P' = P - P_{atm}$. According to Equation 11.3, $F = P'A$, so the force exerted on the bed of the truck can be expressed as $F = (P - P_{atm})A$. If we assume that the plunger remains perpendicular to the floor of the load bed, the torque that the plunger creates about the axis shown in the figure in the text is

$$\tau = F\ell = (P - P_{atm})A\ell = (P - P_{atm})(\pi r^2)\ell$$

SOLUTION   Direct substitution of the numerical data into the expression above gives

$$\tau = \left(3.54 \times 10^6 \text{ Pa} - 1.01 \times 10^5 \text{ Pa}\right)\pi(0.150 \text{ m})^2 (3.50 \text{ m}) = \boxed{8.50 \times 10^5 \text{ N} \cdot \text{m}}$$

---

41. **REASONING**   The buoyant force exerted on the balloon by the air must be equal in magnitude to the weight of the balloon and its contents (load and hydrogen). The magnitude of the buoyant force is given by $\rho_{air} V g$. Therefore,

$$\rho_{air} V g = W_{load} + \rho_{hydrogen} V g$$

where, since the balloon is spherical, $V = (4/3)\pi r^3$. Making this substitution for V and solving for r, we obtain

$$r = \left[\frac{3W_{load}}{4\pi g(\rho_{air} - \rho_{hydrogen})}\right]^{1/3}$$

SOLUTION   Direct substitution of the data given in the problem yields

$$r = \left[\frac{3(5750 \text{ N})}{4\pi(9.80 \text{ m/s}^2)(1.29 \text{ kg/m}^3 - 0.0899 \text{ kg/m}^3)}\right]^{1/3} = \boxed{4.89 \text{ m}}$$

---

45. **REASONING** According to Equation 11.1, the density of the life jacket is its mass divided by its volume. The volume is given. To obtain the mass, we note that the person wearing the life jacket is floating, so that the upward-acting buoyant force balances the downward-acting weight of the person and life jacket. The magnitude of the buoyant force is the weight of the

displaced water, according to Archimedes' principle. We can express each of the weights as mg (Equation 4.5) and then relate the mass of the displaced water to the density of water and the displaced volume by using Equation 11.1.

SOLUTION According to Equation 11.1, the density of the life jacket is

$$\rho_J = \frac{m_J}{V_J} \qquad (1)$$

Since the person wearing the life jacket is floating, the upward-acting buoyant force F_B balances the downward-acting weight W_P of the person and the weight W_J of the life jacket. The buoyant force has a magnitude that equals the weight W_{H_2O} of the displaced water, as stated by Archimedes' principle. Thus, we have

$$F_B = W_{H_2O} = W_P + W_J \qquad (2)$$

In Equation (2), we can use Equation 4.5 to express each weight as mass m times the magnitude g of the acceleration due to gravity. Then, the mass of the water can be expressed as $m_{H_2O} = \rho_{H_2O} V_{H_2O}$ (Equation 11.1). With these substitutions, Equation (2) becomes

$$m_{H_2O} g = m_P g + m_J g \quad \text{or} \quad (\rho_{H_2O} V_{H_2O}) g = m_P g + m_J g$$

Solving this result for m_J shows that

$$m_J = \rho_{H_2O} V_{H_2O} - m_P$$

Substituting this result into Equation (1) and noting that the volume of the displaced water is $V_{H_2O} = 3.1 \times 10^{-2} \text{ m}^3 + 6.2 \times 10^{-2} \text{ m}^3$ gives

$$\rho_J = \frac{\rho_{H_2O} V_{H_2O} - m_P}{V_J} = \frac{\left(1.00 \times 10^3 \text{ kg/m}^3\right)\left(3.1 \times 10^{-2} \text{ m}^3 + 6.2 \times 10^{-2} \text{ m}^3\right) - 81 \text{ kg}}{3.1 \times 10^{-2} \text{ m}^3} = \boxed{390 \text{ kg/m}^3}$$

---

49. **REASONING AND SOLUTION** The weight of the coin in air is equal to the sum of the weights of the silver and copper that make up the coin:

$$W_{air} = m_{coin} g = \left(m_{silver} + m_{copper}\right) g = \left(\rho_{silver} V_{silver} + \rho_{copper} V_{copper}\right) g \qquad (1)$$

The weight of the coin in water is equal to its weight in air minus the buoyant force exerted on the coin by the water. Therefore,

$$W_{water} = W_{air} - \rho_{water}(V_{silver} + V_{copper})g \qquad (2)$$

Solving Equation (2) for the sum of the volumes gives

$$V_{silver} + V_{copper} = \frac{W_{air} - W_{water}}{\rho_{water}\,g} = \frac{m_{coin}\,g - W_{water}}{\rho_{water}\,g}$$

or

$$V_{silver} + V_{copper} = \frac{(1.150 \times 10^{-2}\ kg)(9.80\ m/s^2) - 0.1011\ N}{(1.000 \times 10^3\ kg/m^3)(9.80\ m/s^2)} = 1.184 \times 10^{-6}\ m^3$$

From Equation (1) we have

$$V_{silver} = \frac{m_{coin} - \rho_{copper}V_{copper}}{\rho_{silver}} = \frac{1.150 \times 10^{-2}\ kg - (8890\ kg/m^3)V_{copper}}{10\ 500\ kg/m^3}$$

or

$$V_{silver} = 1.095 \times 10^{-6}\ m^3 - (0.8467)V_{copper} \qquad (3)$$

Substitution of Equation (3) into Equation (2) gives

$$W_{water} = W_{air} - \rho_{water}\left[1.095 \times 10^{-6}\ m^3 - (0.8467)V_{copper} + V_{copper}\right]g$$

Solving for V_{copper} gives $V_{copper} = 5.806 \times 10^{-7}\ m^3$. Substituting into Equation (3) gives

$$V_{silver} = 1.095 \times 10^{-6}\ m^3 - (0.8467)(5.806 \times 10^{-7}\ m^3) = 6.034 \times 10^{-7}\ m^3$$

From Equation 11.1, the mass of the silver is

$$m_{silver} = \rho_{silver}V_{silver} = (10\ 500\ kg/m^3)(6.034 \times 10^{-7}\ m^3) = \boxed{6.3 \times 10^{-3}\ kg}$$

---

51. **REASONING** The height of the cylinder that is in the oil is given by $h_{oil} = V_{oil}/(\pi r^2)$, where V_{oil} is the volume of oil displaced by the cylinder and r is the radius of the cylinder. We must, therefore, find the volume of oil displaced by the cylinder. After the oil is poured in, the buoyant force that acts on the cylinder is equal to the sum of the weight of the water displaced by the cylinder and the weight of the oil displaced by the cylinder. Therefore, the magnitude of the buoyant force is given by $F = \rho_{water}gV_{water} + \rho_{oil}gV_{oil}$. Since the cylinder

floats in the fluid, the net force that acts on the cylinder must be zero. Therefore, the buoyant force that supports the cylinder must be equal to the weight of the cylinder, or

$$\rho_{water} g V_{water} + \rho_{oil} g V_{oil} = mg$$

where m is the mass of the cylinder. Substituting values into the expression above leads to

$$V_{water} + (0.725)V_{oil} = 7.00 \times 10^{-3} \ m^3 \tag{1}$$

From the figure in the text, $V_{cylinder} = V_{water} + V_{oil}$. Substituting values into the expression for $V_{cylinder}$ gives

$$V_{water} + V_{oil} = 8.48 \times 10^{-3} \ m^3 \tag{2}$$

Subtracting Equation (1) from Equation (2) yields $V_{oil} = 5.38 \times 10^{-3} \ m^3$.

SOLUTION The height of the cylinder that is in the oil is, therefore,

$$h_{oil} = \frac{V_{oil}}{\pi r^2} = \frac{5.38 \times 10^{-3} \ m^3}{\pi (0.150 \ m)^2} = \boxed{7.6 \times 10^{-2} \ m}$$

---

55. **REASONING AND SOLUTION** The mass flow rate Q_{mass} is the amount of fluid mass that flows per unit time. Therefore,

$$Q_{mass} = \frac{m}{t} = \frac{\rho V}{t} = \frac{(1030 \ kg/m^3)(9.5 \times 10^{-4} \ m^3)}{6.0 \ h} \left(\frac{1.0 \ h}{3600 \ s} \right) = \boxed{4.5 \times 10^{-5} \ kg/s}$$

---

61. **REASONING AND SOLUTION**
a. Using Equation 11.12, the form of Bernoulli's equation with $y_1 = y_2$, we have

$$P_1 - P_2 = \frac{1}{2}\rho \left(v_2^2 - v_1^2\right) = \frac{1.29 \ kg/m^3}{2}\left[(15 \ m/s)^2 - (0 \ m/s)^2\right] = \boxed{150 \ Pa}$$

b. The pressure inside the roof is greater than the pressure on the outside. Therefore, there is a net outward force on the roof. If the wind speed is sufficiently high, some roofs are "blown outward."

63. **REASONING AND SOLUTION**   Let the speed of the air below and above the wing be given by v_1 and v_2, respectively. According to Equation 11.12, the form of Bernoulli's equation with $y_1 = y_2$, we have

$$P_1 - P_2 = \tfrac{1}{2}\rho\left(v_2^2 - v_1^2\right) = \frac{1.29 \text{ kg} / \text{m}^3}{2}\left[(251 \text{ m}/\text{s})^2 - (225 \text{ m}/\text{s})^2\right] = 7.98 \times 10^3 \text{ Pa}$$

From Equation 11.3, the lifting force is, therefore,

$$F = \left(P_1 - P_2\right)A = (7.98 \times 10^3 \text{ Pa})(24.0 \text{ m}^2) = \boxed{1.92 \times 10^5 \text{ N}}$$

69. **REASONING**   Since the pressure difference is known, Bernoulli's equation can be used to find the speed v_2 of the gas in the pipe. Bernoulli's equation also contains the unknown speed v_1 of the gas in the Venturi meter; therefore, we must first express v_1 in terms of v_2. This can be done by using Equation 11.9, the equation of continuity.

SOLUTION

a.   From the equation of continuity (Equation 11.9) it follows that $v_1 = \left(A_2 / A_1\right)v_2$. Therefore,

$$v_1 = \frac{0.0700 \text{ m}^2}{0.0500 \text{ m}^2}v_2 = (1.40)v_2$$

Substituting this expression into Bernoulli's equation (Equation 11.12), we have

$$P_1 + \tfrac{1}{2}\rho(1.40\, v_2)^2 = P_2 + \tfrac{1}{2}\rho v_2^2$$

Solving for v_2, we obtain

$$v_2 = \sqrt{\frac{2(P_2 - P_1)}{\rho\left[(1.40)^2 - 1\right]}} = \sqrt{\frac{2(120 \text{ Pa})}{(1.30 \text{ kg}/\text{m}^3)\left[(1.40)^2 - 1\right]}} = \boxed{14 \text{ m}/\text{s}}$$

b.   According to Equation 11.10, the volume flow rate is

$$Q = A_2 v_2 = (0.0700 \text{ m}^2)(14 \text{ m}/\text{s}) = \boxed{0.98 \text{ m}^3 / \text{s}}$$

75. ***REASONING AND SOLUTION*** Bernoulli's equation (Equation 11.12) is

$$P_1 + \tfrac{1}{2}\rho v_1^2 = P_2 + \tfrac{1}{2}\rho v_2^2$$

If we let the right hand side refer to the air above the plate, and the left hand side refer to the air below the plate, then $v_1 = 0\,\text{m/s}$, since the air below the plate is stationary. We wish to find v_2 for the situation illustrated in part *b* of the figure shown in the text. Solving the equation above for v_2 (with $v_2 = v_{2b}$ and $v_1 = 0$) gives

$$v_{2b} = \sqrt{\frac{2(P_1 - P_2)}{\rho}} \qquad (1)$$

In Equation (1), P_1 is atmospheric pressure and P_2 must be determined. We must first consider the situation in part *a* of the text figure.

The figure at the right shows the forces that act on the rectangular plate in part *a* of the text drawing. F_1 is the force exerted on the plate from the air below the plate, and F_2 is the force exerted on the plate from the air above the plate. Applying Newton's second law, we have (taking "up" to be the positive direction),

$$F_1 - F_2 - mg = 0$$

$$F_1 - F_2 = mg$$

Thus, the difference in pressures exerted by the air on the plate in part *a* of the drawing is

$$P_1 - P_2 = \frac{F_1 - F_2}{A} = \frac{mg}{A} \qquad (2)$$

where A is the area of the plate. From Bernoulli's equation (Equation 11.12) we have, with $v_2 = v_{2a}$ and $v_{1a} = 0\,\text{m/s}$,

$$P_1 - P_2 = \tfrac{1}{2}\rho v_{2a}^2 \qquad (3)$$

where v_{2a} is the speed of the air along the top of the plate in part *a* of the text drawing. Combining Equations (2) and (3) we have

$$\frac{mg}{A} = \tfrac{1}{2}\rho v_{2a}^2 \qquad (4)$$

The figure below, on the left, shows the forces that act on the plate in part *b* of the text drawing. The notation is the same as that used when the plate was horizontal (part *a* of the text figure). The figure at the right below shows the same forces resolved into components along the plate and perpendicular to the plate.

Applying Newton's second law we have $F_1 - F_2 - mg\sin\theta = 0$, or

$$F_1 - F_2 = mg\sin\theta$$

Thus, the difference in pressures exerted by the air on the plate in part b of the text figure is

$$P_1 - P_2 = \frac{F_1 - F_2}{A} = \frac{mg\sin\theta}{A}$$

Using Equation (4) above,

$$P_1 - P_2 = \tfrac{1}{2}\rho v_{2a}^2 \sin\theta$$

Thus, Equation (1) becomes

$$v_{2b} = \sqrt{\frac{2\left(\tfrac{1}{2}\rho v_{2a}^2 \sin\theta\right)}{\rho}}$$

Therefore,

$$v_{2b} = \sqrt{v_{2a}^2 \sin\theta} = \sqrt{(11.0\ \text{m/s})^2 \sin 30.0^\circ} = \boxed{7.78\ \text{m/s}}$$

---

77. **REASONING AND SOLUTION**

a. If the water behaves as an ideal fluid, and since the pipe is horizontal and has the same radius throughout, the speed and pressure of the water are the same at all points in the pipe. Since the right end of the pipe is open to the atmosphere, the pressure at the right end is atmospheric pressure; therefore, the pressure at the left end is also atmospheric pressure, or $\boxed{1.01\times 10^5\ \text{Pa}}$.

b. If the water is treated as a viscous fluid, the volume flow rate Q is described by Poiseuille's law (Equation 11.14):

$$Q = \frac{\pi R^4 (P_2 - P_1)}{8 \eta L}$$

Let P_1 represent the pressure at the right end of the pipe, and let P_2 represent the pressure at the left end of the pipe. Solving for P_2 (with P_1 equal to atmospheric pressure), we obtain

$$P_2 = \frac{8 \eta L Q}{\pi R^4} + P_1$$

Therefore,

$$P_2 = \frac{8(1.00 \times 10^{-3} \text{ Pa} \cdot \text{s})(1.3 \text{ m})(9.0 \times 10^{-3} \text{ m}^3 / \text{s})}{\pi (6.4 \times 10^{-3} \text{ m})^4} + 1.013 \times 10^5 \text{ Pa} = \boxed{1.19 \times 10^5 \text{ Pa}}$$

---

81. **REASONING** The volume flow rate Q of a viscous fluid flowing through a pipe of radius R is given by Equation 11.14 as $Q = \dfrac{\pi R^4 (P_2 - P_1)}{8 \eta L}$, where $P_2 - P_1$ is the pressure difference between the ends of the pipe, L is the length of the pipe, and η is the viscosity of the fluid. Since all the variables are known except L, we can use this relation to find it.

SOLUTION Solving Equation 11.14 for the pipe length, we have

$$L = \frac{\pi R^4 (P_2 - P_1)}{8 \eta Q} = \frac{\pi (5.1 \times 10^{-3} \text{ m})^4 (1.8 \times 10^3 \text{ Pa})}{8 (1.0 \times 10^{-3} \text{ Pa} \cdot \text{s})(2.8 \times 10^{-4} \text{ m}^3/\text{s})} = \boxed{1.7 \text{ m}}$$

---

85. **REASONING** Since the faucet is closed, the water in the pipe may be treated as a static fluid. The gauge pressure P_2 at the faucet on the first floor is related to the gauge pressure P_1 at the faucet on the second floor by Equation 11.4, $P_2 = P_1 + \rho g h$.

SOLUTION
a. Solving Equation 11.4 for P_1, we find the gauge pressure at the second-floor faucet is

$$P_1 = P_2 - \rho g h = 1.90 \times 10^5 \text{ Pa} - (1.00 \times 10^3 \text{ kg/m}^3)(9.80 \text{ m/s}^2)(6.50 \text{ m}) = \boxed{1.26 \times 10^5 \text{ Pa}}$$

b. If the second faucet were placed at a height h above the first-floor faucet so that the gauge pressure P_1 at the second faucet were zero, then no water would flow from the second faucet, even if it were open. Solving Equation 11.4 for h when P_1 equals zero, we obtain

$$h = \frac{P_2 - P_1}{\rho g} = \frac{1.90 \times 10^5 \ \text{Pa} - 0}{(1.00 \times 10^3 \ \text{kg} / \text{m}^3)(9.80 \ \text{m} / \text{s}^2)} = \boxed{19.4 \ \text{m}}$$

87. **REASONING** According to Archimedes principle, the buoyant force that acts on the block is equal to the weight of the water that is displaced by the block. The block displaces an amount of water V, where V is the volume of the block. Therefore, the weight of the water displaced by the block is $W = mg = (\rho_{\text{water}} V) g$.

SOLUTION The buoyant force that acts on the block is, therefore,

$$F = \rho_{\text{water}} V g = (1.00 \times 10^3 \ \text{kg/m}^3)(0.10 \ \text{m} \times 0.20 \ \text{m} \times 0.30 \ \text{m})(9.80 \ \text{m/s}^2) = \boxed{59 \ \text{N}}$$

95. **REASONING AND SOLUTION**
a. The volume flow rate is given by Equation 11.10. Assuming that the line has a circular cross section, $A = \pi r^2$, we have

$$Q = Av = (\pi r^2) v = \pi \ (0.0065 \ \text{m})^2 (1.2 \ \text{m/s}) = \boxed{1.6 \times 10^{-4} \ \text{m}^3 / \text{s}}$$

b. The volume flow rate calculated in part (a) above is the flow rate for all twelve holes. Therefore, the volume flow rate through one of the twelve holes is

$$Q_{\text{hole}} = \frac{Q}{12} = (\pi r_{\text{hole}}^2) v_{\text{hole}}$$

Solving for v_{hole} we have

$$v_{\text{hole}} = \frac{Q}{12 \pi r_{\text{hole}}^2} = \frac{1.6 \times 10^{-4} \ \text{m}^3/\text{s}}{12 \pi (4.6 \times 10^{-4} \ \text{m})^2} = \boxed{2.0 \times 10^1 \ \text{m/s}}$$

99. **REASONING AND SOLUTION** The pressure at the bottom of the container is

$$P = P_{\text{atm}} + \rho_{\text{w}} g h_{\text{w}} + \rho_{\text{m}} g h_{\text{m}}$$

We want $P = 2P_{\text{atm}}$, and we know $h = h_{\text{w}} + h_{\text{m}} = 1.00$ m. Using the above and rearranging gives

$$h_{\text{m}} = \frac{P_{\text{atm}} - \rho_{\text{w}} g h}{(\rho_{\text{m}} - \rho_{\text{w}}) g} = \frac{1.01 \times 10^5 \ \text{Pa} - (1.00 \times 10^3 \ \text{kg/m}^3)(9.80 \ \text{m/s}^2)(1.00 \ \text{m})}{(13.6 \times 10^3 \ \text{kg/m}^3 - 1.00 \times 10^3 \ \text{kg/m}^3)(9.80 \ \text{m/s}^2)} = \boxed{0.74 \ \text{m}}$$

CHAPTER 12 | *TEMPERATURE AND HEAT*

1. **REASONING AND SOLUTION** The temperature of –273.15 °C is 273.15 Celsius degrees *below* the ice point of 0 °C. This number of Celsius degrees corresponds to

$$(273.15 \text{ C}°) \left(\frac{(9/5) \text{ F}°}{1 \text{ C}°} \right) = 491.67 \text{ F}°$$

Subtracting 491.67 F° from the ice point of 32.00 °F on the Fahrenheit scale gives a Fahrenheit temperature of $\boxed{-459.67 \text{ °F}}$

5. **REASONING AND SOLUTION**
a. The Kelvin temperature and the temperature on the Celsius scale are related by Equation 12.1: $T = T_c + 273.15$, where T is the Kelvin temperature and T_c is the Celsius temperature. Therefore, a temperature of 77 K on the Celsius scale is

$$T_c = T - 273.15 = 77 \text{ K} - 273.15 \text{ K} = \boxed{-196 \text{ °C}}$$

b. The temperature of –196 °C is 196 Celsius degrees *below* the ice point of 0 °C. Since $1 \text{ C}° = \frac{9}{5} \text{ F}°$, this number of Celsius degrees corresponds to

$$196 \text{ C}° \left(\frac{\frac{9}{5} \text{ F}°}{1 \text{ C}°} \right) = 353 \text{ F}°$$

Subtracting 353 Fahrenheit degrees from the ice point of 32.0 °F on the Fahrenheit scale gives a Fahrenheit temperature of $\boxed{-321 \text{ °F}}$.

9. **REASONING AND SOLUTION** The Rankine and Fahrenheit degrees are the same size, since the difference between the steam point and ice point temperatures is the same for both. The difference in the ice points of the two scales is $491.67 - 32.00 = 459.67$. To get Rankine from Fahrenheit this amount must be added, so $\boxed{T_R = T_F + 459.67}$.

11. ***REASONING AND SOLUTION***   The steel in the bridge expands according to Equation 12.2, $\Delta L = \alpha L_0 \Delta T$. Solving for L_0 and using the value for the coefficient of thermal expansion of steel given in Table 12.1, we find that the approximate length of the Golden Gate bridge is

$$L_0 = \frac{\Delta L}{\alpha \Delta T} = \frac{0.53 \text{ m}}{\left[12 \times 10^{-6} \text{ (C}°)^{-1}\right](32\,°\text{C} - 2\,°\text{C})} = \boxed{1500 \text{ m}}$$

15. ***REASONING AND SOLUTION***   The change in the coin's diameter is $\Delta d = \alpha d_0 \Delta T$, according to Equation 12.2. Solving for α gives

$$\alpha = \frac{\Delta d}{d_0 \, \Delta T} = \frac{2.3 \times 10^{-5} \text{ m}}{(1.8 \times 10^{-2} \text{ m})(75\,\text{C}°)} = \boxed{1.7 \times 10^{-5} \text{ (C}°)^{-1}} \tag{12.2}$$

21. ***REASONING AND SOLUTION***   Recall that $\omega = 2\pi / T$ (Equation 10.6), where ω is the angular frequency of the pendulum and T is the period. Using this fact and Equation 10.16, we know that the period of the pendulum before the temperature rise is given by $T_1 = 2\pi\sqrt{L_0/g}$, where L_0 is the length of the pendulum. After the temperature has risen, the period becomes (using Equation 12.2), $T_2 = 2\pi\sqrt{\left[L_0 + \alpha L_0 \Delta T\right]/g}$. Dividing these expressions, solving for T_2, and taking the coefficient of thermal expansion of brass from Table 12.1, we find that

$$T_2 = T_1\sqrt{1 + \alpha \Delta T} = (2.0000 \text{ s})\sqrt{1 + (19 \times 10^{-6}/\text{C}°)(140 \text{ C}°)} = \boxed{2.0027 \text{ s}}$$

25. ***REASONING AND SOLUTION***   Let $L_0 = 0.50$ m and L be the true length of the line at 40.0 °C. The ruler has expanded an amount

$$\Delta L_r = L - L_0 = \alpha_r L_0 \Delta T_r \tag{1}$$

The copper plate must shrink by an amount

$$\Delta L_p = L_0 - L = \alpha_p L \Delta T_p \tag{2}$$

Eliminating L from Equations (1) and (2), solving for ΔT_p, and using values of the coefficients of thermal expansion for copper and steel from Table 12.1, we find that

$$\Delta T_p = \frac{-\alpha_r \Delta T_r}{\alpha_p \left(1 + \alpha_r \Delta T_r\right)}$$

$$= \frac{-\left[12 \times 10^{-6} \ (C°)^{-1}\right](40.0 \ °C - 20.0 \ °C)}{\left[17 \times 10^{-6} \ (C°)^{-1}\right]\left\{1 + \left[12 \times 10^{-6} \ (C°)^{-1}\right](40.0 \ °C - 20.0 \ °C)\right\}} = -14 \ °C$$

Therefore, $T_p = 40.0 \ °C - 14 \ C° = \boxed{26 \ °C}$.

---

31. **REASONING** The increase ΔV in volume is given by Equation 12.3 as $\Delta V = \beta V_0 \Delta T$, where β is the coefficient of volume expansion, V_0 is the initial volume, and ΔT is the increase in temperature. The lead and quartz objects experience the same change in volume. Therefore, we can use Equation 12.3 to express the two volume changes and set them equal. We will solve the resulting equation for ΔT_{Quartz}.

SOLUTION Recognizing that the lead and quartz objects experience the same change in volume and expressing that change with Equation 12.3, we have

$$\underbrace{\beta_{Lead} V_0 \Delta T_{Lead}}_{\Delta V_{Lead}} = \underbrace{\beta_{Quartz} V_0 \Delta T_{Quartz}}_{\Delta V_{Quartz}}$$

In this result V_0 is the initial volume of each object. Solving for ΔT_{Quartz} and taking values for the coefficients of volume expansion for lead and quartz from Table 12.1 gives

$$\Delta T_{Quartz} = \frac{\beta_{Lead} \Delta T_{Lead}}{\beta_{Quartz}} = \frac{\left[87 \times 10^{-6} \ (C°)^{-1}\right](4.0 \ C°)}{1.5 \times 10^{-6} \ (C°)^{-1}} = \boxed{230 \ C°}$$

---

35. **REASONING AND SOLUTION** Both the gasoline and the tank expand as the temperature increases. The coefficients of volumetric expansion β_g and β_s for gasoline and steel are available in Table 12.1. According to Equation 12.3, the volume expansion of the gasoline is

$$\Delta V_g = \beta_g V_0 \Delta T = \left[950 \times 10^{-6} \ (C°)^{-1}\right](20.0 \ gal)(18 \ C°) = 0.34 \ gal$$

while the volume of the steel tank expands by an amount

$$\Delta V_{\rm s} = \beta_{\rm s} V_0 \Delta T = \left[36\times10^{-6}\,({\rm C}°)^{-1}\right](20.0\text{ gal})(18\text{ C}°) = 0.013\text{ gal}$$

The amount of gasoline which spills out is

$$\Delta V_{\rm g} - \Delta V_{\rm s} = \boxed{0.33\text{ gal}}$$

---

39. **REASONING** In order to keep the water from expanding as its temperature increases from 15 to 25 °C, the atmospheric pressure must be increased to compress the water as it tries to expand. The magnitude of the pressure change ΔP needed to compress a substance by an amount ΔV is, according to Equation 10.20, $\Delta P = B(\Delta V/V_0)$. The ratio $\Delta V/V_0$ is, according to Equation 12.3, $\Delta V/V_0 = \beta\,\Delta T$. Combining these two equations yields

$$\Delta P = B\beta\Delta T$$

SOLUTION Taking the value for the coefficient of volumetric expansion β for water from Table 12.1, we find that the change in atmospheric pressure that is required to keep the water from expanding is

$$\Delta P = (2.2\times10^9\text{ N/m}^2)\left[207\times10^{-6}\,(\text{ C}°)^{-1}\right](25\text{ °C}-15\text{ °C})$$

$$= \left(4.6\times10^6\text{ Pa}\right)\left(\frac{1\text{ atm}}{1.01\times10^5\text{ Pa}}\right) = \boxed{45\text{ atm}}$$

---

41. **REASONING** The cavity that contains the liquid in either Pyrex thermometer expands according to Equation 12.3, $\Delta V_{\rm g} = \beta_{\rm g} V_0 \Delta T$. On the other hand, the volume of mercury expands by an amount $\Delta V_{\rm m} = \beta_{\rm m} V_0 \Delta T$, while the volume of alcohol expands by an amount $\Delta V_{\rm a} = \beta_{\rm a} V_0 \Delta T$. Therefore, the net change in volume for the mercury thermometer is

$$\Delta V_{\rm m} - \Delta V_{\rm g} = (\beta_{\rm m} - \beta_{\rm g})V_0 \Delta T$$

while the net change in volume for the alcohol thermometer is

$$\Delta V_{\rm a} - \Delta V_{\rm g} = (\beta_{\rm a} - \beta_{\rm g})V_0 \Delta T$$

In each case, this volume change is related to a movement of the liquid into a cylindrical region of the thermometer with volume $\pi r^2 h$, where r is the radius of the region and h is the height of the region. For the mercury thermometer, therefore,

$$h_m = \frac{(\beta_m - \beta_g)V_0 \Delta T}{\pi r^2}$$

Similarly, for the alcohol thermometer

$$h_a = \frac{(\beta_a - \beta_g)V_0 \Delta T}{\pi r^2}$$

These two expressions can be combined to give the ratio of the heights, h_a/h_m.

SOLUTION Taking the values for the coefficients of volumetric expansion for methyl alcohol, Pyrex glass, and mercury from Table 12.1, we divide the two expressions for the heights of the liquids in the thermometers and find that

$$\frac{h_a}{h_m} = \frac{\beta_a - \beta_g}{\beta_m - \beta_g} = \frac{1200 \times 10^{-6} \ (C°)^{-1} - 9.9 \times 10^{-6} \ (C°)^{-1}}{182 \times 10^{-6} \ (C°)^{-1} - 9.9 \times 10^{-6} \ (C°)^{-1}} = 6.9$$

Therefore, the degree marks are $\boxed{\text{6.9 times further apart}}$ on the alcohol thermometer than on the mercury thermometer.

---

43. **REASONING** Since there is no heat lost or gained by the system, the heat lost by the water in cooling down must be equal to the heat gained by the thermometer in warming up. The heat Q lost or gained by a substance is given by Equation 12.4 as $Q = cm\Delta T$, where c is the specific heat capacity, m is the mass, and ΔT is the change in temperature. Thus, we have that

$$\underbrace{c_{H_2O} \ m_{H_2O} \ \Delta T_{H_2O}}_{\text{Heat lost by water}} = \underbrace{c_{therm} \ m_{therm} \ \Delta T_{therm}}_{\text{Heat gained by thermometer}}$$

We can use this equation to find the temperature of the water before the insertion of the thermometer.

SOLUTION Solving the equation above for ΔT_{H_2O}, and using the value of c_{H_2O} from Table 12.2, we have

$$\Delta T_{H_2O} = \frac{c_{therm} \ m_{therm} \ \Delta T_{therm}}{c_{H_2O} \ m_{H_2O}}$$

$$= \frac{\left[815 \ J/(kg \cdot C°)\right](31.0 \ g)(41.5 \ °C - 12.0 \ °C)}{\left[4186 \ J/(kg \cdot C°)\right](119 \ g)} = 1.50 \ C°$$

The temperature of the water before the insertion of the thermometer was

$$T = 41.5 \ °C + 1.50 \ C° = \boxed{43.0 \ °C}$$

49. ***REASONING*** Let the system be comprised only of the metal forging and the oil. Then, according to the principle of energy conservation, the heat lost by the forging equals the heat gained by the oil, or $Q_{metal} = Q_{oil}$. According to Equation 12.4, the heat lost by the forging is $Q_{metal} = c_{metal} m_{metal} (T_{0metal} - T_{eq})$, where T_{eq} is the final temperature of the system at thermal equilibrium. Similarly, the heat gained by the oil is given by $Q_{oil} = c_{oil} m_{oil} (T_{eq} - T_{0oil})$.

SOLUTION

$$Q_{metal} = Q_{oil}$$

$$c_{metal} m_{metal} (T_{0metal} - T_{eq}) = c_{oil} m_{oil} (T_{eq} - T_{0oil})$$

Solving for T_{0metal}, we have

$$T_{0metal} = \frac{c_{oil} m_{oil} (T_{eq} - T_{0oil})}{c_{metal} m_{metal}} + T_{eq}$$

or

$$T_{0metal} = \frac{[2700 \text{ J/(kg} \cdot \text{C}°)](710 \text{ kg})(47 \text{ °C} - 32 \text{ °C})}{[430 \text{ J/(kg} \cdot \text{C}°)](75 \text{ kg})} + 47 \text{ °C} = \boxed{940 \text{ °C}}$$

---

55. ***REASONING*** When heat Q is supplied to the block, its temperature changes by an amount ΔT. The relation between Q and ΔT is given by

$$Q = c \, m \, \Delta T \qquad (12.4)$$

where c is the specific heat capacity and m is the mass. When the temperature of the block changes by an amount ΔT, the change ΔV in its volume is given by Equation 12.3 as $\Delta V = \beta V_0 \Delta T$, where β is the coefficient of volume expansion and V_0 is the initial volume of the block. Solving for ΔT gives

$$\Delta T = \frac{\Delta V}{\beta V_0}$$

Substituting this expression for ΔT into Equation 12.4 gives

$$Q = c \, m \, \Delta T = c \, m \left(\frac{\Delta V}{\beta V_0} \right)$$

SOLUTION The heat supplied to the block is

$$Q = cm\left(\frac{\Delta V}{\beta V_0}\right) = \frac{\left[750 \text{ J}/(\text{kg} \cdot \text{C}^\circ)\right](130 \text{ kg})(1.2 \times 10^{-5} \text{ m}^3)}{\left[6.4 \times 10^{-5} \ (\text{C}^\circ)^{-1}\right](4.6 \times 10^{-2} \text{ m}^3)} = \boxed{4.0 \times 10^5 \text{ J}}$$

---

57. **REASONING** Heat Q_1 must be added to raise the temperature of the aluminum in its solid phase from 130 °C to its melting point at 660 °C. According to Equation 12.4, $Q_1 = cm\Delta T$. The specific heat c of aluminum is given in Table 12.2. Once the solid aluminum is at its melting point, additional heat Q_2 must be supplied to change its phase from solid to liquid. The additional heat required to melt or liquefy the aluminum is $Q_2 = mL_f$, where L_f is the latent heat of fusion of aluminum. Therefore, the total amount of heat which must be added to the aluminum in its solid phase to liquefy it is

$$Q_{\text{total}} = Q_1 + Q_2 = m(c\Delta T + L_f)$$

SOLUTION Substituting values, we obtain

$$Q_{\text{total}} = (0.45 \text{ kg})\left\{[9.00 \times 10^2 \text{ J}/(\text{kg} \cdot \text{C}^\circ)](660 \text{ °C} - 130 \text{ °C}) + 4.0 \times 10^5 \text{ J/kg}\right\} = \boxed{3.9 \times 10^5 \text{ J}}$$

---

61. **REASONING** From the conservation of energy, the heat lost by the mercury is equal to the heat gained by the water. As the mercury loses heat, its temperature decreases; as the water gains heat, its temperature rises to its boiling point. Any remaining heat gained by the water will then be used to vaporize the water.

According to Equation 12.4, the heat lost by the mercury is $Q_{\text{mercury}} = (cm\Delta T)_{\text{mercury}}$. The heat required to vaporize the water is, from Equation 12.5, $Q_{\text{vap}} = (m_{\text{vap}}L_v)_{\text{water}}$. Thus, the total amount of heat gained by the water is $Q_{\text{water}} = (cm\Delta T)_{\text{water}} + (m_{\text{vap}}L_v)_{\text{water}}$.

SOLUTION

$$Q_{\substack{\text{lost by} \\ \text{mercury}}} = Q_{\substack{\text{gained by} \\ \text{water}}}$$

$$(cm\Delta T)_{\text{mercury}} = (cm\Delta T)_{\text{water}} + (m_{\text{vap}}L_v)_{\text{water}}$$

where $\Delta T_{\text{mercury}} = (205 \text{ °C} - 100.0 \text{ °C})$ and $\Delta T_{\text{water}} = (100.0 \text{ °C} - 80.0 \text{ °C})$. The specific heats of mercury and water are given in Table 12.2, and the latent heat of vaporization of water is given in Table 12.3. Solving for the mass of the water that vaporizes gives

$$m_{vap} = \frac{c_{mercury} m_{mercury} \Delta T_{mercury} - c_{water} m_{water} \Delta T_{water}}{(L_v)_{water}}$$

$$= \frac{[139 \text{ J/(kg} \cdot \text{C}°)](2.10 \text{ kg})(105 \text{ C}°) - [4186 \text{ J/(kg} \cdot \text{C}°)](0.110 \text{ kg})(20.0 \text{ C}°)}{22.6 \times 10^5 \text{ J/kg}}$$

$$= \boxed{9.49 \times 10^{-3} \text{ kg}}$$

---

67. **REASONING** According to the statement of the problem, the initial state of the system is comprised of the ice and the steam. From the principle of energy conservation, the heat lost by the steam equals the heat gained by the ice, or $Q_{steam} = Q_{ice}$. When the ice and the steam are brought together, the steam immediately begins losing heat to the ice. An amount $Q_{1(lost)}$ is released as the temperature of the steam drops from 130 °C to 100 °C, the boiling point of water. Then an amount of heat $Q_{2(lost)}$ is released as the steam condenses into liquid water at 100 °C. The remainder of the heat lost by the "steam" $Q_{3(lost)}$ is the heat that is released as the water at 100 °C cools to the equilibrium temperature of $T_{eq} = 50.0$ °C. According to Equation 12.4, $Q_{1(lost)}$ and $Q_{3(lost)}$ are given by

$$Q_{1(lost)} = c_{steam} m_{steam} (T_{steam} - 100.0 \text{ °C}) \quad \text{and} \quad Q_{3(lost)} = c_{water} m_{steam} (100.0 \text{ °C} - T_{eq})$$

$Q_{2(lost)}$ is given by $Q_{2(lost)} = m_{steam} L_v$, where L_v is the latent heat of vaporization of water. The total heat lost by the steam has three effects on the ice. First, a portion of this heat $Q_{1(gained)}$ is used to raise the temperature of the ice to its melting point at 0.00 °C. Then, an amount of heat $Q_{2(gained)}$ is used to melt the ice completely (we know this because the problem states that after thermal equilibrium is reached the liquid phase is present at 50.0 °C). The remainder of the heat $Q_{3(gained)}$ gained by the "ice" is used to raise the temperature of the resulting liquid at 0.0 °C to the final equilibrium temperature. According to Equation 12.4, $Q_{1(gained)}$ and $Q_{3(gained)}$ are given by

$$Q_{1(gained)} = c_{ice} m_{ice} (0.00 \text{ °C} - T_{ice}) \quad \text{and} \quad Q_{3(gained)} = c_{water} m_{ice} (T_{eq} - 0.00 \text{ °C})$$

$Q_{2(gained)}$ is given by $Q_{2(gained)} = m_{ice} L_f$, where L_f is the latent heat of fusion of ice.

SOLUTION According to the principle of energy conservation, we have

$$Q_{steam} = Q_{ice}$$

$$Q_{1(lost)} + Q_{2(lost)} + Q_{3(lost)} = Q_{1(gained)} + Q_{2(gained)} + Q_{3(gained)}$$

or

$$c_{steam}m_{steam}(T_{steam} - 100.0\ ^\circ C) + m_{steam}L_v + c_{water}m_{steam}(100.0\ ^\circ C - T_{eq})$$

$$= c_{ice}m_{ice}(0.00\ ^\circ C - T_{ice}) + m_{ice}L_f + c_{water}m_{ice}(T_{eq} - 0.00\ ^\circ C)$$

Values for specific heats are given in Table 12.2, and values for the latent heats are given in Table 12.3. Solving for the ratio of the masses gives

$$\frac{m_{steam}}{m_{ice}} = \frac{c_{ice}(0.00\ ^\circ C - T_{ice}) + L_f + c_{water}(T_{eq} - 0.00\ ^\circ C)}{c_{steam}(T_{steam} - 100.0\ ^\circ C) + L_v + c_{water}(100.0\ ^\circ C - T_{eq})}$$

$$= \frac{\left[2.00 \times 10^3\ J/(kg \cdot C^\circ)\right]\left[0.0\ ^\circ C - (-10.0\ ^\circ C)\right] + 33.5 \times 10^4\ J/kg + \left[4186\ J/(kg \cdot C^\circ)\right](50.0\ ^\circ C - 0.0\ ^\circ C)}{\left[2020\ J/(kg \cdot C^\circ)\right](130\ ^\circ C - 100.0\ ^\circ C) + 22.6 \times 10^5\ J/kg + \left[4186\ J/(kg \cdot C^\circ)\right](100.0\ ^\circ C - 50.0\ ^\circ C)}$$

or

$$\frac{m_{steam}}{m_{ice}} = \boxed{0.223}$$

---

69. **REASONING** The system is comprised of the unknown material, the glycerin, and the aluminum calorimeter. From the principle of energy conservation, the heat gained by the unknown material is equal to the heat lost by the glycerin and the calorimeter. The heat gained by the unknown material is used to melt the material and then raise its temperature from the initial value of $-25.0\ ^\circ C$ to the final equilibrium temperature of $T_{eq} = 20.0\ ^\circ C$.

SOLUTION

$$\underset{\substack{\text{gained by} \\ \text{unknown}}}{Q} = \underset{\substack{\text{lost by} \\ \text{glycerine}}}{Q} + \underset{\substack{\text{lost by} \\ \text{calorimeter}}}{Q}$$

$$m_u L_f + c_u m_u \Delta T_u = c_{gl} m_{gl} \Delta T_{gl} + c_{al} m_{al} \Delta T_{al}$$

Taking values for the specific heat capacities of glycerin and aluminum from Table 12.2, we have

$$(0.10\ kg)L_f + [160\ J/(kg \cdot C^\circ)](0.10\ kg)(45.0\ C^\circ) = [2410\ J/(kg \cdot C^\circ)](0.100\ kg)(7.0\ C^\circ)$$
$$+ [9.0 \times 10^2\ J/(kg \cdot C^\circ)](0.150\ kg)(7.0\ C^\circ)$$

Solving for L_f yields,

$$L_f = \boxed{1.9 \times 10^4 \text{ J/kg}}$$

---

71. **REASONING** In order to melt, the bullet must first heat up to 327.3 °C (its melting point) and then undergo a phase change. According to Equation 12.4, the amount of heat necessary to raise the temperature of the bullet to 327.3 °C is $Q = cm(327.3 \text{ °C} - 30.0 \text{ °C})$, where m is the mass of the bullet. The amount of heat required to melt the bullet is given by $Q_{melt} = mL_f$, where L_f is the latent heat of fusion of lead.

The lead bullet melts completely when it comes to a sudden halt; all of the kinetic energy of the bullet is converted into heat; therefore,

$$\text{KE} = Q + Q_{melt}$$

$$\tfrac{1}{2}mv^2 = cm(327.3 \text{ °C} - 30.0 \text{ °C}) + mL_f$$

The value for the specific heat c of lead is given in Table 12.2, and the value for the latent heat of fusion L_f of lead is given in Table 12.3. This expression can be solved for v, the minimum speed of the bullet for such an event to occur.

SOLUTION Solving for v, we find that the minimum speed of the lead bullet is

$$v = \sqrt{2L_f + 2c\,(327.3 \text{ °C} - 30.0 \text{ °C})}$$

$$v = \sqrt{2(2.32 \times 10^4 \text{ J/kg}) + 2[128 \text{ J/(kg} \cdot \text{C°)}](327.3 \text{ °C} - 30.0 \text{ °C})} = \boxed{3.50 \times 10^2 \text{ m/s}}$$

---

75. **REASONING** The definition of percent relative humidity is given by Equation 12.6 as follows:

$$\text{Percent relative humidity} = \frac{\text{Partial pressure of water vapor}}{\text{Equilibrium vapor pressure of water at the existing temperature}} \times 100$$

Using R to denote the percent relative humidity, P to denote the partial pressure of water vapor, and P_V to denote the equilibrium vapor pressure of water at the existing temperature, we can write Equation 12.6 as

$$R = \frac{P}{P_V} \times 100$$

The partial pressure of water vapor P is the same at the two given temperatures. The relative humidity is not the same at the two temperatures, however, because the equilibrium vapor pressure P_V is different at each temperature, with values that are available from the vapor pressure curve given with the problem statement. To determine the ratio R_{10}/R_{40}, we will apply Equation 12.6 at each temperature.

SOLUTION Using Equation 12.6 and reading the values of $P_{V, 10}$ and $P_{V, 40}$ from the vapor pressure curve given with the problem statement, we find

$$\frac{R_{10}}{R_{40}} = \frac{P/P_{V, 10}}{P/P_{V, 40}} = \frac{P_{V, 40}}{P_{V, 10}} = \frac{7200 \text{ Pa}}{1300 \text{ Pa}} = \boxed{5.5}$$

---

83. **REASONING** We must first find the equilibrium temperature T_{eq} of the iced tea. Once this is known, we can use the vapor pressure curve that accompanies Problem 75 to find the partial pressure of water vapor at that temperature and then estimate the relative humidity using Equation 12.6.

According to the principle of energy conservation, when the ice is mixed with the tea, the heat lost by the tea is gained by the ice, or $Q_{tea} = Q_{ice}$. The heat gained by the ice is used to melt the ice at 0.0 °C; the remainder of the heat is used to bring the water at 0.0 °C up to the final equilibrium temperature T_{eq}.

SOLUTION

$$Q_{tea} = Q_{ice}$$

$$c_{water} m_{tea} (30.0 \text{ °C} - T_{eq}) = m_{ice} L_f + c_{water} m_{ice} (T_{eq} - 0.00 \text{ °C})$$

The specific heat capacity of water is given in Table 12.2, and the latent heat of fusion L_f of water is given in Table 12.3. Solving for T_{eq}, we have

$$T_{eq} = \frac{c_{water} m_{tea} (30.0 \text{ °C}) - m_{ice} L_f}{c_{water} (m_{tea} + m_{ice})}$$

$$= \frac{[4186 \text{ J/(kg} \cdot \text{C°)}] (0.300 \text{ kg})(30.0 \text{ °C}) - (0.0670 \text{ kg})(33.5 \times 10^4 \text{ J/kg})}{[4186 \text{ J/(kg} \cdot \text{C°)}] (0.300 \text{ kg} + 0.0670 \text{ kg})} = 9.91 \text{ °C}$$

According to the vapor pressure curve that accompanies Problem 75, at a temperature of 9.91 °C, the equilibrium vapor pressure is approximately 1250 Pa. At 30 °C, the equilibrium vapor pressure is approximately 4400 Pa.   Therefore, according to Equation 12.6, the percent relative humidity is approximately

$$\text{Percent relative humidity} = \left(\frac{1250 \text{ Pa}}{4400 \text{ Pa}} \right) \times 100 = \boxed{28\%}$$

---

87. ***REASONING***   From the principle of conservation of energy, the heat lost by the coin must be equal to the heat gained by the liquid nitrogen. The heat lost by the silver coin is, from Equation 12.4, $Q = c_{\text{coin}} m_{\text{coin}} \Delta T_{\text{coin}}$ (see Table 12.2 for the specific heat capacity of silver). If the liquid nitrogen is at its boiling point, -195.8 °C, then the heat gained by the nitrogen will cause it to change phase from a liquid to a vapor. The heat gained by the liquid nitrogen is $Q = m_{\text{nitrogen}} L_{\text{v}}$, where m_{nitrogen} is the mass of liquid nitrogen that vaporizes, and L_{v} is the latent heat of vaporization for nitrogen (see Table 12.3).

SOLUTION

$$Q_{\substack{\text{lost by} \\ \text{coin}}} = Q_{\substack{\text{gained by} \\ \text{nitrogen}}}$$

$$c_{\text{coin}} m_{\text{coin}} \Delta T_{\text{coin}} = m_{\text{nitrogen}} L_{\text{v}}$$

Solving for the mass of the nitrogen that vaporizes

$$m_{\text{nitrogen}} = \frac{c_{\text{coin}} m_{\text{coin}} \Delta T_{\text{coin}}}{L_{\text{v}}}$$

$$= \frac{[235 \text{ J/(kg} \cdot \text{C}°)](1.5 \times 10^{-2} \text{ kg})[25 \text{ °C} - (-195.8 \text{ °C})]}{2.00 \times 10^{5} \text{ J/kg}} = \boxed{3.9 \times 10^{-3} \text{ kg}}$$

---

91. ***REASONING AND SOLUTION*** The volume V_0 of an object changes by an amount ΔV when its temperature changes by an amount ΔT; the mathematical relationship is given by Equation 12.3: $\Delta V = \beta V_0 \Delta T$. Thus, the volume of the kettle at 24 °C can be found by solving Equation 12.3 for V_0. According to Table 12.1, the coefficient of volumetric expansion for copper is 51×10^{-6} $(\text{C}°)^{-1}$. Solving Equation 12.3 for V_0, we have

$$V_0 = \frac{\Delta V}{\beta \Delta T} = \frac{1.2 \times 10^{-5} \text{ m}^3}{[51 \times 10^{-6} \text{ (C}°)^{-1}](100 \text{ °C} - 24 \text{ °C})} = \boxed{3.1 \times 10^{-3} \text{ m}^3}$$

---

93. ***REASONING AND SOLUTION***   From the vapor pressure curve that accompanies Problem 75, it is seen that the partial pressure of water vapor in the atmosphere at 10 °C is about 1400 Pa, and that the equilibrium vapor pressure at 30 °C is about 4200 Pa. The relative humidity is, from Equation 12.6,

$$\begin{array}{c} \text{Percent} \\ \text{relative} \\ \text{humidity} \end{array} = \left(\frac{1400 \text{ Pa}}{4200 \text{ Pa}} \right) \times 100 = \boxed{33\%}$$

99. ***REASONING AND SOLUTION***   As the rock falls through a distance h, its initial potential energy $m_{\text{rock}}gh$ is converted into kinetic energy. This kinetic energy is then converted into heat when the rock is brought to rest in the pail. If we ignore the heat absorbed by the pail, the principle of conservation of energy indicates that

$$m_{\text{rock}}gh = c_{\text{rock}}m_{\text{rock}}\Delta T + c_{\text{water}}m_{\text{water}}\Delta T$$

where we have used Equation 12.4 to express the heat absorbed by the rock and the water. Table 12.2 gives the specific heat capacity of the water. Solving for ΔT yields

$$\Delta T = \frac{m_{\text{rock}}gh}{c_{\text{rock}}m_{\text{rock}} + c_{\text{water}}m_{\text{water}}}$$

Substituting values yields

$$\Delta T = \frac{(0.20 \text{ kg})(9.80 \text{ m/s}^2)(15 \text{ m})}{[1840 \text{ J/(kg}\cdot\text{C}°)](0.20 \text{ kg}) + [4186 \text{ J/(kg}\cdot\text{C}°)](0.35 \text{ kg})} = \boxed{0.016 \text{ C}°}$$

CHAPTER 13 | *THE TRANSFER OF HEAT*

3. **REASONING AND SOLUTION** According to Equation 13.1, the heat per second lost is

$$\frac{Q}{t} = \frac{kA\,\Delta T}{L} = \frac{[0.040 \text{ J/(s·m·C}^\circ)]\,(1.6 \text{ m}^2)(25 \text{ C}^\circ)}{2.0 \times 10^{-3} \text{ m}} = \boxed{8.0 \times 10^2 \text{ J/s}}$$

where the value for the thermal conductivity k of wool has been taken from Table 13.1.

---

5. **REASONING** The heat transferred in a time t is given by Equation 13.1, $Q = (kA\,\Delta T)t/L$. If the same amount of heat per second is conducted through the two plates, then $(Q/t)_{al} = (Q/t)_{st}$. Using Equation 13.1, this becomes

$$\frac{k_{al}A\,\Delta T}{L_{al}} = \frac{k_{st}A\,\Delta T}{L_{st}}$$

This expression can be solved for L_{st}.

SOLUTION Solving for L_{st} gives

$$L_{st} = \frac{k_{st}}{k_{al}} L_{al} = \frac{14 \text{ J/(s·m·C}^\circ)}{240 \text{ J/(s·m·C}^\circ)}(0.035 \text{ m}) = \boxed{2.0 \times 10^{-3} \text{ m}}$$

---

7. **REASONING AND SOLUTION** Values for the thermal conductivities of Styrofoam and air are given in Table 11.1. The conductance of an 0.080 mm thick sample of Styrofoam of cross-sectional area A is

$$\frac{k_s A}{L_s} = \frac{[0.010 \text{ J/(s·m·C}^\circ)]\,A}{0.080 \times 10^{-3} \text{ m}} = [125 \text{ J/(s·m}^2\text{·C}^\circ)]\,A$$

The conductance of a 3.5 mm thick sample of air of cross-sectional area A is

$$\frac{k_a A}{L_a} = \frac{[0.0256 \text{ J/(s·m·C}^\circ)]\,A}{3.5 \times 10^{-3} \text{ m}} = [7.3 \text{ J/(s·m}^2\text{·C}^\circ)]\,A$$

Dividing the conductance of Styrofoam by the conductance of air for samples of the same cross-sectional area A, gives

$$\frac{[125 \text{ J/(s} \cdot \text{m}^2 \cdot \text{C}°)] A}{[7.3 \text{ J/(s} \cdot \text{m}^2 \cdot \text{C}°)] A} = 17$$

Therefore, the body can adjust the conductance of the tissues beneath the skin by a factor of 17 .

---

13. **REASONING AND SOLUTION** The rate of heat transfer is the same for all three materials so

$$Q/t = k_p A \Delta T_p/L = k_b A \Delta T_b/L = k_w A \Delta T_w/L$$

Let T_i be the inside temperature, T_1 be the temperature at the plasterboard-brick interface, T_2 be the temperature at the brick-wood interface, and T_0 be the outside temperature. Then

$$k_p T_i - k_p T_1 = k_b T_1 - k_b T_2 \tag{1}$$

and

$$k_b T_1 - k_b T_2 = k_w T_2 - k_w T_0 \tag{2}$$

Solving (1) for T_2 gives

$$T_2 = (k_p + k_b) T_1/k_b - (k_p/k_b) T_i$$

a. Substituting this into (2) and solving for T_1 yields

$$T_1 = \frac{\left(k_p/k_b\right)\left(1 + k_w/k_b\right) T_i + \left(k_w/k_b\right) T_0}{\left(1 + k_w/k_b\right)\left(1 + k_p/k_b\right) - 1} = \boxed{21\,°\text{C}}$$

b. Using this value in (1) yields

$$\boxed{T_2 = 18\ °\text{C}}$$

---

19. **REASONING** The rate at which heat is conducted along either rod is given by Equation 13.1, $Q/t = (kA\,\Delta T)/L$. Since both rods conduct the same amount of heat per second, we have

$$\frac{k_s A_s\ \Delta T}{L_s} = \frac{k_i A_i\ \Delta T}{L_i} \tag{1}$$

Since the same temperature difference is maintained across both rods, we can algebraically cancel the ΔT terms. Because both rods have the same mass, $m_s = m_i$; in terms of the

densities of silver and iron, the statement about the equality of the masses becomes
$\rho_s \left(L_s A_s \right) = \rho_i \left(L_i A_i \right)$, or

$$\frac{A_s}{A_i} = \frac{\rho_i L_i}{\rho_s L_s} \qquad (2)$$

Equations (1) and (2) may be combined to find the ratio of the lengths of the rods. Once the ratio of the lengths is known, Equation (2) can be used to find the ratio of the cross-sectional areas of the rods. If we assume that the rods have circular cross sections, then each has an area of $A = \pi r^2$. Hence, the ratio of the cross-sectional areas can be used to find the ratio of the radii of the rods.

SOLUTION
a. Solving Equation (1) for the ratio of the lengths and substituting the right hand side of Equation (2) for the ratio of the areas, we have

$$\frac{L_s}{L_i} = \frac{k_s A_s}{k_i A_i} = \frac{k_s \left(\rho_i L_i \right)}{k_i \left(\rho_s L_s \right)} \qquad \text{or} \qquad \left(\frac{L_s}{L_i} \right)^2 = \frac{k_s \rho_i}{k_i \rho_s}$$

Solving for the ratio of the lengths, we have

$$\frac{L_s}{L_i} = \sqrt{\frac{k_s \rho_i}{k_i \rho_s}} = \sqrt{\frac{[420 \text{ J/(s·m·C°)}](7860 \text{ kg/m}^3)}{[79 \text{ J/(s·m·C°)}](10\,500 \text{ kg/m}^3)}} = \boxed{2.0}$$

b. From Equation (2) we have

$$\frac{\pi r_s^2}{\pi r_i^2} = \frac{\rho_i L_i}{\rho_s L_s} \qquad \text{or} \qquad \left(\frac{r_s}{r_i} \right)^2 = \frac{\rho_i L_i}{\rho_s L_s}$$

Solving for the ratio of the radii, we have

$$\frac{r_s}{r_i} = \sqrt{\frac{\rho_i}{\rho_s} \left(\frac{L_i}{L_s} \right)} = \sqrt{\frac{7860 \text{ kg/m}^3}{10\,500 \text{ kg/m}^3} \left(\frac{1}{2.0} \right)} = \boxed{0.61}$$

---

21. ***REASONING AND SOLUTION*** Solving the Stefan-Boltzmann law, Equation 13.2, for the time t, and using the fact that $Q_{blackbody} = Q_{bulb}$, we have

$$t_{blackbody} = \frac{Q_{blackbody}}{\sigma T^4 A} = \frac{Q_{bulb}}{\sigma T^4 A} = \frac{P_{bulb}\, t_{bulb}}{\sigma T^4 A}$$

where P_{bulb} is the power rating of the light bulb. Therefore,

$$t_{blackbody} = \frac{(100.0 \text{ J/s}) (3600 \text{ s})}{\left[5.67 \times 10^{-8} \text{ J/(s} \cdot \text{m}^2 \cdot \text{K}^4)\right] (303 \text{ K})^4 \left[(6 \text{ sides})(0.0100 \text{ m})^2 / \text{side}\right]}$$

$$\times \left(\frac{1 \text{ h}}{3600 \text{ s}}\right)\left(\frac{1 \text{ d}}{24 \text{ h}}\right) = \boxed{14.5 \text{ d}}$$

---

27. **REASONING AND SOLUTION** The net power generated by the stove is given by Equation 13.3, $P_{net} = e\sigma A(T^4 - T_0^4)$. Solving for T gives

$$T = \left(\frac{P_{net}}{e\sigma A} + T_0^4\right)^{1/4}$$

$$= \left\{\frac{7300 \text{ W}}{(0.900)[5.67 \times 10^{-8} \text{ J}/(\text{s} \cdot \text{m}^2 \cdot \text{K}^4)](2.00 \text{ m}^2)} + (302 \text{ K})^4\right\}^{1/4} = \boxed{532 \text{ K}}$$

---

29. **REASONING AND SOLUTION** The power radiated per square meter by the car when it has reached a temperature T is given by the Stefan-Boltzmann law, Equation 13.2, $P_{radiated} / A = e\sigma T^4$, where $P_{radiated} = Q/t$. Solving for T we have

$$T = \left[\frac{(P_{radiated} / A)}{e\sigma}\right]^{1/4} = \left\{\frac{560 \text{ W/m}^2}{(1.00)\left[5.67 \times 10^{-8} \text{ J/(s} \cdot \text{m}^2 \cdot \text{K}^4)\right]}\right\}^{1/4} = \boxed{320 \text{ K}}$$

---

33. **REASONING** The total radiant power emitted by an object that has a Kelvin temperature T, surface area A, and emissivity e can be found by rearranging Equation 13.2, the Stefan-Boltzmann law: $Q = e\sigma T^4 At$. The emitted power is $P = Q/t = e\sigma T^4 A$. Therefore, when the original cylinder is cut perpendicular to its axis into N smaller cylinders, the ratio of the power radiated by the pieces to that radiated by the original cylinder is

$$\frac{P_{pieces}}{P_{original}} = \frac{e\sigma T^4 A_2}{e\sigma T^4 A_1} \tag{1}$$

where A_1 is the surface area of the original cylinder, and A_2 is the sum of the surface areas of all N smaller cylinders. The surface area of the original cylinder is the sum of the surface area of the ends and the surface area of the cylinder body; therefore, if L and r represent the length and cross-sectional radius of the original cylinder, with $L = 10r$,

$$A_1 = (\text{area of ends}) + (\text{area of cylinder body})$$

$$= 2(\pi r^2) + (2\pi r)L = 2(\pi r^2) + (2\pi r)(10r) = 22\pi r^2$$

When the original cylinder is cut perpendicular to its axis into N smaller cylinders, the total surface area A_2 is

$$A_2 = N2(\pi r^2) + (2\pi r)L = N2(\pi r^2) + (2\pi r)(10r) = (2N + 20)\pi r^2$$

Substituting the expressions for A_1 and A_2 into Equation (1), we obtain the following expression for the ratio of the power radiated by the N pieces to that radiated by the original cylinder

$$\frac{P_{\text{pieces}}}{P_{\text{original}}} = \frac{e\sigma T^4 A_2}{e\sigma T^4 A_1} = \frac{(2N+20)\pi r^2}{22\pi r^2} = \frac{N+10}{11}$$

SOLUTION Since the total radiant power emitted by the N pieces is twice that emitted by the original cylinder, $P_{\text{pieces}} / P_{\text{original}} = 2$, we have $(N + 10)/11 = 2$. Solving this expression for N gives $N = 12$. Therefore, $\boxed{\text{there are 12 smaller cylinders}}$.

---

35. **REASONING** The heat conducted through the iron poker is given by Equation 13.1, $Q = (kA \, \Delta T)t \, / \, L$. If we assume that the poker has a circular cross-section, then its cross-sectional area is $A = \pi r^2$. Table 13.1 gives the thermal conductivity of iron as $79 \, \text{J} / (\text{s} \cdot \text{m} \cdot \text{C}°)$.

SOLUTION The amount of heat conducted from one end of the poker to the other in 5.0 s is, therefore,

$$Q = \frac{(k \, A \Delta T)t}{L} = \frac{\left[79 \, \text{J}/(\text{s} \cdot \text{m} \cdot \text{C}°)\right]\pi\left(5.0 \times 10^{-3} \, \text{m}\right)^2 (502 \, °\text{C} - 26 \, °\text{C})(5.0 \, \text{s})}{1.2 \, \text{m}} = \boxed{12 \, \text{J}}$$

---

43. **REASONING** Heat flows along the rods via conduction, so that Equation 13.1 applies: $Q = \frac{(kA\Delta T)t}{L}$, where Q is the amount of heat that flows in a time t, k is the thermal conductivity of the material from which a rod is made, A is the cross-sectional area of the rod, and ΔT is the difference in temperature between the ends of a rod. In arrangement a,

this expression applies to each rod and ΔT has the same value of $\Delta T = T_W - T_C$. The total heat Q' is the sum of the heats through each rod. In arrangement b, the situation is more complicated. We will use the fact that the same heat flows through each rod to determine the temperature at the interface between the rods and then use this temperature to determine ΔT and the heat flow through either rod.

SOLUTION For arrangement a, we apply Equation 13.1 to each rod and obtain for the total heat that

$$Q' = Q_1 + Q_2 = \frac{k_1 A (T_W - T_C) t}{L} + \frac{k_2 A (T_W - T_C) t}{L} = \frac{(k_1 + k_2) A (T_W - T_C) t}{L} \quad (1)$$

For arrangement b, we use T to denote the temperature at the interface between the rods and note that the same heat flows through each rod. Thus, using Equation 13.1 to express the heat flowing in each rod, we have

$$\underbrace{\frac{k_1 A (T_W - T) t}{L}}_{\substack{\text{Heat flowing} \\ \text{through rod 1}}} = \underbrace{\frac{k_2 A (T - T_C) t}{L}}_{\substack{\text{Heat flowing} \\ \text{through rod 2}}} \quad \text{or} \quad k_1 (T_W - T) = k_2 (T - T_C)$$

Solving this expression for the temperature T gives

$$T = \frac{k_1 T_W + k_2 T_C}{k_1 + k_2} \quad (2)$$

Applying Equation 13.1 to either rod in arrangement b and using Equation (2) for the interface temperature, we can determine the heat Q that is flowing. Choosing rod 2, we find that

$$Q = \frac{k_2 A (T - T_C) t}{L} = \frac{k_2 A \left(\dfrac{k_1 T_W + k_2 T_C}{k_1 + k_2} - T_C \right) t}{L}$$

$$= \frac{k_2 A \left(\dfrac{k_1 T_W - k_1 T_C}{k_1 + k_2} \right) t}{L} = \frac{k_2 A k_1 (T_W - T_C) t}{L (k_1 + k_2)} \quad (3)$$

Using Equations (1) and (3), we obtain for the desired ratio that

$$\frac{Q'}{Q} = \frac{\dfrac{(k_1 + k_2) A (T_W - T_C) t}{L}}{\dfrac{k_2 A k_1 (T_W - T_C) t}{L(k_1 + k_2)}} = \frac{(k_1 + k_2) A \cancel{(T_W - T_C)} \cancel{t} \cancel{L} (k_1 + k_2)}{\cancel{L} k_2 A k_1 \cancel{(T_W - T_C)} \cancel{t}} = \frac{(k_1 + k_2)^2}{k_2 k_1}$$

Using the fact that $k_2 = 2k_1$, we obtain

$$\frac{Q'}{Q} = \frac{(k_1 + k_2)^2}{k_2 k_1} = \frac{(k_1 + 2k_1)^2}{2k_1 k_1} = \boxed{4.5}$$

CHAPTER 14 | *THE IDEAL GAS LAW AND KINETIC THEORY*

1. **REASONING AND SOLUTION** Since hemoglobin has a molecular mass of 64 500 u, the mass per mole of hemoglobin is 64 500 g/mol. The number of hemoglobin molecules per mol is Avogadro's number, or 6.022×10^{23} mol^{-1}. Therefore, one molecule of hemoglobin has a mass (in kg) of

$$\left(\frac{64\ 500 \text{ g/mol}}{6.022 \times 10^{23} \text{ mol}^{-1}} \right) \left(\frac{1 \text{ kg}}{1000 \text{ g}} \right) = \boxed{1.07 \times 10^{-22} \text{ kg}}$$

5. **REASONING** The mass (in grams) of the active ingredient in the standard dosage is the number of molecules in the dosage times the mass per molecule (in grams per molecule). The mass per molecule can be obtained by dividing the molecular mass (in grams per mole) by Avogadro's number. The molecular mass is the sum of the atomic masses of the molecule's atomic constituents.

SOLUTION Using N to denote the number of molecules in the standard dosage and m_{molecule} to denote the mass of one molecule, the mass (in grams) of the active ingredient in the standard dosage can be written as follows:

$$m = N m_{\text{molecule}}$$

Using M to denote the molecular mass (in grams per mole) and recognizing that $m_{\text{molecule}} = \dfrac{M}{N_A}$, where N_A is Avogadro's number and is the number of molecules per mole, we have

$$m = N m_{\text{molecule}} = N \left(\frac{M}{N_A} \right)$$

M (in grams per mole) is equal to the molecular mass in atomic mass units, and we can obtain this quantity by referring to the periodic table on the inside of the back cover of the text to find the molecular masses of the constituent atoms in the active ingredient. Thus, we have

$$\text{Molecular mass} = \underbrace{22(12.011 \text{ u})}_{\text{Carbon}} + \underbrace{23(1.00794 \text{ u})}_{\text{Hydrogen}} + \underbrace{1(35.453 \text{ u})}_{\text{Chlorine}} + \underbrace{2(14.0067 \text{ u})}_{\text{Nitrogen}} + \underbrace{2(15.9994 \text{ u})}_{\text{Oxygen}}$$

$$= 382.89 \text{ u}$$

The mass of the active ingredient in the standard dosage is

$$m = N\left(\frac{M}{N_A}\right) = \left(1.572 \times 10^{19} \text{ molecules}\right)\left(\frac{382.89 \text{ g/mol}}{6.022 \times 10^{23} \text{ molecules/mol}}\right) = \boxed{1.00 \times 10^{-2} \text{ g}}$$

---

9. ***REASONING*** The number n of moles of water molecules in the glass is equal to the mass m of water divided by the mass per mole. According to Equation 11.1, the mass of water is equal to its density ρ times its volume V. Thus, we have

$$n = \frac{m}{\text{Mass per mole}} = \frac{\rho V}{\text{Mass per mole}}$$

The volume of the cylindrical glass is $V = \pi r^2 h$, where r is the radius of the cylinder and h is its height. The number of moles of water can be written as

$$n = \frac{\rho V}{\text{Mass per mole}} = \frac{\rho\left(\pi r^2 h\right)}{\text{Mass per mole}}$$

SOLUTION The molecular mass of water (H_2O) is $2(1.00794 \text{ u}) + (15.9994 \text{ u}) = 18.0 \text{ u}$. The mass per mole of H_2O is 18.0 g/mol. The density of water (see Table 11.1) is $1.00 \times 10^3 \text{ kg/m}^3$ or 1.00 g/cm^3.

$$n = \frac{\rho\left(\pi r^2 h\right)}{\text{Mass per mole}} = \frac{\left(1.00 \text{ g/cm}^3\right)(\pi)(4.50 \text{ cm})^2 (12.0 \text{ cm})}{18.0 \text{ g/mol}} = \boxed{42.4 \text{ mol}}$$

---

11. ***REASONING*** Both gases fill the balloon to the same pressure P, volume V, and temperature T. Assuming that both gases are ideal, we can apply the ideal gas law $PV = nRT$ to each and conclude that the same number of moles n of each gas is needed to fill the balloon. Furthermore, the number of moles can be calculated from the mass m (in grams) and the mass per mole M (in grams per mole), according to $n = \dfrac{m}{M}$. Using this expression in the equation $n_{\text{Helium}} = n_{\text{Nitrogen}}$ will allow us to obtain the desired mass of nitrogen.

SOLUTION Since the number of moles of helium equals the number of moles of nitrogen, we have

$$\underbrace{\frac{m_{\text{Helium}}}{M_{\text{Helium}}}}_{\substack{\text{Number of moles} \\ \text{of helium}}} = \underbrace{\frac{m_{\text{Nitrogen}}}{M_{\text{Nitrogen}}}}_{\substack{\text{Number of moles} \\ \text{of nitrogen}}}$$

Solving for m_{Nitrogen} and taking the values of mass per mole for helium (He) and nitrogen (N_2) from the periodic table on the inside of the back cover of the text, we find

$$m_{\text{Nitrogen}} = \frac{M_{\text{Nitrogen}} m_{\text{Helium}}}{M_{\text{Helium}}} = \frac{(28.0 \text{ g/mol})(0.16 \text{ g})}{4.00 \text{ g/mol}} = \boxed{1.1 \text{ g}}$$

19. **REASONING** According to the ideal gas law (Equation 14.1), $PV = nRT$. Since n, the number of moles, is constant, $n_1 R = n_2 R$. Thus, according to Equation 14.1, we have

$$\frac{P_1 V_1}{T_1} = \frac{P_2 V_2}{T_2}$$

SOLUTION Solving for T_2, we have

$$T_2 = \left(\frac{P_2}{P_1}\right)\left(\frac{V_2}{V_1}\right) T_1 = \left(\frac{48.5 \, P_1}{P_1}\right)\left[\frac{V_1/16}{V_1}\right](305 \text{ K}) = \boxed{925 \text{ K}}$$

23. **REASONING AND SOLUTION**
a. Since the heat gained by the gas in one tank is equal to the heat lost by the gas in the other tank, $Q_1 = Q_2$, or (letting the subscript 1 correspond to the neon in the left tank, and letting 2 correspond to the neon in the right tank) $cm_1 \Delta T_1 = cm_2 \Delta T_2$,

$$cm_1(T - T_1) = cm_2(T_2 - T)$$

$$m_1(T - T_1) = m_2(T_2 - T)$$

Solving for T gives

$$T = \frac{m_2 T_2 + m_1 T_1}{m_2 + m_1} \tag{1}$$

The masses m_1 and m_2 can be found by first finding the number of moles n_1 and n_2. From the ideal gas law, $PV = nRT$, so

$$n_1 = \frac{P_1 V_1}{R T_1} = \frac{(5.0 \times 10^5 \text{ Pa})(2.0 \text{ m}^3)}{[8.31 \text{ J/(mol} \cdot \text{K)}](220 \text{ K})} = 5.5 \times 10^2 \text{ mol}$$

This corresponds to a mass $m_1 = (5.5 \times 10^2 \text{ mol})(20.179 \text{ g/mol}) = 1.1 \times 10^4 \text{ g} = 1.1 \times 10^1 \text{ kg}$. Similarly, $n_2 = 2.4 \times 10^2$ mol and $m_2 = 4.9 \times 10^3$ g = 4.9 kg. Substituting these mass values into Equation (1) yields

$$T = \frac{(4.9 \text{ kg})(580 \text{ K}) + (1.1 \times 10^1 \text{ kg})(220 \text{ K})}{(4.9 \text{ kg}) + (1.1 \times 10^1 \text{ kg})} = \boxed{3.3 \times 10^2 \text{ K}}$$

b. From the ideal gas law,

$$P = \frac{nRT}{V} = \frac{[(5.5 \times 10^2 \text{ mol}) + (2.4 \times 10^2 \text{ mol})][8.31 \text{ J/(mol·K)}](3.3 \times 10^2 \text{ K})}{2.0 \text{ m}^3 + 5.8 \text{ m}^3} = \boxed{2.8 \times 10^5 \text{Pa}}$$

27. **REASONING** The graph that accompanies Problem 75 in Chapter 12 can be used to determine the equilibrium vapor pressure of water in the air when the temperature is 30.0 °C (303 K). Equation 12.6 can then be used to find the partial pressure of water in the air at this temperature. Using this pressure, the ideal gas law can then be used to find the number of moles of water vapor per cubic meter.

SOLUTION According to the graph that accompanies Problem 75 in Chapter 12, the equilibrium vapor pressure of water vapor at 30.0 °C is approximately 4250 Pa. According to Equation 12.6,

$$\begin{pmatrix} \text{Partial} \\ \text{pressure of} \\ \text{water vapor} \end{pmatrix} = \begin{pmatrix} \text{Percent relative} \\ \text{humidity} \end{pmatrix}\begin{pmatrix} \text{Equilibrium vapor pressure of} \\ \text{water at the existing temperature} \end{pmatrix} \times \frac{1}{100}$$

$$= \frac{(55)(4250 \text{ Pa})}{100} = 2.34 \times 10^3 \text{ Pa}$$

The ideal gas law then gives the number of moles of water vapor per cubic meter of air as

$$\frac{n}{V} = \frac{P}{RT} = \frac{(2.34 \times 10^3 \text{ Pa})}{[8.31 \text{ J/(mol·K)}](303 \text{ K})} = \boxed{0.93 \text{ mol/m}^3} \tag{14.1}$$

31. **REASONING** According to the ideal gas law (Equation 14.1), $PV = nRT$. Since n, the number of moles of the gas, is constant, $n_1 R = n_2 R$. Therefore, $P_1 V_1 / T_1 = P_2 V_2 / T_2$, where $T_1 = 273$ K and T_2 is the temperature we seek. Since the beaker is cylindrical, the volume V of the gas is equal to Ad, where A is the cross-sectional area of the cylindrical volume and d is the height of the region occupied by the gas, as measured from the bottom of the beaker.

With this substitution for the volume, the expression obtained from the ideal gas law becomes

$$\frac{P_1 d_1}{T_1} = \frac{P_2 d_2}{T_2} \qquad (1)$$

where the pressures P_1 and P_2 are equal to the sum of the atmospheric pressure and the pressure caused by the mercury in each case. These pressures can be determined using Equation 11.4. Once the pressures are known, Equation (1) can be solved for T_2.

SOLUTION Using Equation 11.4, we obtain the following values for the pressures P_1 and P_2. Note that the initial height of the mercury is $h_1 = \frac{1}{2}(1.520 \text{ m}) = 0.760 \text{ m}$, while the final height of the mercury is $h_2 = \frac{1}{4}(1.520 \text{ m}) = 0.380 \text{ m}$.

$$P_1 = P_0 + \rho g h_1 = \left(1.01 \times 10^5 \text{ Pa}\right) + \left[\left(1.36 \times 10^4 \text{kg/m}^3\right)(9.80 \text{ m/s}^2)(0.760 \text{ m})\right] = 2.02 \times 10^5 \text{ Pa}$$

$$P_2 = P_0 + \rho g h_2 = \left(1.01 \times 10^5 \text{ Pa}\right) + \left[\left(1.36 \times 10^4 \text{kg/m}^3\right)(9.80 \text{ m/s}^2)(0.380 \text{ m})\right] = 1.52 \times 10^5 \text{ Pa}$$

In these pressure calculations, the density of mercury is $\rho = 1.36 \times 10^4$ kg/m^3. In Equation (1) we note that $d_1 = 0.760$ m and $d_2 = 1.14$ m. Solving Equation (1) for T_2 and substituting values, we obtain

$$T_2 = \left(\frac{P_2 d_2}{P_1 d_1}\right) T_1 = \left[\frac{\left(1.52 \times 10^5 \text{ Pa}\right)(1.14 \text{ m})}{\left(2.02 \times 10^5 \text{ Pa}\right)(0.760 \text{ m})}\right](273 \text{ K}) = \boxed{308 \text{ K}}$$

---

37. **REASONING** The behavior of the molecules is described by Equation 14.5: $PV = \frac{2}{3} N(\frac{1}{2} m v_{\text{rms}}^2)$. Since the pressure and volume of the gas are kept constant, while the number of molecules is doubled, we can write $P_1 V_2 = P_2 V_2$, where the subscript 1 refers to the initial condition, and the subscript 2 refers to the conditions after the number of molecules is doubled. Thus,

$$\frac{2}{3} N_1 \left[\frac{1}{2} m (v_{\text{rms}})_1^2\right] = \frac{2}{3} N_2 \left[\frac{1}{2} m (v_{\text{rms}})_2^2\right] \quad \text{or} \quad N_1 (v_{\text{rms}})_1^2 = N_2 (v_{\text{rms}})_2^2$$

The last expression can be solved for $(v_{\text{rms}})_2$, the final translational rms speed.

SOLUTION Since the number of molecules is doubled, $N_2 = 2N_1$. Solving the last expression above for $(v_{rms})_2$, we find

$$(v_{rms})_2 = (v_{rms})_1 \sqrt{\frac{N_1}{N_2}} = (463 \text{ m/s}) \sqrt{\frac{N_1}{2N_1}} = \frac{463 \text{ m/s}}{\sqrt{2}} = \boxed{327 \text{ m/s}}$$

---

41. **REASONING** The rms-speed v_{rms} of the sulfur dioxide molecules is related to the Kelvin temperature T by $\frac{1}{2}mv_{rms}^2 = \frac{3}{2}kT$ (Equation 14.6), where m is the mass of a SO_2 molecule and k is Boltzmann's constant. Solving this equation for the rms-speed gives

$$v_{rms} = \sqrt{\frac{3kT}{m}} \tag{1}$$

The temperature can be found from the ideal gas law, Equation 14.1, as $T = PV/(nR)$, where P is the pressure, V is the volume, n is the number of moles, and R is the universal gas constant. All the variables in this relation are known. Substituting this expression for T into Equation (1) yields

$$v_{rms} = \sqrt{\frac{3k\left(\dfrac{PV}{nR}\right)}{m}} = \sqrt{\frac{3kPV}{nmR}}$$

The mass m of a single SO_2 molecule will be calculated in the **SOLUTION** section.

SOLUTION Using the periodic table on the inside of the text's back cover, we find the molecular mass of a sulfur dioxide molecule (SO_2) to be

$$\underbrace{32.07 \text{ u}}_{\substack{\text{Mass of a single} \\ \text{sulfur atom}}} + \underbrace{2\,(15.9994 \text{ u})}_{\substack{\text{Mass of two} \\ \text{oxygen atoms}}} = 64.07 \text{ u}$$

Since $1 \text{ u} = 1.6605 \times 10^{-27}$ kg (see Section 14.1), the mass of a sulfur dioxide molecule is

$$m = (64.07 \text{ u})\left(\frac{1.6605 \times 10^{-27} \text{ kg}}{1 \text{ u}}\right) = 1.064 \times 10^{-25} \text{ kg}$$

The translational rms-speed of the sulfur dioxide molecules is

$$v_{rms} = \sqrt{\frac{3kPV}{nmR}}$$

$$= \sqrt{\frac{3(1.38 \times 10^{-23} \text{ J/K})(2.12 \times 10^4 \text{ Pa})(50.0 \text{ m}^3)}{(421 \text{ mol})(1.064 \times 10^{-25} \text{ kg})[8.31 \text{ J/(mol} \cdot \text{K)}]}} = \boxed{343 \text{ m/s}}$$

---

43. **REASONING AND SOLUTION**
a. Assuming that the direction of travel of the bullets is positive, the average change in momentum per second is

$$\Delta p/\Delta t = m\Delta v/\Delta t = (200)(0.0050\ \text{kg})[(0\ \text{m/s}) - 1200\ \text{m/s}]/(10.0\ \text{s}) = \boxed{-120\ \text{N}}$$

b. The average force exerted on the bullets is $\overline{F} = \Delta p/\Delta t$. According to Newton's third law, the average force exerted on the wall is $-\overline{F} = \boxed{120\ \text{N}}$.

c. The pressure P is the magnitude of the force on the wall per unit area, so

$$P = (120\ \text{N})/(3.0 \times 10^{-4}\ \text{m}^2) = \boxed{4.0 \times 10^5\ \text{Pa}}$$

---

47. **REASONING AND SOLUTION** According to Fick's law of diffusion (Equation 14.8), the mass of ethanol that diffuses through the cylinder in one hour (3600 s) is

$$m = \frac{(DA\ \Delta C)t}{L} = \frac{(12.4 \times 10^{-10}\ \text{m}^2/\text{s})(4.00 \times 10^{-4}\ \text{m}^2)(1.50\ \text{kg/m}^3)(3600\ \text{s})}{(0.0200\ \text{m})} = \boxed{1.34 \times 10^{-7}\ \text{kg}}$$

---

51. **REASONING AND SOLUTION**
a. The average concentration is $C_{av} = (1/2)(C_1 + C_2) = (1/2)C_2 = m/V = m/(AL)$, so that $C_2 = 2m/(AL)$. Fick's law then becomes $m = DAC_2 t/L = DA(2m/AL)t/L = 2Dmt/L^2$. Solving for t yields

$$\boxed{t = L^2/(2D)}$$

b. Substituting the given data into this expression yields

$$t = (2.5 \times 10^{-2}\ \text{m})^2/[2(1.0 \times 10^{-5}\ \text{m}^2/\text{s})] = \boxed{31\ \text{s}}$$

---

55. **REASONING** According to Equation 11.1, the mass density ρ of a substance is defined as its mass m divided by its volume V: $\rho = m/V$. The mass of nitrogen is equal to the number n of moles of nitrogen times its mass per mole: $m = n$ (Mass per mole). The number of moles can be obtained from the ideal gas law (see Equation 14.1) as $n = (PV)/(RT)$. The mass per mole (in g/mol) of nitrogen has the same numerical value as its molecular mass (which we know).

SOLUTION Substituting $m = n\,(\text{Mass per mole})$ into $\rho = m/V$, we obtain

$$\rho = \frac{m}{V} = \frac{n\,(\text{Mass per mole})}{V} \tag{1}$$

Substituting $n = (PV)/(RT)$ from the ideal gas law into Equation 1 gives the following result:

$$\rho = \frac{n\,(\text{Mass per mole})}{V} = \frac{\left(\dfrac{P\cancel{V}}{RT}\right)(\text{Mass per mole})}{\cancel{V}} = \frac{P\,(\text{Mass per mole})}{RT}$$

The pressure is 2.0 atmospheres, or $P = 2\,(1.013 \times 10^5 \text{ Pa})$. The molecular mass of nitrogen is given as 28 u, which means that its mass per mole is 28 g/mol. Expressed in terms of kilograms per mol, the mass per mole is

$$\text{Mass per mole} = \left(28\,\frac{\cancel{g}}{\text{mol}}\right)\left(\frac{1 \text{ kg}}{10^3 \cancel{g}}\right)$$

The density of the nitrogen gas is

$$\rho = \frac{P\,(\text{Mass per mole})}{RT} = \frac{2\,(1.013 \times 10^5 \text{ Pa})\left(28\,\dfrac{\cancel{g}}{\text{mol}}\right)\left(\dfrac{1 \text{ kg}}{10^3 \cancel{g}}\right)}{[8.31 \text{ J}/(\text{mol} \cdot \text{K})](310 \text{ K})} = \boxed{2.2 \text{ kg/m}^3}$$

---

57. **REASONING AND SOLUTION**   Using the expressions for $\overline{v^2}$ and $(\overline{v})^2$ given in the statement of the problem, we obtain:

a. $\qquad \overline{v^2} = \frac{1}{3}(v_1^2 + v_2^2 + v_3^2) = \frac{1}{3}\left[(3.0 \text{ m/s})^2 + (7.0 \text{ m/s})^2 + (9.0 \text{ m/s})^2\right] = \boxed{46.3 \text{ m}^2/\text{s}^2}$

b. $\qquad (\overline{v})^2 = \left[\frac{1}{3}(v_1 + v_2 + v_3)\right]^2 = \left[\frac{1}{3}(3.0 \text{ m/s} + 7.0 \text{ m/s} + 9.0 \text{ m/s})\right]^2 = \boxed{40.1 \text{ m}^2/\text{s}^2}$

$\overline{v^2}$ and $(\overline{v})^2$ are *not* equal, because they are two different physical quantities.

---

59.   ***REASONING*** The mass m of nitrogen that must be removed from the tank is equal to the number of moles withdrawn times the mass per mole. The number of moles withdrawn is the initial number n_i minus the final number of moles n_f, so we can write

$$m = \underbrace{(n_i - n_f)}_{\substack{\text{Number of} \\ \text{moles withdrawn}}} \text{(Mass per mole)} \tag{1}$$

The final number of moles is related to the initial number by the ideal gas law.

SOLUTION   From the ideal gas law (Equation 14.1), we have

$$n_f = \frac{P_f V}{RT} \quad \text{and} \quad n_i = \frac{P_i V}{RT}$$

Note that the volume V and temperature T do not change. Dividing the first by the second equation gives

$$\frac{n_f}{n_i} = \frac{\dfrac{P_f V}{RT}}{\dfrac{P_i V}{RT}} = \frac{P_f}{P_i} \quad \text{or} \quad n_f = n_i \left(\frac{P_f}{P_i} \right)$$

Substituting this expression for n_f into Equation 1 gives

$$m = (n_i - n_f)(\text{Mass per mole}) = \left[n_i - n_i \left(\frac{P_f}{P_i} \right) \right](\text{Mass per mole})$$

The molecular mass of nitrogen (N_2) is $2\,(14.0067\text{ u}) = 28.0134\text{ u}$. Therefore, the mass per mole is 28.0134 g/mol. The mass of nitrogen that must be removed is

$$m = \left[n_i - n_i \left(\frac{P_f}{P_i} \right) \right](\text{Mass per mole})$$

$$= \left[0.85\text{ mol} - (0.85\text{ mol}) \left(\frac{25\text{ atm}}{38\text{ atm}} \right) \right] \left(28.0134\ \frac{\text{g}}{\text{mol}} \right) = \boxed{8.1\text{ g}}$$

CHAPTER 15 | *THERMODYNAMICS*

3. **REASONING** Energy in the form of work leaves the system, while energy in the form of heat enters. More energy leaves than enters, so we expect the internal energy of the system to decrease, that is, we expect the change ΔU in the internal energy to be negative. The first law of thermodynamics will confirm our expectation. As far as the environment is concerned, we note that when the system loses energy, the environment gains it, and when the system gains energy the environment loses it. Therefore, the change in the internal energy of the environment must be opposite to that of the system.

SOLUTION
a. The system gains heat so Q is positive, according to our convention. The system does work, so W is also positive, according to our convention. Applying the first law of thermodynamics from Equation 15.1, we find for the system that

$$\Delta U = Q - W = (77 \text{ J}) - (164 \text{ J}) = \boxed{-87 \text{ J}}$$

As expected, this value is negative, indicating a decrease.

b. The change in the internal energy of the environment is opposite to that of the system, so that $\boxed{\Delta U_{\text{environment}} = +87 \text{ J}}$.

5. **REASONING** Since the change in the internal energy and the heat released in the process are given, the first law of thermodynamics (Equation 15.1) can be used to find the work done. Since we are told how much work is required to make the car go one mile, we can determine how far the car can travel. When the gasoline burns, its internal energy decreases and heat flows into the surroundings; therefore, both ΔU and Q are negative.

SOLUTION According to the first law of thermodynamics, the work that is done when one gallon of gasoline is burned in the engine is

$$W = Q - \Delta U = -1.00 \times 10^8 \text{ J} - (-1.19 \times 10^8 \text{ J}) = 0.19 \times 10^8 \text{ J}$$

Since 6.0×10^5 J of work is required to make the car go one mile, the car can travel

$$0.19 \times 10^8 \text{ J} \left(\frac{1 \text{ mile}}{6.0 \times 10^5 \text{ J}} \right) = \boxed{32 \text{ miles}}$$

9. **REASONING** According to Equation 15.2, $W = P\Delta V$, the average pressure $\overline{P}$ of the expanding gas is equal to $\overline{P} = W / \Delta V$, where the work W done by the gas on the bullet can be found from the work-energy theorem (Equation 6.3). Assuming that the barrel of the gun is cylindrical with radius r, the volume of the barrel is equal to its length L multiplied by the area (πr^2) of its cross section. Thus, the change in volume of the expanding gas is $\Delta V = L\pi r^2$.

SOLUTION The work done by the gas on the bullet is given by Equation 6.3 as

$$W = \tfrac{1}{2}m(v_{final}^2 - v_{initial}^2) = \tfrac{1}{2}(2.6\times10^{-3} \text{ kg})[(370 \text{ m/s})^2 - 0] = 180 \text{ J}$$

The average pressure of the expanding gas is, therefore,

$$\overline{P} = \frac{W}{\Delta V} = \frac{180 \text{ J}}{(0.61\,\text{m})\pi(2.8\times10^{-3} \text{ m})^2} = \boxed{1.2\times10^7 \text{ Pa}}$$

---

15. **REASONING** The work done in an isobaric process is given by Equation 15.2, $W = P\Delta V$; therefore, the pressure is equal to $P = W / \Delta V$. In order to use this expression, we must first determine a numerical value for the work done; this can be calculated using the first law of thermodynamics (Equation 15.1), $\Delta U = Q - W$.

SOLUTION Solving Equation 15.1 for the work W, we find

$$W = Q - \Delta U = 1500 \text{ J} - (+4500 \text{ J}) = -3.0\times10^3 \text{ J}$$

Therefore, the pressure is

$$P = \frac{W}{\Delta V} = \frac{-3.0\times10^3 \text{ J}}{-0.010 \text{ m}^3} = \boxed{3.0\times10^5 \text{ Pa}}$$

The change in volume ΔV, which is the final volume minus the initial volume, is negative because the final volume is 0.010 m^3 *less* than the initial volume.

---

17. **REASONING AND SOLUTION** The first law of thermodynamics states that $\Delta U = Q - W$. The work W involved in an isobaric process is, according to Equation 15.2, $W = P\Delta V$. Combining these two expressions leads to $\Delta U = Q - P\Delta V$. Solving for Q gives

$$Q = \Delta U + P\Delta V \qquad (1)$$

Since this is an expansion, $\Delta V > 0$, so $P\Delta V > 0$. From the ideal gas law, $PV = nRT$, we have $P\Delta V = nR\,\Delta T$. Since $P\Delta V > 0$, it follows that $nR\,\Delta T > 0$. The internal energy of an ideal gas is directly proportional to its Kelvin temperature T. Therefore, since $nR\,\Delta T > 0$, it follows that $\Delta U > 0$. Since both terms on the right hand side of Equation (1) are positive, the left hand side of Equation (1) must also be positive. Thus, Q is positive. By the convention described in the text, this means that

$$\boxed{\text{heat can only flow into an ideal gas during an isobaric expansion}}$$

21. **REASONING AND SOLUTION**

a. Since the temperature of the gas is kept constant at all times, the process is isothermal; therefore, the internal energy of an ideal gas does not change and $\boxed{\Delta U = 0}$.

b. From the first law of thermodynamics (Equation 15.1), $\Delta U = Q - W$. But $\Delta U = 0$, so that $Q = W$. Since work is done on the gas, the work is negative, and $\boxed{Q = -6.1 \times 10^3 \text{ J}}$.

c. The work done in an isothermal compression is given by Equation 15.3:

$$W = nRT \ln\left(\frac{V_f}{V_i}\right)$$

Therefore, the temperature of the gas is

$$T = \frac{W}{nR \ln\left(V_f / V_i\right)} = \frac{-6.1 \times 10^3 \text{ J}}{(3.0 \text{ mol})[8.31 \text{ J/(mol·K)}] \ln\left[(2.5 \times 10^{-2} \text{ m}^3)/(5.5 \times 10^{-2} \text{ m}^3)\right]} = \boxed{310 \text{ K}}$$

25. **REASONING** When the expansion is isothermal, the work done can be calculated from Equation (15.3): $W = nRT \ln\left(V_f / V_i\right)$. When the expansion is adiabatic, the work done can be calculated from Equation 15.4: $W = \frac{3}{2} nR(T_i - T_f)$.

Since the gas does the same amount of work whether it expands adiabatically or isothermally, we can equate the right hand sides of these two equations. We also note that since the initial temperature is the same for both cases, the temperature T in the isothermal expansion is the same as the initial temperature T_i for the adiabatic expansion. We then have

$$nRT_i \ln\left(\frac{V_f}{V_i}\right) = \frac{3}{2} nR(T_i - T_f) \quad \text{or} \quad \ln\left(\frac{V_f}{V_i}\right) = \frac{\frac{3}{2}(T_i - T_f)}{T_i}$$

SOLUTION   Solving for the ratio of the volumes gives

$$\frac{V_f}{V_i} = e^{\frac{3}{2}(T_i - T_f)/T_i} = e^{\frac{3}{2}(405 \text{ K} - 245 \text{ K})/(405 \text{ K})} = \boxed{1.81}$$

---

29. ***REASONING* AND *SOLUTION***

Step A → B

The internal energy of a monatomic ideal gas is $U = (3/2)nRT$. Thus, the change is

$$\Delta U = \tfrac{3}{2} nR \,\Delta T = \tfrac{3}{2}(1.00 \text{ mol})\big[8.31 \text{ J}/(\text{mol}\cdot\text{K})\big](800.0 \text{ K} - 400.0 \text{ K}) = \boxed{4990 \text{ J}}$$

The work for this constant pressure step is $W = P\Delta V$. But the ideal gas law applies, so

$$W = P\Delta V = nR \,\Delta T = (1.00 \text{ mol})\big[8.31 \text{ J}/(\text{mol}\cdot\text{K})\big](800.0 \text{ K} - 400.0 \text{ K}) = \boxed{3320 \text{ J}}$$

The first law of thermodynamics indicates that the heat is

$$Q = \Delta U + W = \tfrac{3}{2} nR \,\Delta T + nR \,\Delta T$$

$$= \tfrac{5}{2}(1.00 \text{ mol})\big[8.31 \text{ J}/(\text{mol}\cdot\text{K})\big](800.0 \text{ K} - 400.0 \text{ K}) = \boxed{8310 \text{ J}}$$

Step B → C

The internal energy of a monatomic ideal gas is $U = (3/2)nRT$. Thus, the change is

$$\Delta U = \tfrac{3}{2} nR \,\Delta T = \tfrac{3}{2}(1.00 \text{ mol})\big[8.31 \text{ J}/(\text{mol}\cdot\text{K})\big](400.0 \text{ K} - 800.0 \text{ K}) = \boxed{-4990 \text{ J}}$$

The volume is constant in this step, so the work done by the gas is $\boxed{W = 0 \text{ J}}$.

The first law of thermodynamics indicates that the heat is

$$Q = \Delta U + W = \Delta U = \boxed{-4990 \text{ J}}$$

Step C → D

The internal energy of a monatomic ideal gas is $U = (3/2)nRT$. Thus, the change is

$$\Delta U = \tfrac{3}{2} nR \,\Delta T = \tfrac{3}{2}(1.00 \text{ mol})\big[8.31 \text{ J}/(\text{mol}\cdot\text{K})\big](200.0 \text{ K} - 400.0 \text{ K}) = \boxed{-2490 \text{ J}}$$

The work for this constant pressure step is $W = P\Delta V$. But the ideal gas law applies, so

$$W = P\Delta V = nR \,\Delta T = (1.00 \text{ mol})\big[8.31 \text{ J}/(\text{mol}\cdot\text{K})\big](200.0 \text{ K} - 400.0 \text{ K}) = \boxed{-1660 \text{ J}}$$

The first law of thermodynamics indicates that the heat is

$$Q = \Delta U + W = \tfrac{3}{2} nR\,\Delta T + nR\,\Delta T$$

$$= \tfrac{5}{2}(1.00 \text{ mol})\big[8.31 \text{ J/}(\text{mol}\cdot\text{K})\big](200.0 \text{ K} - 400.0 \text{ K}) = \boxed{-4150 \text{ J}}$$

Step D $\rightarrow$ A
The internal energy of a monatomic ideal gas is $U = (3/2)nRT$. Thus, the change is

$$\Delta U = \tfrac{3}{2} nR\,\Delta T = \tfrac{3}{2}(1.00 \text{ mol})\big[8.31 \text{ J/}(\text{mol}\cdot\text{K})\big](400.0 \text{ K} - 200.0 \text{ K}) = \boxed{2490 \text{ J}}$$

The volume is constant in this step, so the work done by the gas is $\boxed{W = 0 \text{ J}}$

The first law of thermodynamics indicates that the heat is

$$Q = \Delta U + W = \Delta U = \boxed{2490 \text{ J}}$$

---

33. ***REASONING AND SOLUTION*** Let the left be side 1 and the right be side 2. Since the partition moves to the right, side 1 does work on side 2, so that the work values involved satisfy the relation $W_1 = -W_2$. Using Equation 15.4 for each work value, we find that

$$\tfrac{3}{2} nR\big(T_{1i} - T_{1f}\big) = -\tfrac{3}{2} nR\big(T_{2i} - T_{2f}\big) \qquad \text{or}$$

$$T_{1f} + T_{2f} = T_{1i} + T_{2i} = 525 \text{ K} + 275 \text{ K} = 8.00 \times 10^2 \text{ K}$$

We now seek a second equation for the two unknowns T_{1f} and T_{2f}. Equation 15.5 for an adiabatic process indicates that $P_{1i}V_{1i}^{\gamma} = P_{1f}V_{1f}^{\gamma}$ and $P_{2i}V_{2i}^{\gamma} = P_{2f}V_{2f}^{\gamma}$. Dividing these two equations and using the facts that $V_{1i} = V_{2i}$ and $P_{1f} = P_{2f}$, gives

$$\frac{P_{1i}V_{1i}^{\gamma}}{P_{2i}V_{2i}^{\gamma}} = \frac{P_{1f}V_{1f}^{\gamma}}{P_{2f}V_{2f}^{\gamma}} \qquad \text{or} \qquad \frac{P_{1i}}{P_{2i}} = \left(\frac{V_{1f}}{V_{2f}}\right)^{\gamma}$$

Using the ideal gas law, we find that

$$\frac{P_{1i}}{P_{2i}} = \left(\frac{V_{1f}}{V_{2f}}\right)^{\gamma} \qquad \text{becomes} \qquad \frac{nRT_{1i}/V_{1i}}{nRT_{2i}/V_{2i}} = \left(\frac{nRT_{1f}/P_{1f}}{nRT_{2f}/P_{2f}}\right)^{\gamma}$$

Since $V_{1i} = V_{2i}$ and $P_{1f} = P_{2f}$, the result above reduces to

$$\frac{T_{1i}}{T_{2i}} = \left(\frac{T_{1f}}{T_{2f}}\right)^{\gamma} \quad \text{or} \quad \frac{T_{1f}}{T_{2f}} = \left(\frac{T_{1i}}{T_{2i}}\right)^{1/\gamma} = \left(\frac{525 \text{ K}}{275 \text{ K}}\right)^{1/\gamma} = 1.474$$

Using this expression for the ratio of the final temperatures in $T_{1f} + T_{2f} = 8.00 \times 10^2$ K, we find that

a. $\boxed{T_{1f} = 477 \text{ K}}$ and b. $\boxed{T_{2f} = 323 \text{ K}}$

---

35. **_REASONING AND SOLUTION_** According to the first law of thermodynamics (Equation 15.1), $\Delta U = U_f - U_i = Q - W$. Since the internal energy of this gas is doubled by the addition of heat, the initial and final internal energies are U and $2U$, respectively. Therefore,

$$\Delta U = U_f - U_i = 2U - U = U$$

Equation 15.1 for this situation then becomes $U = Q - W$. Solving for Q gives

$$Q = U + W \tag{1}$$

The initial internal energy of the gas can be calculated from Equation 14.7:

$$U = \frac{3}{2}nRT = \frac{3}{2}(2.5 \text{ mol})\left[8.31 \text{ J}/(\text{mol} \cdot \text{K})\right](350 \text{ K}) = 1.1 \times 10^4 \text{ J}$$

a. If the process is carried out isochorically (i.e., at constant volume), then W = 0, and the heat required to double the internal energy is

$$Q = U + W = U + 0 = \boxed{1.1 \times 10^4 \text{ J}}$$

b. If the process is carried out isobarically (i.e., at constant pressure), then $W = P\Delta V$, and Equation (1) above becomes

$$Q = U + W = U + P\Delta V \tag{2}$$

From the ideal gas law, $PV = nRT$, we have that $P\Delta V = nR\Delta T$, and Equation (2) becomes

$$Q = U + nR\Delta T \tag{3}$$

The internal energy of an ideal gas is directly proportional to its Kelvin temperature. Since the internal energy of the gas is doubled, the final Kelvin temperature will be twice the initial Kelvin temperature, or $\Delta T = 350$ K. Substituting values into Equation (3) gives

$$Q = 1.1 \times 10^4 \text{ J} + (2.5 \text{ mol})[8.31 \text{ J/(mol·K)}](350 \text{ K}) = \boxed{1.8 \times 10^4 \text{ J}}$$

---

37. ***REASONING*** When the temperature of a gas changes as a result of heat Q being added, the change ΔT in temperature is related to the amount of heat according to $Q = Cn\Delta T$ (Equation 15.6), where C is the molar specific heat capacity, and n is the number of moles. The heat Q_V added under conditions of constant volume is $Q_V = C_V n \Delta T_V$, where C_V is the specific heat capacity at constant volume and is given by $C_V = \frac{3}{2}R$ (Equation 15.8) and R is the universal gas constant. The heat Q_P added under conditions of constant pressure is $Q_P = C_P n \Delta T_P$, where C_P is the specific heat capacity at constant pressure and is given by $C_P = \frac{5}{2}R$ (Equation 15.7). It is given that $Q_V = Q_P$, and this fact will allow us to find the change in temperature of the gas whose pressure remains constant.

SOLUTION Setting $Q_V = Q_P$, gives

$$\underbrace{C_V n \Delta T_V}_{Q_V} = \underbrace{C_P n \Delta T_P}_{Q_P}$$

Algebraically eliminating n and solving for ΔT_P, we obtain

$$\Delta T_P = \left(\frac{C_V}{C_P}\right)\Delta T_V = \left(\frac{\frac{3}{2}R}{\frac{5}{2}R}\right)(75 \text{ K}) = \boxed{45 \text{ K}}$$

---

45. ***REASONING AND SOLUTION*** The efficiency of a heat engine is defined by Equation 15.11 as $e = |W|/|Q_H|$, where $|W|$ is the magnitude of the work done and $|Q_H|$ is the magnitude of the heat input. The principle of energy conservation requires that $|Q_H| = |W| + |Q_C|$, where $|Q_C|$ is the magnitude of the heat rejected to the cold reservoir (Equation 15.12). Combining Equations 15.11 and 15.12 gives

$$e = \frac{|W|}{|W| + |Q_C|} = \frac{16\ 600 \text{ J}}{16\ 600 \text{ J} + 9700 \text{ J}} = \boxed{0.631}$$

49. **_REASONING_** The efficiency e of an engine can be expressed as (see Equation 15.13) $e = 1 - |Q_C| / |Q_H|$, where $|Q_C|$ is the magnitude of the heat delivered to the cold reservoir and $|Q_H|$ is the magnitude of the heat supplied to the engine from the hot reservoir. Solving this equation for $|Q_C|$ gives $|Q_C| = (1-e) |Q_H|$. We will use this expression twice, once for the improved engine and once for the original engine. Taking the ratio of these expressions will give us the answer that we seek.

SOLUTION Taking the ratio of the heat rejected to the cold reservoir by the improved engine to that for the original engine gives

$$\frac{|Q_{C, \text{ improved}}|}{|Q_{C, \text{ original}}|} = \frac{(1 - e_{\text{improved}}) |Q_{H, \text{ improved}}|}{(1 - e_{\text{original}}) |Q_{H, \text{ original}}|}$$

But the input heat to both engines is the same, so $|Q_{H, \text{ improved}}| = |Q_{H, \text{ original}}|$. Thus, the ratio becomes

$$\frac{|Q_{C, \text{ improved}}|}{|Q_{C, \text{ original}}|} = \frac{1 - e_{\text{improved}}}{1 - e_{\text{original}}} = \frac{1 - 0.42}{1 - 0.23} = \boxed{0.75}$$

---

53. **_REASONING_** The efficiency e of a Carnot engine is given by Equation 15.15, $e = 1 - (T_C / T_H)$, where, according to Equation 15.14, $|Q_C| / |Q_H| = T_C / T_H$. Since the efficiency is given along with T_C and $|Q_C|$, Equation 15.15 can be used to calculate T_H. Once T_H is known, the ratio T_C / T_H is thus known, and Equation 15.14 can be used to calculate $|Q_H|$.

SOLUTION
a. Solving Equation 15.15 for T_H gives

$$T_H = \frac{T_C}{1 - e} = \frac{378 \text{ K}}{1 - 0.700} = \boxed{1260 \text{ K}}$$

b. Solving Equation 15.14 for $|Q_H|$ gives

$$|Q_H| = |Q_C| \left(\frac{T_H}{T_C} \right) = (5230 \text{ J}) \left(\frac{1260 \text{ K}}{378 \text{ K}} \right) = \boxed{1.74 \times 10^4 \text{ J}}$$

59. **REASONING** The maximum efficiency e at which the power plant can operate is given by Equation 15.15, $e = 1 - (T_C / T_H)$. The power output is given; it can be used to find the magnitude $|W|$ of the work output for a 24 hour period. With the efficiency and $|W|$ known, Equation 15.11, $e = |W|/|Q_H|$, can be used to find $|Q_H|$, the magnitude of the input heat. The magnitude $|Q_C|$ of the exhaust heat can then be found from Equation 15.12, $|Q_H| = |W| + |Q_C|$.

SOLUTION
a. The maximum efficiency is

$$e = 1 - \frac{T_C}{T_H} = 1 - \frac{323 \text{ K}}{505 \text{ K}} = \boxed{0.360}$$

b. Since the power output of the power plant is $P = 84\,000$ kW, the required heat input $|Q_H|$ for a 24 hour period is

$$|Q_H| = \frac{|W|}{e} = \frac{Pt}{e} = \frac{(8.4 \times 10^7 \text{ J/s})(24 \text{ h})}{0.360} \left(\frac{3600 \text{ s}}{1 \text{ h}} \right) = 2.02 \times 10^{13} \text{ J}$$

Therefore, solving Equation 15.12 for $|Q_C|$, we have

$$|Q_C| = |Q_H| - |W| = 2.02 \times 10^{13} \text{ J} - 7.3 \times 10^{12} \text{ J} = \boxed{1.3 \times 10^{13} \text{ J}}$$

61. **REASONING** The expansion from point a to point b and the compression from point c to point d occur isothermally, and we will apply the first law of thermodynamics to these parts of the cycle in order to obtain expressions for the input and rejected heats, magnitudes $|Q_H|$ and $|Q_C|$, respectively. In order to simplify the resulting expression for $|Q_C|/|Q_H|$, we will then use the fact that the expansion from point b to point c and the compression from point d to point a are adiabatic.

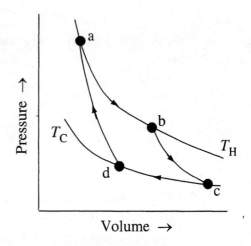

SOLUTION According to the first law of thermodynamics, the change in internal energy ΔU is given by $\Delta U = Q - W$ (Equation 15.1), where Q is the heat and W is the work. Since the internal energy of an ideal gas is proportional to the Kelvin temperature and the temperature is constant for an isothermal process, it follows that $\Delta U = 0$ J for such a case. The work of isothermal expansion or compression for an ideal gas is $W = nRT \ln\left(V_f / V_i\right)$ (Equation 15.3), where n is the number of moles, R is the universal gas constant, T is the Kelvin temperature, V_f is the final volume of the gas, and V_i is the initial volume. We have, then, that

$$\Delta U = Q - W \quad \text{or} \quad 0 = Q - nRT \ln\left(\frac{V_f}{V_i}\right) \quad \text{or} \quad Q = nRT \ln\left(\frac{V_f}{V_i}\right)$$

Applying this result for Q to the isothermal expansion (temperature $= T_H$) from point a to point b and the isothermal compression (temperature $= T_C$) from point c to point d, we have

$$Q_H = nRT_H \ln\left(\frac{V_b}{V_a}\right) \quad \text{and} \quad Q_C = nRT_C \ln\left(\frac{V_d}{V_c}\right)$$

where $V_f = V_b$ and $V_i = V_a$ for the isotherm at T_H and $V_f = V_d$ and $V_i = V_c$ for the isotherm at T_C. In this problem, we are interested in the magnitude of the heats. For Q_H, this poses no problem, since $V_b > V_a$, $\ln\left(V_b / V_a\right)$ is positive, and we have

$$\left|Q_H\right| = nRT_H \ln\left(\frac{V_b}{V_a}\right) \tag{1}$$

However, for Q_C, we need to be careful, because $V_c > V_d$ and $\ln\left(V_d / V_c\right)$ is negative. Thus, we write for the magnitude of Q_C that

$$\left|Q_C\right| = -nRT_C \ln\left(\frac{V_d}{V_c}\right) = nRT_C \ln\left(\frac{V_c}{V_d}\right) \tag{2}$$

According to Equations (1) and (2), the ratio of the magnitudes of the rejected and input heats is

$$\frac{\left|Q_C\right|}{\left|Q_H\right|} = \frac{nRT_C \ln\left(\dfrac{V_c}{V_d}\right)}{nRT_H \ln\left(\dfrac{V_b}{V_a}\right)} = \frac{T_C \ln\left(\dfrac{V_c}{V_d}\right)}{T_H \ln\left(\dfrac{V_b}{V_a}\right)} \tag{3}$$

We now consider the adiabatic parts of the Carnot cycle. For the adiabatic expansion or compression of an ideal gas the initial pressure and volume (P_i and V_i) are related to the final pressure and volume (P_f and V_f) according to

$$P_i V_i^{\gamma} = P_f V_f^{\gamma} \tag{15.5}$$

where γ is the ratio of the specific heats at constant pressure and constant volume. It is also true that $P = nRT/V$ (Equation 14.1), according to the ideal gas law. Substituting this expression for the pressure into Equation 15.5 gives

$$\left(\frac{nRT_i}{V_i}\right)V_i^{\gamma} = \left(\frac{nRT_f}{V_f}\right)V_f^{\gamma} \quad \text{or} \quad T_iV_i^{\gamma-1} = T_fV_f^{\gamma-1}$$

Applying this result to the adiabatic expansion from point b to point c and to the adiabatic compression from point d to point a, we obtain

$$T_HV_b^{\gamma-1} = T_CV_c^{\gamma-1} \quad \text{and} \quad T_CV_d^{\gamma-1} = T_HV_a^{\gamma-1}$$

Dividing the first of these equations by the second shows that

$$\frac{T_HV_b^{\gamma-1}}{T_HV_a^{\gamma-1}} = \frac{T_CV_c^{\gamma-1}}{T_CV_d^{\gamma-1}} \quad \text{or} \quad \frac{V_b^{\gamma-1}}{V_a^{\gamma-1}} = \frac{V_c^{\gamma-1}}{V_d^{\gamma-1}} \quad \text{or} \quad \frac{V_b}{V_a} = \frac{V_c}{V_d}$$

With this result, Equation (3) becomes

$$\frac{|Q_C|}{|Q_H|} = \frac{T_C \ln\left(\dfrac{V_c}{V_d}\right)}{T_H \ln\left(\dfrac{V_b}{V_a}\right)} = \boxed{\frac{T_C}{T_H}}$$

---

63. **_REASONING AND SOLUTION_** Equation 15.14 holds for a Carnot air conditioner as well as a Carnot engine. Therefore, solving Equation 15.14 for $|Q_C|$, we have

$$|Q_C| = |Q_H|\left(\frac{T_C}{T_H}\right) = \left(6.12\times10^5 \text{ J}\right)\left(\frac{299 \text{ K}}{312 \text{ K}}\right) = \boxed{5.86\times10^5 \text{ J}}$$

---

71. **_REASONING_** The conservation of energy applies to the air conditioner, so that $|Q_H| = |W| + |Q_C|$, where $|Q_H|$ is the amount of heat put into the room by the unit, $|Q_C|$ is the amount of heat removed from the room by the unit, and $|W|$ is the amount of work needed to operate the unit. Therefore, a net heat of $|Q_H| - |Q_C| = |W|$ is added to and heats up the room. To find the temperature rise of the room, we will use the COP to determine $|W|$ and then use the given molar specific heat capacity.

SOLUTION Let COP denote the coefficient of performance. By definition (Equation 15.16), $COP = |Q_C|/|W|$, so that

$$|W| = \frac{|Q_C|}{COP} = \frac{7.6 \times 10^4 \text{ J}}{2.0} = 3.8 \times 10^4 \text{ J}$$

The temperature rise in the room can be found as follows: $|Q_H| - |Q_C| = |W| = C_V n \Delta T$. Solving for ΔT gives

$$\Delta T = \frac{|W|}{C_V n} = \frac{|W|}{\left(\frac{5}{2}R\right)n} = \frac{3.8 \times 10^4 \text{ J}}{\frac{5}{2}\left[8.31 \text{ J/(mol·K)}\right](3800 \text{ mol})} = \boxed{0.48 \text{ K}}$$

---

73. **REASONING** Let the coefficient of performance be represented by the symbol COP. Then according to Equation 15.16, $COP = |Q_C|/|W|$. From the statement of energy conservation for a Carnot refrigerator (Equation 15.12), $|W| = |Q_H| - |Q_C|$. Combining Equations 15.16 and 15.12 leads to

$$COP = \frac{|Q_C|}{|Q_H| - |Q_C|} = \frac{|Q_C|/|Q_C|}{(|Q_H| - |Q_C|)/|Q_C|} = \frac{1}{(|Q_H|/|Q_C|) - 1}$$

Replacing the ratio of the heats with the ratio of the Kelvin temperatures, according to Equation 15.14, leads to

$$COP = \frac{1}{T_H/T_C - 1} \tag{1}$$

The heat $|Q_C|$ that must be removed from the refrigerator when the water is cooled can be calculated using Equation 12.4, $|Q_C| = cm\Delta T$; therefore,

$$|W| = \frac{|Q_C|}{COP} = \frac{cm\,\Delta T}{COP} \tag{2}$$

SOLUTION
a. Substituting values into Equation (1) gives

$$COP = \frac{1}{\dfrac{T_H}{T_C} - 1} = \frac{1}{\dfrac{(20.0 + 273.15) \text{ K}}{(6.0 + 273.15) \text{ K}} - 1} = \boxed{2.0 \times 10^1}$$

b. Substituting values into Equation (2) gives

$$|W| = \frac{cm\,\Delta T}{\text{COP}} = \frac{\left[4186\ \text{J/(kg} \cdot \text{C}°)\right](5.00\ \text{kg})(20.0\ °\text{C} - 6.0\ °\text{C})}{2.0 \times 10^1} = \boxed{1.5 \times 10^4\ \text{J}}$$

---

77. **REASONING AND SOLUTION** The change in entropy ΔS of a system for a process in which heat Q enters or leaves the system reversibly at a constant temperature T is given by Equation 15.18, $\Delta S = (Q/T)_\text{R}$. For a phase change, $Q = mL$, where L is the latent heat (see Section 12.8).

a. If we imagine a reversible process in which 3.00 kg of ice melts into water at 273 K, the change in entropy of the water molecules is

$$\Delta S = \left(\frac{Q}{T}\right)_\text{R} = \left(\frac{mL_\text{f}}{T}\right)_\text{R} = \frac{(3.00\ \text{kg})\left(3.35 \times 10^5\ \text{J/kg}\right)}{273\ \text{K}} = \boxed{3.68 \times 10^3\ \text{J/K}}$$

b. Similarly, if we imagine a reversible process in which 3.00 kg of water changes into steam at 373 K, the change in entropy of the water molecules is

$$\Delta S = \left(\frac{Q}{T}\right)_\text{R} = \left(\frac{mL_\text{v}}{T}\right)_\text{R} = \frac{(3.00\ \text{kg})\left(2.26 \times 10^6\ \text{J/kg}\right)}{373\ \text{K}} = \boxed{1.82 \times 10^4\ \text{J/K}}$$

c. Since the change in entropy is greater for the vaporization process than for the fusion process, the $\boxed{\text{vaporization process creates more disorder}}$ in the collection of water molecules.

---

83. **REASONING** According to the first law of thermodynamics (Equation 15.1), $\Delta U = Q - W$. For a monatomic ideal gas (Equation 14.7), $U = \frac{3}{2}nRT$. Therefore, for the process in question, the change in the internal energy is $\Delta U = \frac{3}{2}nR\Delta T$. Combining the last expression for ΔU with Equation 15.1 yields

$$\tfrac{3}{2}nR\Delta T = Q - W$$

This expression can be solved for ΔT.

SOLUTION

a. The heat is $Q = +1200\text{ J}$, since it is absorbed by the system. The work is $W = +2500\text{ J}$, since it is done *by* the system. Solving the above expression for ΔT and substituting the values for the data given in the problem statement, we have

$$\Delta T = \frac{Q - W}{\frac{3}{2}nR} = \frac{1200\text{ J} - 2500\text{ J}}{\frac{3}{2}(0.50\text{ mol})[8.31\text{ J}/(\text{mol}\cdot\text{K})]} = \boxed{-2.1 \times 10^2\text{ K}}$$

b. Since $\Delta T = T_{\text{final}} - T_{\text{initial}}$ is negative, T_{initial} must be greater than T_{final}; this change represents a $\boxed{\text{decrease}}$ in temperature.

Alternatively, one could deduce that the temperature decreases from the following physical argument. Since the system loses more energy in doing work than it gains in the form of heat, the internal energy of the system decreases. Since the internal energy of an ideal gas depends only on the temperature, a decrease in the internal energy must correspond to a decrease in the temperature.

---

87. *REASONING AND SOLUTION*

a. Since the energy that becomes unavailable for doing work is zero for the process, we have from Equation 15.19, $W_{\text{unavailable}} = T_0\Delta S_{\text{universe}} = 0$. Therefore, $\Delta S_{\text{universe}} = 0$ and according to the discussion in Section 15.11, the process is $\boxed{\text{reversible}}$.

b. Since the process is reversible, we have (see Section 15.11)

$$\Delta S_{\text{universe}} = \Delta S_{\text{system}} + \Delta S_{\text{surroundings}} = 0$$

Therefore,

$$\Delta S_{\text{surroundings}} = -\Delta S_{\text{system}} = \boxed{-125\text{ J/K}}$$

---

91. *REASONING AND SOLUTION*

a. Starting at point *A*, the work done during the first (vertical) straight-line segment is

$$W_1 = P_1\Delta V_1 = P_1(0\text{ m}^3) = 0\text{ J}$$

For the second (horizontal) straight-line segment, the work is

$$W_2 = P_2\Delta V_2 = 10(1.0 \times 10^4\text{ Pa})6(2.0 \times 10^{-3}\text{ m}^3) = 1200\text{ J}$$

For the third (vertical) straight-line segment the work is

$$W_3 = P_3 \Delta V_3 = P_3(0 \text{ m}^3) = 0 \text{ J}$$

For the fourth (horizontal) straight-line segment the work is

$$W_4 = P_4 \Delta V_4 = 15(1.0 \times 10^4 \text{ Pa})6(2.0 \times 10^{-3} \text{ m}^3) = 1800 \text{ J}$$

The total work done is

$$W = W_1 + W_2 + W_3 + W_4 = \boxed{+3.0 \times 10^3 \text{ J}}$$

b. Since the total work is positive, work is done $\boxed{\text{by the system}}$.

---

93. **REASONING AND SOLUTION**   The change in volume is $\Delta V = -sA$, where s is the distance through which the piston drops and A is the piston area. The minus sign is included because the volume decreases. Thus,

$$s = \frac{-\Delta V}{A}$$

The ideal gas law states that $\Delta V = nR\Delta T/P$. But $Q = C_p n \Delta T = \frac{5}{2} R n \Delta T$. Thus, $\Delta T = Q/\left(\frac{5}{2} Rn\right)$. Using these expressions for ΔV and ΔT, we find that

$$s = \frac{-nR\,\Delta T / P}{A} = \frac{-nR\left[Q/\left(\frac{5}{2} Rn\right)\right]}{PA} = \frac{-Q}{\frac{5}{2} PA}$$

$$= \frac{-(-2093 \text{ J})}{\frac{5}{2}\left(1.01 \times 10^5 \text{ Pa}\right)\left(3.14 \times 10^{-2} \text{ m}^2\right)} = \boxed{0.264 \text{ m}}$$

---

99. **REASONING**   The efficiency of either engine is given by Equation 15.13, $e = 1 - \left(|Q_C|/|Q_H|\right)$. Since engine A receives three times more input heat, produces five times more work, and rejects two times more heat than engine B, it follows that $|Q_{HA}| = 3|Q_{HB}|$, $|W_A| = 5|W_B|$, and $|Q_{CA}| = 2|Q_{CB}|$. As required by the principle of energy conservation for engine A (Equation 15.12),

$$\underbrace{|Q_{HA}|}_{3|Q_{HB}|} = \underbrace{|Q_{CA}|}_{2|Q_{CB}|} + \underbrace{|W_A|}_{5|W_B|}$$

Thus,

$$3\left|Q_{HB}\right| = 2\left|Q_{CB}\right| + 5\left|W_B\right| \qquad (1)$$

Since engine B also obeys the principle of energy conservation (Equation 15.12),

$$\left|Q_{HB}\right| = \left|Q_{CB}\right| + \left|W_B\right| \qquad (2)$$

Substituting $\left|Q_{HB}\right|$ from Equation (2) into Equation (1) yields

$$3(\left|Q_{CB}\right| + \left|W_B\right|) = 2\left|Q_{CB}\right| + 5\left|W_B\right|$$

Solving for $\left|W_B\right|$ gives

$$\left|W_B\right| = \frac{1}{2}\left|Q_{CB}\right|$$

Therefore, Equation (2) predicts for engine B that

$$\left|Q_{HB}\right| = \left|Q_{CB}\right| + \left|W_B\right| = \frac{3}{2}\left|Q_{CB}\right|$$

SOLUTION

a. Substituting $\left|Q_{CA}\right| = 2\left|Q_{CB}\right|$ and $\left|Q_{HA}\right| = 3\left|Q_{HB}\right|$ into Equation 15.13 for engine A, we have

$$e_A = 1 - \frac{\left|Q_{CA}\right|}{\left|Q_{HA}\right|} = 1 - \frac{2\left|Q_{CB}\right|}{3\left|Q_{HB}\right|} = 1 - \frac{2\left|Q_{CB}\right|}{3\left(\frac{3}{2}\left|Q_{CB}\right|\right)} = 1 - \frac{4}{9} = \boxed{\frac{5}{9}}$$

b. Substituting $\left|Q_{HB}\right| = \frac{3}{2}\left|Q_{CB}\right|$ into Equation 15.13 for engine B, we have

$$e_B = 1 - \frac{\left|Q_{CB}\right|}{\left|Q_{HB}\right|} = 1 - \frac{\left|Q_{CB}\right|}{\frac{3}{2}\left|Q_{CB}\right|} = 1 - \frac{2}{3} = \boxed{\frac{1}{3}}$$

CHAPTER 16 | *WAVES AND SOUND*

1. **REASONING** Since light behaves as a wave, its speed v, frequency f, and wavelength λ are related to according to $v = f\lambda$ (Equation 16.1). We can solve this equation for the frequency in terms of the speed and the wavelength.

SOLUTION Solving Equation 16.1 for the frequency, we find that

$$f = \frac{v}{\lambda} = \frac{3.00 \times 10^8 \text{ m/s}}{5.45 \times 10^{-7} \text{ m}} = \boxed{5.50 \times 10^{14} \text{ Hz}}$$

5. **REASONING** As the transverse wave propagates, the colored dot moves up and down in simple harmonic motion with a frequency of 5.0 Hz. The amplitude (1.3 cm) is the magnitude of the maximum displacement of the dot from its equilibrium position.

SOLUTION The period T of the simple harmonic motion of the dot is $T = 1/f = 1/(5.0 \text{ Hz}) = 0.20 \text{ s}$. In one period the dot travels through one complete cycle of its motion, and covers a vertical distance of $4 \times (1.3 \text{ cm}) = 5.2 \text{ cm}$. Therefore, in 3.0 s the dot will have traveled a *total vertical distance* of

$$\left(\frac{3.0 \text{ s}}{0.20 \text{ s}} \right) (5.2 \text{ cm}) = \boxed{78 \text{ cm}}$$

13. **REASONING** The tension F in the violin string can be found by solving Equation 16.2 for F to obtain $F = mv^2/L$, where v is the speed of waves on the string and can be found from Equation 16.1 as $v = f\lambda$.

SOLUTION Combining Equations 16.2 and 16.1 and using the given data, we obtain

$$F = \frac{mv^2}{L} = (m/L)f^2\lambda^2 = \left(7.8 \times 10^{-4} \text{ kg/m}\right)(440 \text{ Hz})^2 \left(65 \times 10^{-2} \text{ m}\right)^2 = \boxed{64 \text{ N}}$$

19. **REASONING** Newton's second law can be used to analyze the motion of the blocks using the methods developed in Chapter 4. We can thus determine an expression that relates the magnitude P of the pulling force to the magnitude F of the tension in the wire. Equation 16.2 $[v = \sqrt{F/(m/L)}]$ can then be used to find the tension in the wire.

SOLUTION The following drawings show a schematic of the situation described in the problem and the free-body diagrams for each block, where $m_1 = 42.0$ kg and $m_2 = 19.0$ kg. The pulling force is **P**, and the tension in the wire gives rise to the forces **F** and **–F**, which act on m_1 and m_2, respectively.

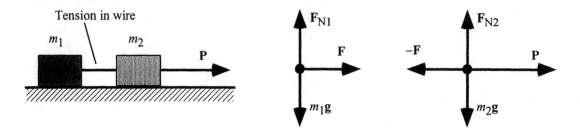

Newton's second law for block 1 is, taking forces that point to the right as positive, $F = m_1 a$, or $a = F/m_1$. For block 2, we obtain $P - F = m_2 a$. Using the expression for a obtained from the equation for block 1, we have

$$P - F = F\left(\frac{m_2}{m_1}\right) \qquad \text{or} \qquad P = F\left(\frac{m_2}{m_1}\right) + F = F\left(\frac{m_2}{m_1} + 1\right)$$

According to Equation 16.2, $F = v^2(m/L)$, where m/L is the mass per unit length of the wire. Combining this expression for F with the expression for P, we have

$$P = v^2(m/L)\left(\frac{m_2}{m_1} + 1\right) = (352 \text{ m/s})^2 (8.50 \times 10^{-4} \text{ kg/m})\left(\frac{19.0 \text{ kg}}{42.0 \text{ kg}} + 1\right) = \boxed{153 \text{ N}}$$

---

25. **REASONING** Since the wave is traveling in the +x direction, its form is given by Equation 16.3 as

$$y = A\sin\left(2\pi ft - \frac{2\pi x}{\lambda}\right)$$

We are given that the amplitude is $A = 0.35$ m. However, we need to evaluate $2\pi f$ and $\dfrac{2\pi}{\lambda}$.

Although the wavelength λ is not stated directly, it can be obtained from the values for the speed v and the frequency f, since we know that $v = f\lambda$ (Equation 16.1).

SOLUTION Since the frequency is $f = 14$ Hz, we have

$$2\pi f = 2\pi(14 \text{ Hz}) = 88 \text{ rad/s}$$

It follows from Equation 16.1 that

$$\frac{2\pi}{\lambda} = \frac{2\pi f}{v} = \frac{2\pi \left(14 \text{ Hz}\right)}{5.2 \text{ m/s}} = 17 \text{ m}^{-1}$$

Using these values for $2\pi f$ and $\dfrac{2\pi}{\lambda}$ in Equation 16.3, we have

$$y = A \sin\left(2\pi ft - \frac{2\pi x}{\lambda}\right)$$

$$\boxed{y = \left(0.35 \text{ m}\right)\sin\left[\left(88 \text{ rad/s}\right)t - \left(17 \text{ m}^{-1}\right)x\right]}$$

---

29. **REASONING** The speed of a wave on the string is given by Equation 16.2 as $v = \sqrt{\dfrac{F}{m/L}}$, where F is the tension in the string and m/L is the mass per unit length (or linear density) of the string. The wavelength λ is the speed of the wave divided by its frequency f (Equation 16.1).

SOLUTION
a. The speed of the wave on the string is

$$v = \sqrt{\frac{F}{(m/L)}} = \sqrt{\frac{15 \text{ N}}{0.85 \text{ kg/m}}} = \boxed{4.2 \text{ m/s}}$$

b. The wavelength is

$$\lambda = \frac{v}{f} = \frac{4.2 \text{ m/s}}{12 \text{ Hz}} = \boxed{0.35 \text{ m}}$$

c. The amplitude of the wave is $A = 3.6 \text{ cm} = 3.6 \times 10^{-2}$ m. Since the wave is moving along the $-x$ direction, the mathematical expression for the wave is given by Equation 16.4 as

$$y = A \sin\left(2\pi ft + \frac{2\pi x}{\lambda}\right)$$

Substituting in the numbers for A, f, and λ, we have

$$y = A \sin\left(2\pi f t + \frac{2\pi x}{\lambda}\right) = \left(3.6 \times 10^{-2} \text{ m}\right) \sin\left[2\pi\left(12 \text{ Hz}\right)t + \frac{2\pi x}{0.35 \text{ m}}\right]$$

$$= \boxed{\left(3.6 \times 10^{-2} \text{ m}\right) \sin\left[\left(75 \text{ rad/s}\right)t + \left(18 \text{ m}^{-1}\right)x\right]}$$

---

31. ***REASONING AND SOLUTION***   The speed of sound in an ideal gas is given by text Equation 16.5

$$v = \sqrt{\frac{\gamma k T}{m}}$$

where m is the mass of a single gas particle (atom or molecule). Solving for T gives

$$T = \frac{mv^2}{\gamma k} \tag{1}$$

The mass of a single helium atom is

$$\frac{4.003 \text{ g/mol}}{6.022 \times 10^{23} \text{ mol}^{-1}}\left(\frac{1 \text{ kg}}{1000 \text{ g}}\right) = 6.650 \times 10^{-27} \text{ kg}$$

The speed of sound in oxygen at 0 °C is 316 m/s (see Table 16.1). Since helium is a monatomic gas, $\gamma = 1.67$. Then, substituting into Equation (1) gives

$$T = \frac{\left(6.65 \times 10^{-27} \text{ kg}\right)\left(316 \text{ m/s}\right)^2}{1.67\left(1.38 \times 10^{-23} \text{ J/K}\right)} = \boxed{28.8 \text{ K}}$$

---

35. ***REASONING***   If we treat the sample of argon atoms like an ideal monatomic gas $(\gamma = 1.67)$ at 298 K, Equation 14.6 $\left(\frac{1}{2}mv_{\text{rms}}^2 = \frac{3}{2}kT\right)$ can be solved for the root-mean-square speed v_{rms} of the argon atoms. The speed of sound in argon can be found from Equation 16.5: $v = \sqrt{\gamma k T / m}$.

SOLUTION   We first find the mass of an argon atom. Since the molecular mass of argon is 39.9 u, argon has a mass per mole of 39.9×10^{-3} kg/mol. Thus, the mass of a single argon atom is

$$m = \frac{39.9 \times 10^{-3} \text{ kg/mol}}{6.022 \times 10^{23} \text{ mol}^{-1}} = 6.63 \times 10^{-26} \text{ kg}$$

a. Solving Equation 14.6 for v_{rms} and substituting the data given in the problem statement, we find

$$v_{\text{rms}} = \sqrt{\frac{3kT}{m}} = \sqrt{\frac{3\left(1.38 \times 10^{-23} \text{ J/K}\right)\left(298 \text{ K}\right)}{6.63 \times 10^{-26} \text{ kg}}} = \boxed{431 \text{ m/s}}$$

b. The speed of sound in argon is, according to Equation 16.5,

$$v = \sqrt{\frac{\gamma kT}{m}} = \sqrt{\frac{\left(1.67\right)\left(1.38 \times 10^{-23} \text{ J/K}\right)\left(298 \text{ K}\right)}{6.63 \times 10^{-26} \text{ kg}}} = \boxed{322 \text{ m/s}}$$

---

43. **REASONING** The sound will spread out uniformly in all directions. For the purposes of counting the echoes, we will consider only the sound that travels in a straight line parallel to the ground and reflects from the vertical walls of the cliff. Let the distance between the hunter and the closer cliff be x_1 and the distance from the hunter to the further cliff be x_2.

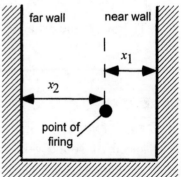

The first echo arrives at the location of the hunter after traveling a total distance $2x_1$ in a time t_1, so that, if v_s is the speed of sound, $t_1 = 2x_1/v_s$. Similarly, the second echo arrives after reflection from the far wall and in an amount of time t_2 after the firing of the gun. The quantity t_2 is related to the distance x_2 and the speed of sound v_s according to $t_2 = 2x_2/v_s$. The time difference between the first and second echo is, therefore

$$\Delta t = t_2 - t_1 = \frac{2}{v_s}\left(x_2 - x_1\right) \tag{1}$$

The third echo arrives in a time t_3 after the second echo. It arises from the sound of the second echo that is reflected from the closer cliff wall. Thus, $t_3 = 2x_1/v_s$, or, solving for x_1, we have

$$x_1 = \frac{v_s t_3}{2} \tag{2}$$

Combining Equations (1) and (2), we obtain

$$\Delta t = t_2 - t_1 = \frac{2}{v_s}\left(x_2 - \frac{v_s t_3}{2}\right)$$

Solving for x_2, we have

$$x_2 = \frac{v_s}{2}(\Delta t + t_3) \tag{3}$$

The distance between the cliffs can be found from $d = x_1 + x_2$, where x_1 and x_2 can be determined from Equations (2) and (3), respectively.

SOLUTION According to Equation (2), the distance x_1 is

$$x_1 = \frac{(343 \text{ m/s})(1.1 \text{ s})}{2} = 190 \text{ m}$$

According to Equation (3), the distance x_2 is

$$x_2 = \frac{(343 \text{ m/s})}{2}(1.6 \text{ s} + 1.1 \text{ s}) = 460 \text{ m}$$

Therefore, the distance between the cliffs is

$$d = x_1 + x_2 = 190 \text{ m} + 460 \text{ m} = \boxed{650 \text{ m}}$$

---

45. **REASONING** Equation 16.7 relates the Young's modulus Y, the mass density ρ, and the speed of sound v in a long slender solid bar. According to Equation 16.7, the Young's modulus is given by $Y = \rho v^2$. The data given in the problem can be used to compute values for both ρ and v.

SOLUTION Using the values of the data given in the problem statement, we find that the speed of sound in the bar is

$$v = \frac{L}{t} = \frac{0.83 \text{ m}}{1.9 \times 10^{-4} \text{ s}} = 4.4 \times 10^3 \text{ m/s}$$

where L is the length of the rod and t is the time required for the wave to travel the length of the rod. The mass density of the bar is, from Equation 11.1, $\rho = m/V = m/(LA)$, where m and A are, respectively, the mass and the cross-sectional area of the rod. The density of the rod is, therefore,

$$\rho = \frac{m}{LA} = \frac{2.1 \text{ kg}}{(0.83 \text{ m})(1.3 \times 10^{-4} \text{ m}^2)} = 1.9 \times 10^4 \text{ kg/m}^3$$

Using these values, we find that Young's modulus for the material of the rod is

$$Y = \rho v^2 = \left(1.9 \times 10^4 \text{ kg/m}^3\right)\left(4.4 \times 10^3 \text{ m/s}\right)^2 = 3.7 \times 10^{11} \text{ N/m}^2$$

Comparing this value to those given in Table 10.1, we conclude that the bar is most likely made of $\boxed{\text{tungsten}}$.

---

51. **REASONING** We must determine the time t for the warning to travel the vertical distance $h = 10.0 \text{ m}$ from the prankster to the ears of the man when he is just under the window. The desired distance above the man's ears is the distance that the balloon would travel in this time and can be found with the aid of the equations of kinematics.

SOLUTION Since sound travels with constant speed v_s, the distance h and the time t are related by $h = v_s t$. Therefore, the time t required for the warning to reach the ground is

$$t = \frac{h}{v_s} = \frac{10.0 \text{ m}}{343 \text{ m/s}} = 0.0292 \text{ s}$$

We now proceed to find the distance that the balloon travels in this time. To this end, we must find the balloon's final speed v_y after falling from rest for 10.0 m. Since the balloon is dropped from rest, we use Equation 3.6b ($v_y^2 = v_{0y}^2 + 2a_y y$) with $v_{0y} = 0$ m/s:

$$v_y = \sqrt{v_{0y}^2 + 2a_y y} = \sqrt{(0 \text{ m/s})^2 + 2(9.80 \text{ m/s}^2)(10.0 \text{ m})} = 14.0 \text{ m/s}$$

Using this result, we can find the balloon's speed 0.0292 seconds before it hits the man by solving Equation 3.3b ($v_y = v_{0y} + a_y t$) for v_{0y}:

$$v_{0y} = v_y - a_y t = \left[(14.0 \text{ m/s}) - (9.80 \text{ m/s}^2)(0.0292 \text{ s})\right] = 13.7 \text{ m/s}$$

Finally, we can find the desired distance y above the man's head from Equation 3.5b:

$$y = v_{0y} t + \frac{1}{2} a_y t^2 = (13.7 \text{ m/s})(0.0292 \text{ s}) + \frac{1}{2}(9.80 \text{ m/s}^2)(0.0292 \text{ s})^2 = \boxed{0.404 \text{ m}}$$

---

53. **_REASONING AND SOLUTION_** Since the sound radiates uniformly in all directions, at a distance r from the source, the energy of the sound wave is distributed over the area of a sphere of radius r. Therefore, according to $I = \dfrac{P}{4\pi r^2}$ (Equation 16.9) with $r = 3.8$ m, the power radiated from the source is

$$P = 4\pi I r^2 = 4\pi(3.6 \times 10^{-2} \text{ W/m}^2)(3.8 \text{ m})^2 = \boxed{6.5 \text{ W}}$$

---

57. **_REASONING AND SOLUTION_** According to Equation 16.8, the power radiated by the speaker is $P = IA = I\pi r^2$, where r is the radius of the circular opening. Thus, the radiated power is

$$P = (17.5 \text{ W/m}^2)(\pi)(0.0950 \text{ m})^2 = 0.496 \text{ W}$$

As a percentage of the electrical power, this is

$$\frac{0.496 \text{ W}}{25.0 \text{ W}} \times 100 \% = \boxed{1.98 \%}$$

---

61. **_REASONING_** Intensity I is power P divided by the area A, or $I = \dfrac{P}{A}$, according to Equation 16.8. The area is given directly, but the power is not. Therefore, we need to recast this expression in terms of the data given in the problem. Power is the change in energy per unit time, according to Equation 6.10b. In this case the energy is the heat Q that causes the temperature of the lasagna to increase. Thus, the power is $P = \dfrac{Q}{t}$, where t denotes the time. As a result, Equation 16.8 for the intensity becomes

$$I = \frac{P}{A} = \frac{Q}{tA} \qquad (1)$$

According to Equation 12.4, the heat that must be supplied to increase the temperature of a substance of mass m by an amount ΔT is $Q = cm\Delta T$, where c is the specific heat capacity. Substituting this expression into Equation (1) gives

$$I = \frac{Q}{tA} = \frac{cm\Delta T}{tA} \qquad (2)$$

SOLUTION Equation (2) reveals that the intensity of the microwaves is

$$I = \frac{cm\Delta T}{t\,A} = \frac{\left[3200 \text{ J/}\left(\text{kg}\cdot\text{C}°\right)\right]\left(0.35 \text{ kg}\right)\left(72 \text{ C}°\right)}{\left(480 \text{ s}\right)\left(2.2\times10^{-2} \text{ m}^2\right)} = \boxed{7.6\times10^3 \text{ W/m}^2}$$

---

65. **REASONING AND SOLUTION** The intensity level β in decibels (dB) is related to the sound intensity I according to Equation 16.10:

$$\beta = \left(10 \text{ dB}\right) \log \left(\frac{I}{I_0}\right)$$

where the quantity I_0 is the reference intensity. Therefore, we have

$$\beta_2 - \beta_1 = \left(10 \text{ dB}\right) \log \left(\frac{I_2}{I_0}\right) - \left(10 \text{ dB}\right) \log \left(\frac{I_1}{I_0}\right) = \left(10 \text{ dB}\right) \log \left(\frac{I_2/I_0}{I_1/I_0}\right) = \left(10 \text{ dB}\right) \log \left(\frac{I_2}{I_1}\right)$$

Solving for the ratio I_2 / I_1, we find

$$30.0 \text{ dB} = \left(10 \text{ dB}\right) \log \left(\frac{I_2}{I_1}\right) \qquad \text{or} \qquad \frac{I_2}{I_1} = 10^{3.0} = 1000$$

Thus, we conclude that the sound intensity $\boxed{\text{increases by a factor of 1000}}$.

---

71. **REASONING** If I_1 is the sound intensity produced by a single person, then NI_1 is the sound intensity generated by N people. The sound intensity level generated by N people is given by Equation 16.10 as

$$\beta_{\text{N}} = \left(10 \text{ dB}\right)\log\left(\frac{NI_1}{I_0}\right)$$

where I_0 is the threshold of hearing. Solving this equation for N yields

$$N = \left(\frac{I_0}{I_1}\right)10^{\frac{\beta_{\text{N}}}{10 \text{ dB}}} \tag{1}$$

We also know that the sound intensity level for one person is

$$\beta_1 = (10 \text{ dB}) \log\left(\frac{I_1}{I_0}\right) \quad \text{or} \quad I_1 = I_0 \, 10^{\frac{\beta_1}{10 \text{ dB}}} \tag{2}$$

Equations (1) and (2) are all that we need in order to find the number of people at the football game.

SOLUTION Substituting the expression for I_1 from Equation (2) into Equation (1) gives the desired result.

$$N = \frac{I_0 \, 10^{\frac{\beta_N}{10 \text{ dB}}}}{I_0 \, 10^{\frac{\beta_1}{10 \text{ dB}}}} = \frac{10^{\frac{109 \text{ dB}}{10 \text{ dB}}}}{10^{\frac{60.0 \text{ dB}}{10 \text{ dB}}}} = \boxed{79\,400}$$

---

75. **REASONING AND SOLUTION** The sound intensity level β in decibels (dB) is related to the sound intensity I according to Equation 16.10, $\beta = (10 \text{ dB}) \log (I/I_0)$, where the quantity I_0 is the reference intensity. According to the problem statement, when the sound intensity level triples, the sound intensity also triples; therefore,

$$3\beta = (10 \text{ dB}) \log\left(\frac{3I}{I_0}\right)$$

Then,

$$3\beta - \beta = (10 \text{ dB}) \log\left(\frac{3I}{I_0}\right) - (10 \text{ dB}) \log\left(\frac{I}{I_0}\right) = (10 \text{ dB}) \log\left(\frac{3I/I_0}{I/I_0}\right)$$

Thus, $2\beta = (10 \text{ dB}) \log 3$ and

$$\beta = (5 \text{ dB}) \log 3 = \boxed{2.39 \text{ dB}}$$

---

77. **REASONING** You hear a frequency f_0 that is 1.0% lower than the frequency f_s emitted by the source. This means that the frequency you observe is 99.0% of the emitted frequency, so that $f_0 = 0.990 \, f_s$. You are an observer who is moving away from a stationary source of sound. Therefore, the Doppler-shifted frequency that you observe is specified by Equation 16.14, which can be solved for the bicycle speed v_0.

SOLUTION Equation 16.14, in which v denotes the speed of sound, states that

$$f_0 = f_s\left(1 - \frac{v_0}{v}\right)$$

Solving for v_0 and using the fact that $f_0 = 0.990\,f_s$ reveal that

$$v_0 = v\left(1 - \frac{f_0}{f_s}\right) = (343 \text{ m/s})\left(1 - \frac{0.990\,f_s}{f_s}\right) = \boxed{3.4 \text{ m/s}}$$

---

81. **REASONING AND SOLUTION** The speed v_s of the Bungee jumper after she has fallen a distance y can be obtained from Equation 2.9:

$$v_s^2 = v_0^2 + 2ay$$

Since she falls from rest, it follows that $v_0 = 0$ m/s. Therefore, taking upward as the positive direction, we have

$$v_s = \sqrt{2ay} = \sqrt{2\left(-9.80 \text{ m/s}^2\right)\left(-11.0 \text{ m}\right)} = 14.7 \text{ m/s}$$

Then, from Equation 16.11

$$f_0 = f_s\left(\frac{1}{1 - v_s/v}\right) = (589 \text{ Hz})\left[\frac{1}{1 - (14.7 \text{ m/s})/(343 \text{ m/s})}\right] = \boxed{615 \text{ Hz}}.$$

---

87. **REASONING**
a. Since the two submarines are approaching each other head on, the frequency f_0 detected by the observer (sub B) is related to the frequency f_s emitted by the source (sub A) by

$$f_0 = f_s\left(\frac{1 + \dfrac{v_0}{v}}{1 - \dfrac{v_s}{v}}\right) \qquad\qquad (16.15)$$

where v_0 and v_s are the speed of the observer and source, respectively, and v is the speed of the underwater sound

b. The sound reflected from submarine B has the same frequency that it detects, namely, f_o. Now sub B becomes the source of sound and sub A is the observer. We can still use Equation 16.15 to find the frequency detected by sub A.

SOLUTION

a. The frequency f_o detected by sub B is

$$f_o = f_s \left(\frac{1 + \dfrac{v_o}{v}}{1 - \dfrac{v_s}{v}} \right) = (1550 \text{ Hz}) \left(\frac{1 + \dfrac{8 \text{ m/s}}{1522 \text{ m/s}}}{1 - \dfrac{12 \text{ m/s}}{1522 \text{ m/s}}} \right) = \boxed{1570 \text{ Hz}}$$

b. The sound reflected from submarine B has the same frequency that it detects, namely, 1570 Hz. Now sub B is the source of sound whose frequency is $f_s = 1570$ Hz. The speed of sub B is $v_s = 8$ m/s. The frequency detected by sub A (whose speed is $v_o = 12$ m/s) is

$$f_o = f_s \left(\frac{1 + \dfrac{v_o}{v}}{1 - \dfrac{v_s}{v}} \right) = (1570 \text{ Hz}) \left(\frac{1 + \dfrac{12 \text{ m/s}}{1522 \text{ m/s}}}{1 - \dfrac{8 \text{ m/s}}{1522 \text{ m/s}}} \right) = \boxed{1590 \text{ Hz}}$$

---

91. **REASONING AND SOLUTION** The speed of sound in a liquid is given by Equation 16.6, $v = \sqrt{B_{ad} / \rho}$, where B_{ad} is the adiabatic bulk modulus and ρ is the density of the liquid. Solving for B_{ad}, we obtain $B_{ad} = v^2 \rho$. Values for the speed of sound in fresh water and in ethyl alcohol are given in Table 16.1. The ratio of the adiabatic bulk modulus of fresh water to that of ethyl alcohol at 20°C is, therefore,

$$\frac{(B_{ad})_{\text{water}}}{(B_{ad})_{\text{ethyl alcohol}}} = \frac{v^2_{\text{water}} \rho_{\text{water}}}{v^2_{\text{ethyl alcohol}} \rho_{\text{ethyl alcohol}}} = \frac{(1482 \text{ m/s})^2 (998 \text{ kg/m}^3)}{(1162 \text{ m/s})^2 (789 \text{ kg/m}^3)} = \boxed{2.06}$$

---

93. **REASONING** According to Equation 16.2, the linear density of the string is given by $(m/L) = F/v^2$, where the speed v of waves on the middle C string is given by Equation 16.1, $v = f\lambda = \left(\dfrac{1}{T} \right) \lambda$, where T is the period.

SOLUTION Combining Equations 16.2 and 16.1 and using the given data, we obtain

$$m/L = \frac{F}{v^2} = \frac{FT^2}{\lambda^2} = \frac{(944 \text{ N})(3.82 \times 10^{-3} \text{ s})^2}{(1.26 \text{ m})^2} = \boxed{8.68 \times 10^{-3} \text{ kg/m}}$$

95. **REASONING** This is a situation in which the intensities I_{man} and I_{woman} (in watts per square meter) detected by the man and the woman are compared using the intensity level β, expressed in decibels. This comparison is based on Equation 16.10, which we rewrite as follows:

$$\beta = (10 \text{ dB})\log\left(\frac{I_{man}}{I_{woman}}\right)$$

SOLUTION Using Equation 16.10, we have

$$\beta = 7.8 \text{ dB} = (10 \text{ dB})\log\left(\frac{I_{man}}{I_{woman}}\right) \quad \text{or} \quad \log\left(\frac{I_{man}}{I_{woman}}\right) = \frac{7.8 \text{ dB}}{10 \text{ dB}} = 0.78$$

Solving for the intensity ratio gives

$$\frac{I_{man}}{I_{woman}} = 10^{0.78} = \boxed{6.0}$$

97. **REASONING** When the end of the Slinky is moved up and down continuously, a transverse wave is produced. The distance between two adjacent crests on the wave, is, by definition, one wavelength. The wavelength λ is related to the speed and frequency of a periodic wave by $\lambda = v/f$ (Equation 16.1). In order to use Equation 16.1, we must first determine the frequency of the wave. The wave on the Slinky will have the same frequency as the simple harmonic motion of the hand. According to the data given in the problem statement, the frequency is $f = (2.00 \text{ cycles})/(1 \text{ s}) = 2.00 \text{ Hz}$.

SOLUTION Substituting the values for λ and f, we find that the distance between crests is

$$\lambda = \frac{v}{f} = \frac{0.50 \text{ m/s}}{2.00 \text{ Hz}} = \boxed{0.25 \text{ m}}$$

99. **REASONING** Since you detect a frequency that is smaller than that emitted by the car when the car is stationary, the car must be moving away from you. Therefore, according to Equation 16.12, the frequency f_o heard by a stationary observer from a source moving away from the observer is given by

$$f_o = f_s\left(\frac{1}{1 + \dfrac{v_s}{v}}\right)$$

where f_s is the frequency emitted from the source when it is stationary with respect to the observer, v is the speed of sound, and v_s is the speed of the moving source. This expression can be solved for v_s.

SOLUTION We proceed to solve for v_s and substitute the data given in the problem statement. Rearrangement gives

$$\frac{v_s}{v} = \frac{f_s}{f_o} - 1$$

Solving for v_s and noting that $f_o / f_s = 0.86$ yields

$$v_s = v\left(\frac{f_s}{f_o} - 1\right) = (343 \text{ m/s})\left(\frac{1}{0.86} - 1\right) = \boxed{56 \text{ m/s}}$$

---

101. **REASONING** According to Equation 16.10, the sound intensity level β in decibels (dB) is related to the sound intensity I according to $\beta = (10\text{ dB}) \log (I/I_0)$, where the quantity I_0 is the reference intensity. Since the sound is emitted uniformly in all directions, the intensity, or power per unit area, is given by $I = P/(4\pi r^2)$. Thus, the sound intensity at position 1 can be written as $I_1 = P/(4\pi r_1^2)$, while the sound intensity at position 2 can be written as $I_2 = P/(4\pi r_2^2)$. We can obtain the sound intensity levels from Equation 16.10 for these two positions by using these expressions for the intensities.

SOLUTION Using Equation 16.10 and the expressions for the intensities at the two positions, we can write the difference in the sound intensity levels β_{21} between the two positions as follows:

$$\beta_{21} = \beta_2 - \beta_1 = (10\text{ dB}) \log \left(\frac{I_2}{I_0}\right) - (10\text{ dB}) \log \left(\frac{I_1}{I_0}\right)$$

$$= (10\text{ dB}) \log \left(\frac{I_2/I_0}{I_1/I_0}\right) = (10\text{ dB}) \log \left(\frac{I_2}{I_1}\right)$$

$$\beta_{21} = (10\text{ dB}) \log \left[\frac{P/(4\pi r_2^2)}{P/(4\pi r_1^2)}\right] = (10\text{ dB}) \log \left(\frac{r_1^2}{r_2^2}\right) = (10\text{ dB}) \log \left(\frac{r_1}{r_2}\right)^2$$

$$= (20\text{ dB}) \log \left(\frac{r_1}{r_2}\right) = (20\text{ dB}) \log \left(\frac{r_1}{2r_1}\right) = (20\text{ dB}) \log (1/2) = \boxed{-6.0 \text{ dB}}$$

The negative sign indicates that the sound intensity level decreases.

105. **REASONING** Using the procedures developed in Chapter 4 for using Newton's second law to analyze the motion of bodies and neglecting the weight of the wire relative to the tension in the wire lead to the following equations of motion for the two blocks:

$$\sum F_x = F - m_1 g \ (\sin \ 30.0°) = 0 \tag{1}$$

$$\sum F_y = F - m_2 g = 0 \tag{2}$$

where F is the tension in the wire. In Equation (1) we have taken the direction of the $+x$ axis for block 1 to be parallel to and up the incline. In Equation (2) we have taken the direction of the $+y$ axis to be upward for block 2. This set of equations consists of two equations in three unknowns, m_1, m_2, and F. Thus, a third equation is needed in order to solve for any of the unknowns. A useful third equation can be obtained by solving Equation 16.2 for F:

$$F = (m/L) v^2 \tag{3}$$

Combining Equation (3) with Equations (1) and (2) leads to

$$(m/L) v^2 - m_1 g \ \sin \ 30.0° = 0 \tag{4}$$

$$(m/L) v^2 - m_2 g = 0 \tag{5}$$

Equations (4) and (5) can be solved directly for the masses m_1 and m_2.

SOLUTION Substituting values into Equation (4), we obtain

$$m_1 = \frac{(m/L) v^2}{g \ \sin \ 30.0°} = \frac{(0.0250 \ \text{kg/m})(75.0 \ \text{m/s})^2}{(9.80 \ \text{m/s}^2) \ \sin \ 30.0°} = \boxed{28.7 \ \text{kg}}$$

Similarly, substituting values into Equation (5), we obtain

$$m_2 = \frac{(m/L) v^2}{g} = \frac{(0.0250 \ \text{kg/m})(75.0 \ \text{m/s})^2}{(9.80 \ \text{m/s}^2)} = \boxed{14.3 \ \text{kg}}$$

3. **REASONING AND SOLUTION** The shape of the string looks like

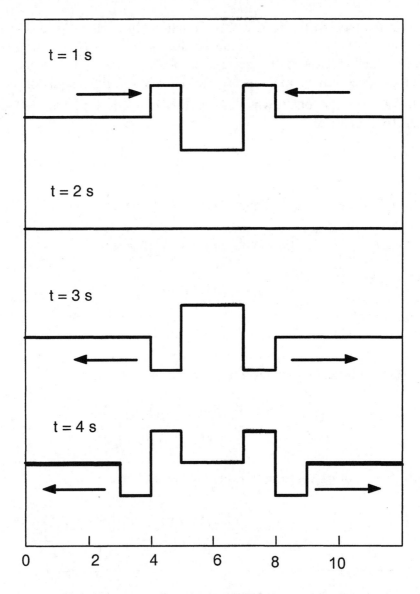

5. ***REASONING*** The tones from the two speakers will produce destructive interference with the smallest frequency when the path length difference at C is one-half of a wavelength. From Figure 17.7, we see that the path length difference is $\Delta s = s_{AC} - s_{BC}$. From Example 1, we know that $s_{AC} = 4.00\ \text{m}$, and from Figure 17.7, $s_{BC} = 2.40\ \text{m}$. Therefore, the path length difference is $\Delta s = 4.00\ \text{m} - 2.40\ \text{m} = 1.60\ \text{m}$.

SOLUTION Thus, destructive interference will occur when

$$\frac{\lambda}{2} = 1.60\ \text{m} \qquad \text{or} \qquad \lambda = 3.20\ \text{m}$$

This corresponds to a frequency of

$$f = \frac{v}{\lambda} = \frac{343\ \text{m/s}}{3.20\ \text{m}} = \boxed{107\ \text{Hz}}$$

7. ***REASONING*** The geometry of the positions of the loudspeakers and the listener is shown in the following drawing.

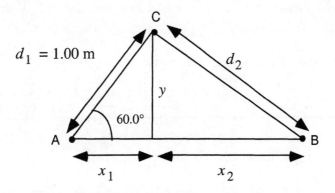

The listener at C will hear either a loud sound or no sound, depending upon whether the interference occurring at C is constructive or destructive. If the listener hears no sound, destructive interference occurs, so

$$d_2 - d_1 = \frac{n\lambda}{2} \qquad n = 1, 3, 5, \ldots \qquad (1)$$

SOLUTION Since $v = \lambda f$, according to Equation 16.1, the wavelength of the tone is

$$\lambda = \frac{v}{f} = \frac{343\ \text{m/s}}{68.6\ \text{Hz}} = 5.00\ \text{m}$$

Speaker B will be closest to Speaker A when $n = 1$ in Equation (1) above, so

$$d_2 = \frac{n\lambda}{2} + d_1 = \frac{5.00 \text{ m}}{2} + 1.00 \text{ m} = 3.50 \text{ m}$$

From the figure above we have that,

$$x_1 = (1.00 \text{ m}) \cos 60.0° = 0.500 \text{ m}$$

$$y = (1.00 \text{ m}) \sin 60.0° = 0.866 \text{ m}$$

Then

$$x_2^2 + y^2 = d_2^2 = (3.50 \text{ m})^2 \qquad \text{or} \qquad x_2 = \sqrt{(3.50 \text{ m})^2 - (0.866 \text{ m})^2} = 3.39 \text{ m}$$

Therefore, the closest that speaker A can be to speaker B so that the listener hears no sound is $x_1 + x_2 = 0.500 \text{ m} + 3.39 \text{ m} = \boxed{3.89 \text{ m}}$.

---

13. ***REASONING*** The diffraction angle for the first minimum for a circular opening is given by Equation 17.2: $\sin\theta = 1.22\lambda / D$, where D is the diameter of the opening.

SOLUTION
a. Using Equation 16.1, we must first find the wavelength of the 2.0-kHz tone:

$$\lambda = \frac{v}{f} = \frac{343 \text{ m/s}}{2.0 \times 10^3 \text{ Hz}} = 0.17 \text{ m}$$

The diffraction angle for a 2.0-kHz tone is, therefore,

$$\theta = \sin^{-1}\left(1.22 \times \frac{0.17 \text{ m}}{0.30 \text{ m}}\right) = \boxed{44°}$$

b. The wavelength of a 6.0-kHz tone is

$$\lambda = \frac{v}{f} = \frac{343 \text{ m/s}}{6.0 \times 10^3 \text{ Hz}} = 0.057 \text{ m}$$

Therefore, if we wish to generate a 6.0-kHz tone whose diffraction angle is as wide as that for the 2.0-kHz tone in part (a), we will need a speaker of diameter D, where

$$D = \frac{1.22 \lambda}{\sin\theta} = \frac{(1.22)(0.057 \text{ m})}{\sin 44°} = \boxed{0.10 \text{ m}}$$

19. *REASONING* The beat frequency of two sound waves is the difference between the two sound frequencies. From the graphs, we see that the period of the wave in the upper text figure is 0.020 s, so its frequency is $f_1 = 1/T_1 = 1/(0.020 \text{ s}) = 5.0 \times 10^1 \text{ Hz}$. The frequency of the wave in the lower figure is $f_2 = 1/(0.024 \text{ s}) = 4.2 \times 10^1 \text{ Hz}$.

SOLUTION The beat frequency of the two sound waves is

$$f_{\text{beat}} = f_1 - f_2 = 5.0 \times 10^1 \text{ Hz} - 4.2 \times 10^1 \text{ Hz} = \boxed{8 \text{ Hz}}$$

22. *REASONING AND SOLUTION* The first case requires that the frequency be either

$$440 \text{ Hz} - 5 \text{ Hz} = 435 \text{ Hz} \quad \text{or} \quad 440 \text{ Hz} + 5 \text{ Hz} = 445 \text{ Hz}$$

The second case requires that the frequency be either

$$436 \text{ Hz} - 9 \text{ Hz} = 427 \text{ Hz} \quad \text{or} \quad 436 \text{ Hz} + 9 \text{ Hz} = 445 \text{ Hz}$$

The frequency of the tuning fork is $\boxed{445 \text{ Hz}}$.

29. *REASONING* The fundamental frequency f_1 is given by Equation 17.3 with $n = 1$: $f_1 = v/(2L)$. Since values for f_1 and L are given in the problem statement, we can use this expression to find the speed of the waves on the cello string. Once the speed is known, the tension F in the cello string can be found by using Equation 16.2, $v = \sqrt{F/(m/L)}$.

SOLUTION Combining Equations 17.3 and 16.2 yields

$$2Lf_1 = \sqrt{\frac{F}{m/L}}$$

Solving for F, we find that the tension in the cello string is

$$F = 4L^2 f_1^2 (m/L) = 4(0.800 \text{ m})^2 (65.4 \text{ Hz})^2 (1.56 \times 10^{-2} \text{ kg/m}) = \boxed{171 \text{ N}}$$

31. **REASONING** For standing waves on a string that is clamped at both ends, Equations 17.3 and 16.2 indicate that the standing wave frequencies are

$$f_n = n\left(\frac{v}{2L}\right) \qquad \text{where} \qquad v = \sqrt{\frac{F}{m/L}}$$

Combining these two expressions, we have, with $n = 1$ for the fundamental frequency,

$$f_1 = \frac{1}{2L}\sqrt{\frac{F}{m/L}}$$

This expression can be used to find the ratio of the two fundamental frequencies.

SOLUTION The ratio of the two fundamental frequencies is

$$\frac{f_{old}}{f_{new}} = \frac{\dfrac{1}{2L}\sqrt{\dfrac{F_{old}}{m/L}}}{\dfrac{1}{2L}\sqrt{\dfrac{F_{new}}{m/L}}} = \sqrt{\frac{F_{old}}{F_{new}}}$$

Since $F_{new} = 4F_{old}$, we have

$$f_{new} = f_{old}\sqrt{\frac{F_{new}}{F_{old}}} = f_{old}\sqrt{\frac{4F_{old}}{F_{old}}} = f_{old}\sqrt{4} = (55.0\ \text{Hz})\,(2) = \boxed{1.10 \times 10^2\ \text{Hz}}$$

---

37. **REASONING**   We can find the extra length that the D-tuner adds to the E-string by calculating the length of the D-string and then subtracting from it the length of the E string. For standing waves on a string that is fixed at both ends, Equation 17.3 gives the frequencies as $f_n = n(v/2L)$. The ratio of the fundamental frequency of the D-string to that of the E-string is

$$\frac{f_D}{f_E} = \frac{v/(2L_D)}{v/(2L_E)} = \frac{L_E}{L_D}$$

This expression can be solved for the length L_D of the D-string in terms of quantities given in the problem statement.

SOLUTION The length of the D-string is

$$L_D = L_E \left(\frac{f_E}{f_D} \right) = (0.628 \text{ m}) \left(\frac{41.2 \text{ Hz}}{36.7 \text{ Hz}} \right) = 0.705 \text{ m}$$

The length of the E-string is extended by the D-tuner by an amount

$$L_D - L_E = 0.705 \text{ m} - 0.628 \text{ m} = \boxed{0.077 \text{ m}}$$

39. *REASONING* The natural frequencies of the cord are, according to Equation 17.3, $f_n = nv/(2L)$, where $n = 1, 2, 3, \ldots$. The speed v of the waves on the cord is, according to Equation 16.2, $v = \sqrt{F/(m/L)}$, where F is the tension in the cord. Combining these two expressions, we have

$$f_n = \frac{nv}{2L} = \frac{n}{2L} \sqrt{\frac{F}{m/L}} \quad \text{or} \quad \left(\frac{f_n 2L}{n} \right)^2 = \frac{F}{m/L}$$

Applying Newton's second law of motion, $\Sigma F = ma$, to the forces that act on the block and are parallel to the incline gives

$$F - Mg \sin\theta = Ma = 0 \quad \text{or} \quad F = Mg \sin\theta$$

where $Mg \sin\theta$ is the component of the block's weight that is parallel to the incline. Substituting this value for the tension into the equation above gives

$$\left(\frac{f_n 2L}{n} \right)^2 = \frac{Mg \sin\theta}{m/L}$$

This expression can be solved for the angle θ and evaluated at the various harmonics. The answer can be chosen from the resulting choices.

SOLUTION Solving this result for $\sin\theta$ shows that

$$\sin\theta = \frac{(m/L)}{Mg} \left(\frac{f_n 2L}{n} \right)^2 = \frac{1.20 \times 10^{-2} \text{ kg/m}}{(15.0 \text{ kg})(9.80 \text{ m/s}^2)} \left[\frac{(165 \text{ Hz})2(0.600 \text{ m})}{n} \right]^2 = \frac{3.20}{n^2}$$

Thus, we have

$$\theta = \sin^{-1} \left(\frac{3.20}{n^2} \right)$$

Evaluating this for the harmonics corresponding to the range of n from $n = 2$ to $n = 4$, we have

$$\theta = \sin^{-1}\left(\frac{3.20}{2^2}\right) = 53.1° \text{ for } n = 2$$

$$\theta = \sin^{-1}\left(\frac{3.20}{3^2}\right) = 20.8° \text{ for } n = 3$$

$$\theta = \sin^{-1}\left(\frac{3.20}{4^2}\right) = 11.5° \text{ for } n = 4$$

The angles between 15.0° and 90.0° are $\boxed{\theta = 20.8°}$ and $\boxed{\theta = 53.1°}$.

---

41. ***REASONING AND SOLUTION*** The distance between one node and an adjacent antinode is $\lambda/4$. Thus, we must first determine the wavelength of the standing wave. A tube open at only one end can develop standing waves only at the odd harmonic frequencies. Thus, for a tube of length L producing sound at the third harmonic ($n = 3$), $L = 3(\lambda/4)$. Therefore, the wavelength of the standing wave is

$$\lambda = \tfrac{4}{3}L = \tfrac{4}{3}(1.5 \text{ m}) = 2.0 \text{ m}$$

and the distance between one node and the adjacent antinode is $\lambda/4 = \boxed{0.50 \text{ m}}$.

---

47. ***REASONING AND SOLUTION***
  a. For a string fixed at both ends the fundamental frequency is $f_1 = v/(2L)$ so $f_n = nf_1$.

$$\boxed{f_2 = 800 \text{ Hz}, \quad f_3 = 1200 \text{ Hz}, \quad f_4 = 1600 \text{ Hz}}$$

  b. For a pipe with both ends open the fundamental frequency is $f_1 = v/(2L)$ so $f_n = nf_1$.

$$\boxed{f_2 = 800 \text{ Hz}, \quad f_3 = 1200 \text{ Hz}, \quad f_4 = 1600 \text{ Hz}}$$

  c. For a pipe open at one end only the fundamental frequency is $f_1 = v/(4L)$ so $f_n = nf_1$ with n odd.

$$\boxed{f_3 = 1200 \text{ Hz}, \quad f_5 = 2000 \text{ Hz}, \quad f_7 = 2800 \text{ Hz}}$$

50. **REASONING** The pressure P_2 at a depth h in a static fluid such as the mercury column is given by $P_2 = P_{atm} + \rho g h$ (Equation 11.4), where $P_{atm} = 1.01 \times 10^5$ Pa is the air pressure at the surface of the fluid, ρ is the density of the fluid, and g is the magnitude of the acceleration due to gravity. Because the air-filled portion of the tube is open at one end, it can have standing waves with natural frequencies given by $f_n = n\left(\dfrac{v}{4L}\right)$ (Equation 17.5), where n can take on only *odd* integral values ($n = 1, 3, 5, \ldots$), and v is the speed of sound in air. The mercury decreases the effective length of the air-filled portion of the tube from its initial length $L_0 = 0.75$ m to its final length L. The third harmonic $f_{3,0}$ of the original tube is found by choosing $n = 3$ in Equation 17.5, and the fundamental frequency f_1 of the shortened tube is found by choosing $n = 1$. We note that the height h of the mercury column is equal to the difference between the original and final lengths of the air in the tube: $h = L_0 - L$.

SOLUTION Substituting $h = L_0 - L$ into $P_2 = P_{atm} + \rho g h$ (Equation 11.4), we obtain

$$P_2 = P_{atm} + \rho g (L_0 - L) \tag{1}$$

The third harmonic frequency of the tube at its initial length L_0 and the fundamental frequency f_1 of the tube at its final length L are equal, so from $f_n = n\left(\dfrac{v}{4L}\right)$ (Equation 17.5), we find that

$$f_{3,0} = 3\left(\frac{\cancel{v}}{4L_0}\right) = 1\left(\frac{\cancel{v}}{4L}\right) = f_1 \quad \text{or} \quad \frac{3}{L_0} = \frac{1}{L} \quad \text{or} \quad L = \frac{L_0}{3} \tag{2}$$

Substituting Equation (2) into Equation (1), we obtain

$$P_2 = P_{atm} + \rho g (L_0 - L) = P_{atm} + \rho g\left(L_0 - \tfrac{1}{3}L_0\right) = P_{atm} + \tfrac{2}{3}\rho g L_0$$

Therefore, the pressure at the bottom of the mercury column is

$$P_2 = 1.01 \times 10^5 \text{ Pa} + \tfrac{2}{3}\left(13\,600 \text{ kg/m}^3\right)\left(9.80 \text{ m/s}^2\right)(0.75 \text{ m}) = \boxed{1.68 \times 10^5 \text{ Pa}}$$

---

55. **REASONING** When constructive interference occurs again at point C, the path length difference is two wavelengths, or $\Delta s = 2\lambda = 3.20$ m. Therefore, we can write the expression for the path length difference as

$$s_{AC} - s_{BC} = \sqrt{s_{AB}^2 + s_{BC}^2} - s_{BC} = 3.20 \text{ m}$$

This expression can be solved for s_{AB}.

SOLUTION Solving for s_{AB}, we find that

$$s_{AB} = \sqrt{(3.20 \text{ m} + 2.40 \text{ m})^2 - (2.40 \text{ m})^2} = \boxed{5.06 \text{ m}}$$

---

59. **REASONING** According to Equation 17.3, the fundamental ($n = 1$) frequency of a string fixed at both ends is related to the wave speed v by $f_1 = v/2L$, where L is the length of the string. Thus, the speed of the wave is $v = 2Lf_1$. Combining this with Equation 16.2, $v = \sqrt{F/(m/L)}$, we have, after some rearranging,

$$\frac{F}{L^2} = 4f_1^2(m/L)$$

Since the strings have the same tension and the same lengths between their fixed ends, we have

$$f_{1E}^2(m/L)_E = f_{1G}^2(m/L)_G$$

where the symbols "E" and "G" represent the E and G strings on the violin. This equation can be solved for the linear density of the G string.

SOLUTION The linear density of the string is

$$(m/L)_G = \frac{f_{1E}^2}{f_{1G}^2}(m/L)_E = \left(\frac{f_{1E}}{f_{1G}}\right)^2(m/L)_E$$

$$= \left(\frac{659.3 \text{ Hz}}{196.0 \text{ Hz}}\right)^2 \left(3.47\times10^{-4} \text{ kg/m}\right) = \boxed{3.93\times10^{-3} \text{ kg/m}}$$

---

63. **REASONING** The beat frequency produced when the piano and the other instrument sound the note (three octaves higher than middle C) is $f_{\text{beat}} = f - f_0$, where f is the frequency of the piano and f_0 is the frequency of the other instrument ($f_0 = 2093 \text{ Hz}$). We can find f by considering the temperature effects and the mechanical effects that occur when the temperature drops from 25.0 °C to 20.0 °C.

SOLUTION The fundamental frequency f_0 of the wire at 25.0 °C is related to the tension F_0 in the wire by

$$f_0 = \frac{v}{2L_0} = \frac{\sqrt{F_0/(m/L)}}{2L_0} \tag{1}$$

where Equations 17.3 and 16.2 have been combined.

The amount ΔL by which the piano wire attempts to contract is (see Equation 12.2) $\Delta L = \alpha L_0 \Delta T$, where α is the coefficient of linear expansion of the wire, L_0 is its length at 25.0 °C, and ΔT is the amount by which the temperature drops. Since the wire is prevented from contracting, there must be a stretching force exerted at each end of the wire. According to Equation 10.17, the magnitude of this force is

$$\Delta F = Y \left(\frac{\Delta L}{L_0} \right) A$$

where Y is the Young's modulus of the wire, and A is its cross-sectional area. Combining this relation with Equation 12.2, we have

$$\Delta F = Y \left(\frac{\alpha L_0 \Delta T}{L_0} \right) A = \alpha (\Delta T) Y A$$

Thus, the frequency f at the lower temperature is

$$f = \frac{v}{2L_0} = \frac{\sqrt{(F_0 + \Delta F)/(m/L)}}{2L_0} = \frac{\sqrt{[F_0 + \alpha(\Delta T)YA]/(m/L)}}{2L_0} \tag{2}$$

Using Equations (1) and (2), we find that the frequency f is

$$f = f_0 \frac{\sqrt{[F_0 + \alpha(\Delta T)YA]/(m/L)}}{\sqrt{F_0/(m/L)}} = f_0 \sqrt{\frac{F_0 + \alpha(\Delta T)YA}{F_0}}$$

$$f = (2093 \text{ Hz}) \sqrt{\frac{818.0 \text{ N} + (12 \times 10^{-6}/\text{C°})(5.0 \text{ C°})(2.0 \times 10^{11} \text{ N/m}^2)(7.85 \times 10^{-7} \text{ m}^2)}{818.0 \text{ N}}}$$

$$= 2105 \text{ Hz}$$

Therefore, the beat frequency is $2105 \text{ Hz} - 2093 \text{ Hz} = \boxed{12 \text{ Hz}}$.

CHAPTER 18 | *ELECTRIC FORCES AND ELECTRIC FIELDS*

1. ***REASONING AND SOLUTION*** The total charge of the electrons is

$$q = N(-e) = (6.0 \times 10^{13})(-1.60 \times 10^{-19} \text{ C})$$

$$q = -9.6 \times 10^{-6} \text{ C} = -9.6 \text{ } \mu\text{C}$$

The net charge on the sphere is, therefore,

$$q_{net} = +8.0 \text{ } \mu\text{C} - 9.6 \text{ } \mu\text{C} = \boxed{-1.6 \text{ } \mu\text{C}}$$

---

5. ***REASONING*** Identical conducting spheres equalize their charge upon touching. When spheres A and B touch, an amount of charge $+q$, flows from A and instantaneously neutralizes the $-q$ charge on B leaving B momentarily neutral. Then, the remaining amount of charge, equal to $+4q$, is equally split between A and B, leaving A and B each with equal amounts of charge $+2q$. Sphere C is initially neutral, so when A and C touch, the $+2q$ on A splits equally to give $+q$ on A and $+q$ on C. When B and C touch, the $+2q$ on B and the $+q$ on C combine to give a total charge of $+3q$, which is then equally divided between the spheres B and C; thus, B and C are each left with an amount of charge $+1.5q$.

SOLUTION Taking note of the initial values given in the problem statement, and summarizing the final results determined in the ***REASONING*** above, we conclude the following:

a. Sphere C ends up with an amount of charge equal to $\boxed{+1.5q}$.

b. The charges on the three spheres before they were touched, are, according to the problem statement, $+5q$ on sphere A, $-q$ on sphere B, and zero charge on sphere C. Thus, the total charge on the spheres is $+5q - q + 0 = \boxed{+4q}$.

c. The charges on the spheres after they are touched are $+q$ on sphere A, $+1.5q$ on sphere B, and $+1.5q$ on sphere C. Thus, the total charge on the spheres is $+q + 1.5q + 1.5q = \boxed{+4q}$.

---

9. ***REASONING*** The number N of excess electrons on one of the objects is equal to the charge q on it divided by the charge of an electron $(-e)$, or $N = q/(-e)$. Since the charge on the object is negative, we can write $q = -|q|$, where $|q|$ is the magnitude of the charge. The magnitude of the charge can be found from Coulomb's law (Equation 18.1), which states

that the magnitude F of the electrostatic force exerted on each object is given by $F = k|q||q|/r^2$, where r is the distance between them.

SOLUTION The number N of excess electrons on one of the objects is

$$N = \frac{q}{-e} = \frac{-|q|}{-e} = \frac{|q|}{e} \qquad (1)$$

To find the magnitude of the charge, we solve Coulomb's law, $F = k|q||q|/r^2$, for $|q|$:

$$|q| = \sqrt{\frac{Fr^2}{k}}$$

Substituting this result into Equation (1) gives

$$N = \frac{|q|}{e} = \frac{\sqrt{\dfrac{Fr^2}{k}}}{e} = \frac{\sqrt{\dfrac{\left(4.55\times10^{-21}\ \text{N}\right)\left(1.80\times10^{-3}\ \text{m}\right)^2}{8.99\times10^9\ \text{N}\cdot\text{m}^2/\text{C}^2}}}{1.60\times10^{-19}\ \text{C}} = \boxed{8}$$

---

11. ***REASONING*** Initially, the two spheres are neutral. Since negative charge is removed from the sphere which loses electrons, it then carries a net positive charge. Furthermore, the neutral sphere to which the electrons are added is then negatively charged. Once the charge is transferred, there exists an electrostatic force on each of the two spheres, the magnitude of which is given by Coulomb's law (Equation 18.1), $F = k|q_1||q_2|/r^2$.

SOLUTION
a. Since each electron carries a charge of -1.60×10^{-19} C, the amount of negative charge removed from the first sphere is

$$\left(3.0\times10^{13}\ \text{electrons}\right)\left(\frac{1.60\times10^{-19}\ \text{C}}{1\ \text{electron}}\right) = 4.8\times10^{-6}\ \text{C}$$

Thus, the first sphere carries a charge $+4.8\times10^{-6}$ C, while the second sphere carries a charge -4.8×10^{-6} C. The magnitude of the electrostatic force that acts on each sphere is, therefore,

$$F = \frac{k|q_1||q_2|}{r^2} = \frac{\left(8.99\times10^9\ \text{N}\cdot\text{m}^2/\text{C}^2\right)\left(4.8\times10^{-6}\ \text{C}\right)^2}{\left(0.50\ \text{m}\right)^2} = \boxed{0.83\ \text{N}}$$

b. Since the spheres carry charges of opposite sign, the force is $\boxed{\text{attractive}}$.

---

13. ***REASONING   AND   SOLUTION***   The   net electrostatic force on charge 3 at $x = +3.0$ m is the vector sum of the forces on charge 3 due to the other two charges, 1 and 2. According to Coulomb's law (Equation 18.1), the magnitude of the force on charge 3 due to charge 1 is

$$F_{13} = \frac{k|q_1||q_3|}{r_{13}^2}$$

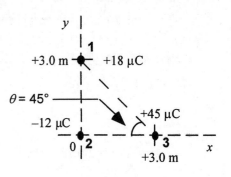

Figure 1

where the distance between charges 1 and 3 is r_{13}.

According to the Pythagorean theorem, $r_{13}^2 = x^2 + y^2$. Therefore,

$$F_{13} = \frac{\left(8.99 \times 10^9 \text{ N} \cdot \text{m}^2 / \text{C}^2\right)\left(18 \times 10^{-6} \text{ C}\right)\left(45 \times 10^{-6} \text{ C}\right)}{\left(3.0 \text{ m}\right)^2 + \left(3.0 \text{ m}\right)^2} = 0.405 \text{ N}$$

Charges 1 and 3 are equidistant from the origin, so that $\theta = 45°$ (see Figure 1). Since charges 1 and 3 are both positive, the force on charge 3 due to charge 1 is repulsive and along the line that connects them, as shown in Figure 2. The components of F_{13} are:

$$F_{13x} = F_{13} \cos 45° = 0.286 \text{ N} \qquad \text{and} \qquad F_{13y} = -F_{13} \sin 45° = -0.286 \text{ N}$$

The second force on charge 3 is the attractive force (opposite signs) due to its interaction with charge 2 located at the origin. The magnitude of the force on charge 3 due to charge 2 is, according to Coulomb's law ,

$$F_{23} = \frac{k|q_2||q_3|}{r_{23}^2} = \frac{k|q_2||q_3|}{x^2}$$

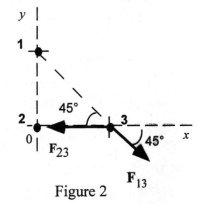

$$= \frac{\left(8.99 \times 10^9 \text{ N} \cdot \text{m}^2 / \text{C}^2\right)\left(12 \times 10^{-6} \text{ C}\right)\left(45 \times 10^{-6} \text{ C}\right)}{\left(3.0 \text{ m}\right)^2}$$

Figure 2

$$= 0.539 \text{ N}$$

Since charges 2 and 3 have opposite signs, they attract each other, and charge 3 experiences a force to the left as shown in Figure 2. Taking up and to the right as the positive directions, we have

$$F_{3x} = F_{13x} + F_{23x} = +0.286 \text{ N} - 0.539 \text{ N} = -0.253 \text{ N}$$

$$F_{3y} = F_{13y} = -0.286 \text{ N}$$

Using the Pythagorean theorem, we find the magnitude of F_3 to be

$$F_3 = \sqrt{F_{3x}^2 + F_{3y}^2} = \sqrt{(-0.253 \text{ N})^2 + (-0.286 \text{ N})^2} = \boxed{0.38 \text{ N}}$$

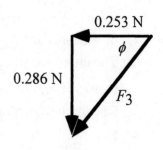

The direction of F_3 relative to the $-x$ axis is specified by the angle ϕ (see Figure 3), where

Figure 3

$$\phi = \tan^{-1}\left(\frac{0.286 \text{ N}}{0.253 \text{ N}}\right) = \boxed{49° \text{ below the } -x \text{ axis}}$$

---

15. **REASONING AND SOLUTION**
a. Since the gravitational force between the spheres is one of attraction and the electrostatic force must balance it, the electric force must be one of repulsion. Therefore, the charges must have $\boxed{\text{the same algebraic signs, both positive or both negative}}$.

b. There are two forces that act on each sphere; they are the gravitational attraction F_G of one sphere for the other, and the repulsive electric force F_E of one sphere on the other. From the problem statement, we know that these two forces balance each other, so that $F_G = F_E$. The magnitude of F_G is given by Newton's law of gravitation (Equation 4.3: $F_G = Gm_1m_2/r^2$), while the magnitude of F_E is given by Coulomb's law (Equation 18.1: $F_E = k|q_1||q_2|/r^2$). Therefore, we have

$$\frac{Gm_1m_2}{r^2} = \frac{k|q_1||q_2|}{r^2} \quad \text{or} \quad Gm^2 = k|q|^2$$

since the spheres have the same mass m and carry charges of the same magnitude $|q|$. Solving for $|q|$, we find

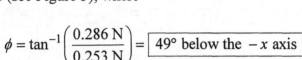

$$|q| = m\sqrt{\frac{G}{k}} = (2.0 \times 10^{-6} \text{ kg})\sqrt{\frac{6.67 \times 10^{-11} \text{ N} \cdot \text{m}^2/\text{kg}^2}{8.99 \times 10^9 \text{ N} \cdot \text{m}^2/\text{C}^2}} = \boxed{1.7 \times 10^{-16} \text{ C}}$$

25.   ***REASONING*** Consider the drawing at the right. It is given that the charges q_A, q_1, and q_2 are each positive. Therefore, the charges q_1 and q_2 each exert a repulsive force on the charge q_A. As the drawing shows, these forces have magnitudes F_{A1} (vertically downward) and F_{A2} (horizontally to the left). The unknown charge placed at the empty

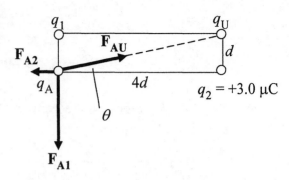

corner of the rectangle is q_U, and it exerts a force on q_A that has a magnitude F_{AU}. In order that the net force acting on q_A point in the vertical direction, the horizontal component of $\mathbf{F_{AU}}$ must cancel out the horizontal force $\mathbf{F_{A2}}$. Therefore, $\mathbf{F_{AU}}$ must point as shown in the drawing, which means that it is an attractive force and q_U must be negative, since q_A is positive.

SOLUTION The basis for our solution is the fact that the horizontal component of $\mathbf{F_{AU}}$ must cancel out the horizontal force $\mathbf{F_{A2}}$. The magnitudes of these forces can be expressed using Coulomb's law $F = k|q||q'|/r^2$, where r is the distance between the charges q and q'. Thus, we have

$$F_{AU} = \frac{k|q_A||q_U|}{(4d)^2 + d^2} \quad \text{and} \quad F_{A2} = \frac{k|q_A||q_2|}{(4d)^2}$$

where we have used the fact that the distance between the charges q_A and q_U is the diagonal of the rectangle, which is $\sqrt{(4d)^2 + d^2}$ according to the Pythagorean theorem, and the fact that the distance between the charges q_A and q_2 is $4d$. The horizontal component of $\mathbf{F_{AU}}$ is $F_{AU} \cos\theta$, which must be equal to F_{A2}, so that we have

$$\frac{k|q_A||q_U|}{(4d)^2 + d^2} \cos\theta = \frac{k|q_A||q_2|}{(4d)^2} \quad \text{or} \quad \frac{|q_U|}{17} \cos\theta = \frac{|q_2|}{16}$$

The drawing in the ***REASONING***, reveals that $\cos\theta = (4d)/\sqrt{(4d)^2 + d^2} = 4/\sqrt{17}$. Therefore, we find that

$$\frac{|q_U|}{17}\left(\frac{4}{\sqrt{17}}\right) = \frac{|q_2|}{16} \quad \text{or} \quad |q_U| = \frac{17\sqrt{17}}{64}|q_2| = \frac{17\sqrt{17}}{64}\left(3.0\times10^{-6} \text{ C}\right) = \boxed{3.3\times10^{-6} \text{ C}}$$

As discussed in the ***REASONING***, the algebraic sign of the charge q_U is $\boxed{\text{negative}}$.

27. **REASONING** The charged insulator experiences an electric force due to the presence of the charged sphere shown in the drawing in the text. The forces acting on the insulator are the downward force of gravity (i.e., its weight, $W = mg$), the electrostatic force $F = k|q_1||q_2|/r^2$ (see Coulomb's law, Equation 18.1) pulling to the right, and the tension T in the wire pulling up and to the left at an angle θ with respect to the vertical as shown in the drawing in the problem statement. We can analyze the forces to determine the desired quantities θ and T.

SOLUTION.
a. We can see from the diagram given with the problem statement that

$$T_x = F \qquad \text{which gives} \qquad T\sin\theta = k|q_1||q_2|/r^2$$

and

$$T_y = W \qquad \text{which gives} \qquad T\cos\theta = mg$$

Dividing the first equation by the second yields

$$\frac{T\sin\theta}{T\cos\theta} = \tan\theta = \frac{k|q_1||q_2|/r^2}{mg}$$

Solving for θ, we find that

$$\theta = \tan^{-1}\left(\frac{k|q_1||q_2|}{mgr^2}\right)$$

$$= \tan^{-1}\left[\frac{(8.99\times10^9 \text{ N}\cdot\text{m}^2/\text{C}^2)(0.600\times10^{-6} \text{ C})(0.900\times10^{-6} \text{ C})}{(8.00\times10^{-2} \text{ kg})(9.80 \text{ m/s}^2)(0.150 \text{ m})^2}\right] = \boxed{15.4°}$$

b. Since $T\cos\theta = mg$, the tension can be obtained as follows:

$$T = \frac{mg}{\cos\theta} = \frac{(8.00\times10^{-2} \text{ kg}) (9.80 \text{ m/s}^2)}{\cos 15.4°} = \boxed{0.813 \text{ N}}$$

31. ***REASONING*** The electric field created by a point charge is inversely proportional to the square of the distance from the charge, according to Equation 18.3. Therefore, we expect the distance r_2 to be greater than the distance r_1, since the field is smaller at r_2 than at r_1. The ratio r_2/r_1, then, should be greater than one.

SOLUTION Applying Equation 18.3 to each position relative to the charge, we have

$$E_1 = \frac{k|q|}{r_1^2} \quad \text{and} \quad E_2 = \frac{k|q|}{r_2^2}$$

Dividing the expression for E_1 by the expression for E_2 gives

$$\frac{E_1}{E_2} = \frac{k|q|/r_1^2}{k|q|/r_2^2} = \frac{r_2^2}{r_1^2}$$

Solving for the ratio r_2/r_1 gives

$$\frac{r_2}{r_1} = \sqrt{\frac{E_1}{E_2}} = \sqrt{\frac{248 \text{ N/C}}{132 \text{ N/C}}} = \boxed{1.37}$$

As expected, this ratio is greater than one.

---

37. ***REASONING***
a. The drawing shows the two point charges q_1 and q_2. Point A is located at $x = 0$ cm, and point B is at $x = +6.0$ cm.

Since q_1 is positive, the electric field points away from it. At point A, the electric field E_1 points to the left, in the $-x$ direction. Since q_2 is negative, the electric field points toward it. At point A, the electric field E_2 points to the right, in the $+x$ direction. The net electric field is $E = -E_1 + E_2$. We can use Equation 18.3, $E = k|q|/r^2$, to find the magnitude of the electric field due to each point charge.

b. The drawing shows the electric fields produced by the charges q_1 and q_2 at point B, which is located at $x = +6.0$ cm.

A   3.0 cm   3.0 cm   B   3.0 cm q_2
q_1
E_1
E_2

Since q_1 is positive, the electric field points away from it. At point B, the electric field points to the right, in the +x direction. Since q_2 is negative, the electric field points toward it. At point B, the electric field points to the right, in the +x direction. The net electric field is $E = +E_1 + E_2$.

SOLUTION
a. The net electric field at the origin (point A) is $E = -E_1 + E_2$:

$$E = -E_1 + E_2 = \frac{-k|q_1|}{r_1^2} + \frac{k|q_2|}{r_2^2}$$

$$= \frac{-\left(8.99 \times 10^9 \ \text{N} \cdot \text{m}^2/\text{C}^2\right)\left(8.5 \times 10^{-6} \ \text{C}\right)}{\left(3.0 \times 10^{-2} \ \text{m}\right)^2} + \frac{\left(8.99 \times 10^9 \ \text{N} \cdot \text{m}^2/\text{C}^2\right)\left(21 \times 10^{-6} \ \text{C}\right)}{\left(9.0 \times 10^{-2} \ \text{m}\right)^2}$$

$$= \boxed{-6.2 \times 10^7 \ \text{N/C}}$$

The minus sign tells us that the net electric field points along the –x axis.

b. The net electric field at $x = +6.0$ cm (point B) is $E = E_1 + E_2$:

$$E = E_1 + E_2 = \frac{k|q_1|}{r_1^2} + \frac{k|q_2|}{r_2^2}$$

$$= \frac{\left(8.99 \times 10^9 \ \text{N} \cdot \text{m}^2/\text{C}^2\right)\left(8.5 \times 10^{-6} \ \text{C}\right)}{\left(3.0 \times 10^{-2} \ \text{m}\right)^2} + \frac{\left(8.99 \times 10^9 \ \text{N} \cdot \text{m}^2/\text{C}^2\right)\left(21 \times 10^{-6} \ \text{C}\right)}{\left(3.0 \times 10^{-2} \ \text{m}\right)^2}$$

$$= \boxed{+2.9 \times 10^8 \ \text{N/C}}$$

The plus sign tells us that the net electric field points along the +x axis.

39. ***REASONING***   Since the charged droplet (charge $= q$) is suspended motionless in the electric field **E**, the net force on the droplet must be zero.   There are two forces that act on the droplet, the force of gravity **W** $= m\mathbf{g}$, and the electric force **F** $= q\mathbf{E}$ due to the electric field.   Since the net force on the droplet is zero, we conclude that $mg = |q|E$.   We can use this reasoning to determine the sign and the magnitude of the charge on the droplet.

SOLUTION

a. Since the net force on the droplet is zero, and the weight of magnitude W points downward, the electric force of magnitude $F = |q|E$ must point upward.   Since the electric field points upward, the excess charge on the droplet must be $\boxed{\text{positive}}$ in order for the force **F** to point upward.

b. Using the expression $mg = |q|E$, we find that the magnitude of the excess charge on the droplet is

$$|q| = \frac{mg}{E} = \frac{(3.50 \times 10^{-9}\ \text{kg})(9.80\ \text{m/s}^2)}{8480\ \text{N/C}} = 4.04 \times 10^{-12}\ \text{C}$$

The charge on a proton is 1.60×10^{-19} C, so the excess number of protons is

$$\left(4.04 \times 10^{-12}\ \text{C}\right)\left(\frac{1\ \text{proton}}{1.60 \times 10^{-19}\ \text{C}}\right) = \boxed{2.53 \times 10^7\ \text{protons}}$$

---

45. ***REASONING***   The drawing shows the arrangement of the three charges. Let $\mathbf{E}_q$ represent the electric field at the empty corner due to the $-q$ charge. Furthermore, let $\mathbf{E}_1$ and $\mathbf{E}_2$ be the electric fields at the empty corner due to charges $+q_1$ and $+q_2$, respectively.

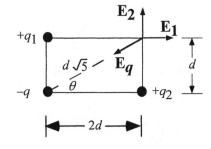

According to the Pythagorean theorem, the distance from the charge $-q$ to the empty corner along the diagonal is given by $\sqrt{(2d)^2 + d^2} = \sqrt{5d^2} = d\sqrt{5}$. The magnitude of each electric field is given by Equation 18.3, $E = k|q|/r^2$. Thus, the magnitudes of each of the electric fields at the empty corner are given as follows:

$$E_q = \frac{k|q|}{r^2} = \frac{k|q|}{\left(d\sqrt{5}\right)^2} = \frac{k|q|}{5d^2}$$

$$E_1 = \frac{k|q_1|}{(2d)^2} = \frac{k|q_1|}{4d^2} \quad \text{and} \quad E_2 = \frac{k|q_2|}{d^2}$$

The angle θ that the diagonal makes with the horizontal is $\theta = \tan^{-1}(d/2d) = 26.57°$. Since the net electric field E_{net} at the empty corner is zero, the horizontal component of the net field must be zero, and we have

$$E_1 - E_q \cos 26.57° = 0 \quad \text{or} \quad \frac{k|q_1|}{4d^2} - \frac{k|q|\cos 26.57°}{5d^2} = 0$$

Similarly, the vertical component of the net field must be zero, and we have

$$E_2 - E_q \sin 26.57° = 0 \quad \text{or} \quad \frac{k|q_2|}{d^2} - \frac{k|q|\sin 26.57°}{5d^2} = 0$$

These last two expressions can be solved for the charge magnitudes $|q_1|$ and $|q_2|$.

SOLUTION Solving the last two expressions for $|q_1|$ and $|q_2|$, we find that

$$|q_1| = \frac{4}{5}q \cos 26.57° = \boxed{0.716\,q}$$

$$|q_2| = \frac{1}{5}q \sin 26.57° = \boxed{0.0895\,q}$$

---

51. **REASONING AND SOLUTION** The net electric field at point P in Figure 1 is the vector sum of the fields $\mathbf{E}_+$ and $\mathbf{E}_-$, which are due, respectively, to the charges $+q$ and $-q$. These fields are shown in Figure 2.

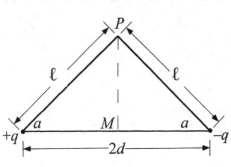

Figure 1

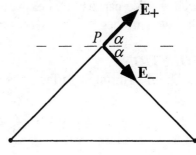

Figure 2

According to Equation 18.3, the magnitudes of the fields E_+ and E_- are the same, since the triangle is an isosceles triangle with equal sides of length ℓ. Therefore, $E_+ = E_- = k|q|/\ell^2$. The vertical components of these two fields cancel, while the horizontal components reinforce, leading to a total field at point P that is horizontal and has a magnitude of

$$E_P = E_+ \cos \alpha + E_- \cos \alpha = 2\left(\frac{k|q|}{\ell^2}\right)\cos \alpha$$

At point M in Figure 1, both $\mathbf{E_+}$ and $\mathbf{E_-}$ are horizontal and point to the right. Again using Equation 18.3, we find

$$E_M = E_+ + E_- = \frac{k|q|}{d^2} + \frac{k|q|}{d^2} = \frac{2k|q|}{d^2}$$

Since $E_M/E_P = 9.0$, we have

$$\frac{E_M}{E_P} = \frac{2k|q|/d^2}{2k|q|(\cos \alpha)/\ell^2} = \frac{1}{(\cos \alpha)d^2/\ell^2} = 9.0$$

But from Figure 1, we can see that $d/\ell = \cos \alpha$. Thus, it follows that

$$\frac{1}{\cos^3\alpha} = 9.0 \qquad \text{or} \qquad \cos \alpha = \sqrt[3]{1/9.0} = 0.48$$

The value for α is, then, $\alpha = \cos^{-1}(0.48) = \boxed{61°}$.

---

55. **REASONING** As discussed in Section 18.9, the electric flux Φ_E through a surface is equal to the component of the electric field that is normal to the surface multiplied by the area of the surface, $\Phi_E = E_\perp A$, where $E_\perp$ is the component of $\mathbf{E}$ that is normal to the surface of area A. We can use this expression and the figure in the text to determine the flux through the two surfaces.

SOLUTION
a. The flux through surface 1 is

$$\left(\Phi_E\right)_1 = (E \cos 35°)A_1 = (250 \text{ N/C})(\cos 35°)(1.7 \text{ m}^2) = \boxed{350 \text{ N} \cdot \text{m}^2/\text{C}}$$

b. Similarly, the flux through surface 2 is

$$\left(\Phi_E\right)_2 = (E \cos 55°)A_2 = (250 \text{ N/C})(\cos 55°)(3.2 \text{ m}^2) = \boxed{460 \text{ N} \cdot \text{m}^2/\text{C}}$$

59. **REASONING** The electric flux through each face of the cube is given by $\Phi_E = (E \cos\phi)A$ (see Section 18.9) where E is the magnitude of the electric field at the face, A is the area of the face, and ϕ is the angle between the electric field and the outward normal of that face. We can use this expression to calculate the electric flux Φ_E through each of the six faces of the cube.

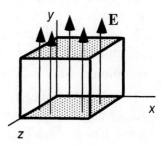

SOLUTION

a. On the bottom face of the cube, the outward normal points parallel to the $-y$ axis, in the opposite direction to the electric field, and $\phi = 180°$. Therefore,

$$\left(\Phi_E\right)_{\text{bottom}} = (1500 \text{ N/C})(\cos 180°)(0.20 \text{ m})^2 = \boxed{-6.0 \times 10^1 \text{ N} \cdot \text{m}^2/\text{C}}$$

On the top face of the cube, the outward normal points parallel to the $+y$ axis, and $\phi = 0.0°$. The electric flux is, therefore,

$$\left(\Phi_E\right)_{\text{top}} = (1500 \text{ N/C})(\cos 0.0°)(0.20 \text{ m})^2 = \boxed{+6.0 \times 10^1 \text{ N} \cdot \text{m}^2/\text{C}}$$

On each of the other four faces, the outward normals are perpendicular to the direction of the electric field, so $\phi = 90°$. So for each of the four side faces,

$$\left(\Phi_E\right)_{\text{sides}} = (1500 \text{ N/C})(\cos 90°)(0.20 \text{ m})^2 = \boxed{0 \text{ N} \cdot \text{m}^2/\text{C}}$$

b. The total flux through the cube is

$$\left(\Phi_E\right)_{\text{total}} = \left(\Phi_E\right)_{\text{top}} + \left(\Phi_E\right)_{\text{bottom}} + \left(\Phi_E\right)_{\text{side 1}} + \left(\Phi_E\right)_{\text{side 2}} + \left(\Phi_E\right)_{\text{side 3}} + \left(\Phi_E\right)_{\text{side 4}}$$

Therefore,

$$\left(\Phi_E\right)_{\text{total}} = (+6.0 \times 10^1 \text{ N} \cdot \text{m}^2/\text{C}) + (-6.0 \times 10^1 \text{ N} \cdot \text{m}^2/\text{C}) + 0 + 0 + 0 + 0 = \boxed{0 \text{ N} \cdot \text{m}^2/\text{C}}$$

67. ***REASONING*** The drawing at the right shows the set-up. Here, the electric field **E** points along the +y axis and applies a force of +**F** to the +q charge and a force of −**F** to the −q charge, where q = 8.0 μC denotes the magnitude of each charge. Each force has the same magnitude of $F = E\,|q|$, according to Equation 18.2.

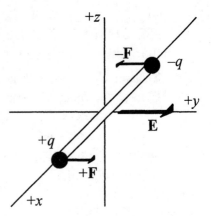

The torque is measured as discussed in Section 9.1. According to Equation 9.1, the torque produced by each force has a magnitude given by the magnitude of the force times the lever arm, which is the perpendicular distance between the point of application of the force and the axis of rotation. In the drawing the z axis is the axis of rotation and is midway between the ends of the rod. Thus, the lever arm for each force is half the length L of the rod or L/2, and the magnitude of the torque produced by each force is $(E\,|q|)(L/2)$.

SOLUTION The +**F** and the −**F** force each cause the rod to rotate in the same sense about the z axis. Therefore, the torques from these forces reinforce one another. Using the expression $(E\,|q|)(L/2)$ for the magnitude of each torque, we find that the magnitude of the net torque is

$$\text{Magnitude of net torque} = E\,|q|\left(\frac{L}{2}\right) + E\,|q|\left(\frac{L}{2}\right) = E\,|q|\,L$$

$$= \left(5.0 \times 10^3 \ \text{N/C}\right)\left(8.0 \times 10^{-6} \ \text{C}\right)\left(4.0 \ \text{m}\right) = \boxed{0.16 \ \text{N} \cdot \text{m}}$$

---

69. ***REASONING*** Each particle will experience an electrostatic force due to the presence of the other charge. According to Coulomb's law (Equation 18.1), the magnitude of the force felt by each particle can be calculated from $F = k\,|q_1||q_2|/r^2$, where $|q_1|$ and $|q_2|$ are the respective charges on particles 1 and 2 and r is the distance between them. According to Newton's second law, the magnitude of the force experienced by each particle is given by $F = ma$, where a is the acceleration of the particle and we have assumed that the electrostatic force is the only force acting.

SOLUTION

a. Since the two particles have identical positive charges, $|q_1| = |q_2| = |q|$, and we have, using the data for particle 1,

$$\frac{k\,|q|^2}{r^2} = m_1 a_1$$

Solving for $|q|$, we find that

$$|q| = \sqrt{\frac{m_1 a_1 r^2}{k}} = \sqrt{\frac{(6.00 \times 10^{-6} \text{ kg}) (4.60 \times 10^3 \text{ m/s}^2) (2.60 \times 10^{-2} \text{ m})^2}{8.99 \times 10^9 \text{ N} \cdot \text{m}^2/\text{C}^2}} = \boxed{4.56 \times 10^{-8} \text{ C}}$$

b. Since each particle experiences a force of the same magnitude (From Newton's third law), we can write $F_1 = F_2$, or $m_1 a_1 = m_2 a_2$. Solving this expression for the mass m_2 of particle 2, we have

$$m_2 = \frac{m_1 a_1}{a_2} = \frac{(6.00 \times 10^{-6} \text{ kg})(4.60 \times 10^3 \text{ m/s}^2)}{8.50 \times 10^3 \text{ m/s}^2} = \boxed{3.25 \times 10^{-6} \text{ kg}}$$

---

75. ***REASONING AND SOLUTION***   Before the spheres have been charged, they exert no forces on each other. After the spheres are charged, each sphere experiences a repulsive force F due to the charge on the other sphere, according to Coulomb's law (Equation 18.1). Therefore, since each sphere has the same charge, the magnitude F of this force is

$$F = \frac{k|q_1||q_2|}{r^2} = \frac{(8.99 \times 10^9 \text{ N} \cdot \text{m}^2/\text{C}^2)(1.60 \times 10^{-6} \text{ C})^2}{(0.100 \text{ m})^2} = 2.30 \text{ N}$$

The repulsive force on each sphere compresses the spring to which it is attached. The magnitude of this repulsive force is related to the amount of compression by Equation 10.1: $F_x^{\text{Applied}} = kx$. Setting $F_x^{\text{Appled}} = F$ and solving for k, we find that

$$k = \frac{F}{x} = \frac{2.30 \text{ N}}{0.0250 \text{ m}} = \boxed{92.0 \text{ N/m}}$$

CHAPTER 19 | *ELECTRIC POTENTIAL ENERGY AND THE ELECTRIC POTENTIAL*

---

3. **REASONING** The work W_{AB} done by an electric force in moving a charge q_0 from point A to point B and the electric potential difference $V_B - V_A$ are related according to

$V_B - V_A = \dfrac{-W_{AB}}{q_0}$ (Equation 19.4). This expression can be solved directly for q_0.

SOLUTION Solving Equation 19.4 for the charge q_0 gives

$$q_0 = \frac{-W_{AB}}{V_B - V_A} = \frac{W_{AB}}{V_A - V_B} = \frac{2.70 \times 10^{-3}\ \text{J}}{50.0\ \text{V}} = \boxed{5.40 \times 10^{-5}\ \text{C}}$$

---

7. **REASONING** The translational speed of the particle is related to the particle's translational kinetic energy, which forms one part of the total mechanical energy that the particle has. The total mechanical energy is conserved, because only the gravitational force and an electrostatic force, both of which are conservative forces, act on the particle (see Section 6.5). Thus, we will determine the speed at point A by utilizing the principle of conservation of mechanical energy.

SOLUTION The particle's total mechanical energy E is

$$E = \underbrace{\tfrac{1}{2}mv^2}_{\substack{\text{Translational} \\ \text{kinetic} \\ \text{energy}}} + \underbrace{\tfrac{1}{2}I\omega^2}_{\substack{\text{Rotational} \\ \text{kinetic} \\ \text{energy}}} + \underbrace{mgh}_{\substack{\text{Gravitational} \\ \text{potential} \\ \text{energy}}} + \underbrace{\tfrac{1}{2}kx^2}_{\substack{\text{Elastic} \\ \text{potential} \\ \text{energy}}} + \underbrace{\text{EPE}}_{\substack{\text{Electric} \\ \text{potential} \\ \text{energy}}}$$

Since the particle does not rotate, the angular speed ω is always zero and since there is no elastic force, we may omit the terms $\tfrac{1}{2}I\omega^2$ and $\tfrac{1}{2}kx^2$ from this expression. With this in mind, we express the fact that $E_B = E_A$ (energy is conserved) as follows:

$$\tfrac{1}{2}mv_B^2 + mgh_B + \text{EPE}_B = \tfrac{1}{2}mv_A^2 + mgh_A + \text{EPE}_A$$

This equation can be simplified further, since the particle travels horizontally, so that $h_B = h_A$, with the result that

$$\tfrac{1}{2}mv_B^2 + \text{EPE}_B = \tfrac{1}{2}mv_A^2 + \text{EPE}_A$$

Solving for v_A gives

$$v_A = \sqrt{v_B^2 + \frac{2\left(\text{EPE}_B - \text{EPE}_A\right)}{m}}.$$

According to Equation 19.4, the difference in electric potential energies $\text{EPE}_B - \text{EPE}_A$ is related to the electric potential difference $V_B - V_A$:

$$\text{EPE}_B - \text{EPE}_A = q_0\left(V_B - V_A\right)$$

Substituting this expression into the expression for v_A gives

$$v_A = \sqrt{v_B^2 + \frac{2q_0\left(V_B - V_A\right)}{m}} = \sqrt{\left(0 \text{ m/s}\right)^2 + \frac{2\left(-2.0\times10^{-5} \text{ C}\right)\left(-36 \text{ V}\right)}{4.0\times10^{-6} \text{ kg}}} = \boxed{19 \text{ m/s}}$$

---

11. **REASONING** The only force acting on the moving charge is the conservative electric force. Therefore, the total energy of the charge remains constant. Applying the principle of conservation of energy between locations A and B, we obtain

$$\tfrac{1}{2} mv_A^2 + \text{EPE}_A = \tfrac{1}{2} mv_B^2 + \text{EPE}_B$$

Since the charged particle starts from rest, $v_A = 0$. The difference in potential energies is related to the difference in potentials by Equation 19.4, $\text{EPE}_B - \text{EPE}_A = q(V_B - V_A)$. Thus, we have

$$q(V_A - V_B) = \tfrac{1}{2} mv_B^2 \tag{1}$$

Similarly, applying the conservation of energy between locations C and B gives

$$q(V_C - V_B) = \tfrac{1}{2} m(2v_B)^2 \tag{2}$$

Dividing Equation (1) by Equation (2) yields

$$\frac{V_A - V_B}{V_C - V_B} = \frac{1}{4}$$

This expression can be solved for V_B.

SOLUTION Solving for V_B, we find that

$$V_B = \frac{4V_A - V_C}{3} = \frac{4(452 \text{ V}) - 791 \text{ V}}{3} = \boxed{339 \text{ V}}$$

---

15. *REASONING* The potential of each charge q at a distance r away is given by Equation 19.6 as $V = kq/r$. By applying this expression to each charge, we will be able to find the desired ratio, because the distances are given for each charge.

SOLUTION According to Equation 19.6, the potentials of each charge are

$$V_A = \frac{kq_A}{r_A} \quad \text{and} \quad V_B = \frac{kq_B}{r_B}$$

Since we know that $V_A = V_B$, it follows that

$$\frac{kq_A}{r_A} = \frac{kq_B}{r_B} \quad \text{or} \quad \frac{q_B}{q_A} = \frac{r_B}{r_A} = \frac{0.43 \text{ m}}{0.18 \text{ m}} = \boxed{2.4}$$

---

17. *REASONING* Initially, suppose that one charge is at C and the other charge is held fixed at B. The charge at C is then moved to position A. According to Equation 19.4, the work W_{CA} done by the electric force as the charge moves from C to A is $W_{CA} = q(V_C - V_A)$, where, from Equation 19.6, $V_C = kq/d$ and $V_A = kq/r$. From the figure at the right we see that $d = \sqrt{r^2 + r^2} = \sqrt{2}r$. Therefore, we find that

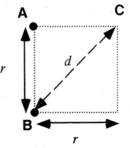

$$W_{CA} = q\left(\frac{kq}{\sqrt{2}r} - \frac{kq}{r}\right) = \frac{kq^2}{r}\left(\frac{1}{\sqrt{2}} - 1\right)$$

SOLUTION Substituting values, we obtain

$$W_{CA} = \frac{(8.99 \times 10^9 \text{ N} \cdot \text{m}^2/\text{C}^2)(3.0 \times 10^{-6} \text{ C})^2}{0.500 \text{ m}}\left(\frac{1}{\sqrt{2}} - 1\right) = \boxed{-4.7 \times 10^{-2} \text{ J}}$$

23. **REASONING** The only force acting on the moving charge is the conservative electric force. Therefore, the sum of the kinetic energy KE and the electric potential energy EPE is the same at points A and B:

$$\tfrac{1}{2}mv_A^2 + \text{EPE}_A = \tfrac{1}{2}mv_B^2 + \text{EPE}_B$$

Since the particle comes to rest at B, $v_B = 0$. Combining Equations 19.3 and 19.6, we have

$$\text{EPE}_A = qV_A = q\left(\frac{kq_1}{d}\right)$$

and

$$\text{EPE}_B = qV_B = q\left(\frac{kq_1}{r}\right)$$

where d is the initial distance between the fixed charge and the moving charged particle, and r is the distance between the charged particles after the moving charge has stopped. Therefore, the expression for the conservation of energy becomes

$$\tfrac{1}{2}mv_A^2 + \frac{kqq_1}{d} = \frac{kqq_1}{r}$$

This expression can be solved for r. Once r is known, the distance that the charged particle moves can be determined.

SOLUTION Solving the expression above for r gives

$$r = \frac{kqq_1}{\tfrac{1}{2}mv_A^2 + \dfrac{kqq_1}{d}}$$

$$= \frac{(8.99\times10^9\,\text{N}\cdot\text{m}^2/\text{C}^2)(-8.00\times10^{-6}\,\text{C})(-3.00\times10^{-6}\,\text{C})}{\tfrac{1}{2}(7.20\times10^{-3}\,\text{kg})(65.0\,\text{m/s})^2 + \dfrac{(8.99\times10^9\,\text{N}\cdot\text{m}^2/\text{C}^2)(-8.00\times10^{-6}\,\text{C})(-3.00\times10^{-6}\,\text{C})}{0.0450\,\text{m}}}$$

$$= 0.0108\,\text{m}$$

Therefore, the charge moves a distance of $0.0450\,\text{m} - 0.0108\,\text{m} = \boxed{0.0342\,\text{m}}$.

27. **REASONING** Initially, the three charges are infinitely far apart. We will proceed as in Example 8 by adding charges to the triangle, one at a time, and determining the electric potential energy at each step. According to Equation 19.3, the electric potential energy EPE is the product of the charge q and the electric potential V at the spot where the charge is

placed, EPE = qV. The total electric potential energy of the group is the sum of the energies of each step in assembling the group.

SOLUTION   Let the corners of the triangle be numbered clockwise as 1, 2 and 3, starting with the top corner. When the first charge (q_1 = 8.00 μC) is placed at a corner 1, the charge has no electric potential energy, EPE_1 = 0. This is because the electric potential V_1 produced by the other two charges at corner 1 is zero, since they are infinitely far away.

Once the 8.00-μC charge is in place, the electric potential V_2 that it creates at corner 2 is

$$V_2 = \frac{kq_1}{r_{21}}$$

where r_{21} = 5.00 m is the distance between corners 1 and 2, and q_1 = 8.00 μC. When the 20.0-μC charge is placed at corner 2, its electric potential energy EPE_2 is

$$\text{EPE}_2 = q_2 V_2 = q_2\left(\frac{kq_1}{r_{21}}\right)$$

$$= \left(20.0 \times 10^{-6}\ \text{C}\right)\left[\frac{\left(8.99 \times 10^9\ \text{N} \cdot \text{m}^2/\text{C}^2\right)\left(8.00 \times 10^{-6}\ \text{C}\right)}{5.00\ \text{m}}\right] = 0.288\ \text{J}$$

The electric potential V_3 at the remaining empty corner is the sum of the potentials due to the two charges that are already in place on corners 1 and 2:

$$V_3 = \frac{kq_1}{r_{31}} + \frac{kq_2}{r_{32}}$$

where q_1 = 8.00 μC, r_{31} = 3.00 m, q_2 = 20.0 μC, and r_{32} = 4.00 m. When the third charge (q_3 = –15.0 μC) is placed at corner 3, its electric potential energy EPE_3 is

$$\text{EPE}_3 = q_3 V_3 = q_3\left(\frac{kq_1}{r_{31}} + \frac{kq_2}{r_{32}}\right) = q_3 k\left(\frac{q_1}{r_{31}} + \frac{q_2}{r_{32}}\right)$$

$$= \left(-15.0 \times 10^{-6}\ \text{C}\right)\left(8.99 \times 10^9\ \text{N} \cdot \text{m}^2/\text{C}^2\right)\left(\frac{8.00 \times 10^{-6}\ \text{C}}{3.00\ \text{m}} + \frac{20.0 \times 10^{-6}\ \text{C}}{4.00\ \text{m}}\right) = -1.034\ \text{J}$$

The electric potential energy of the entire array is given by

$$EPE = EPE_1 + EPE_2 + EPE_3 = 0 + 0.288 \text{ J} + (-1.034 \text{ J}) = \boxed{-0.746 \text{ J}}$$

---

33. **_REASONING_** The magnitude E of the electric field is given by Equation 19.7a (without the minus sign) as $E = \dfrac{\Delta V}{\Delta s}$, where ΔV is the potential difference between the two metal conductors of the spark plug, and Δs is the distance between the two conductors. We can use this relation to find ΔV.

SOLUTION The potential difference between the conductors is

$$\Delta V = E \Delta s = \left(4.7 \times 10^7 \text{ V/m}\right)\left(0.75 \times 10^{-3} \text{ m}\right) = \boxed{3.5 \times 10^4 \text{ V}}$$

---

35. **_REASONING AND SOLUTION_** From Equation 19.7a we know that $E = -\dfrac{\Delta V}{\Delta s}$, where ΔV is the potential difference between the two surfaces of the membrane, and Δs is the distance between them. If A is a point on the positive surface and B is a point on the negative surface, then $\Delta V = V_A - V_B = 0.070 \text{ V}$. The electric field between the surfaces is

$$E = -\frac{\Delta V}{\Delta s} = -\frac{V_B - V_A}{\Delta s} = \frac{V_A - V_B}{\Delta s} = \frac{0.070 \text{ V}}{8.0 \times 10^{-9} \text{ m}} = \boxed{8.8 \times 10^6 \text{ V/m}}$$

---

39. **_REASONING AND SOLUTION_** Let point A be on the x-axis where the potential is 515 V. Let point B be on the x-axis where the potential is 495 V. From Equation 19.7a, the electric field is

$$E = -\frac{\Delta V}{\Delta s} = -\frac{V_B - V_A}{\Delta s} = -\frac{495 \text{ V} - 515 \text{ V}}{-2\left(6.0 \times 10^{-3} \text{ m}\right)} = -1.7 \times 10^3 \text{ V/m}$$

The magnitude of the electric field is $\boxed{1.7 \times 10^3 \text{ V/m}}$. Since the electric field is negative, it points to the $\boxed{\text{left}}$, from the high toward the low potential.

43. **REASONING**   According to Equation 19.11b, the energy stored in a capacitor with capacitance C and potential V across its plates is Energy $= \frac{1}{2}CV^2$.

SOLUTION Therefore, solving Equation 19.11b for V, we have

$$V = \sqrt{\frac{2(\text{Energy})}{C}} = \sqrt{\frac{2(73\ \text{J})}{120 \times 10^{-6}\ \text{F}}} = \boxed{1.1 \times 10^3\ \text{V}}$$

47. **REASONING AND SOLUTION**   Equation 19.10 gives the capacitance for a parallel plate capacitor filled with a dielectric of constant κ: $C = \kappa \varepsilon_0 A / d$. Solving for κ, we have

$$\kappa = \frac{Cd}{\varepsilon_0 A} = \frac{(7.0 \times 10^{-6}\ \text{F})(1.0 \times 10^{-5}\ \text{m})}{(8.85 \times 10^{-12}\ \text{F/m})(1.5\ \text{m}^2)} = \boxed{5.3}$$

51. **REASONING** According to Equation 19.11b, the energy stored in a capacitor with a capacitance C and potential V across its plates is Energy $= \frac{1}{2}CV^2$. Once we determine how much energy is required to operate a 75-W light bulb for one minute, we can then use the expression for the energy to solve for V.

SOLUTION   The energy stored in the capacitor, which is equal to the energy required to operate a 75-W bulb for one minute (= 60 s), is

$$\text{Energy} = Pt = (75\ \text{W})(60\ \text{s}) = 4500\ \text{J}$$

Therefore, solving Equation 19.11b for V, we have

$$V = \sqrt{\frac{2\,(\text{Energy})}{C}} = \sqrt{\frac{2(4500\ \text{J})}{3.3\ \text{F}}} = \boxed{52\ \text{V}}$$

55. **REASONING**   According to Equation 19.10, the capacitance of a parallel plate capacitor filled with a dielectric is $C = \kappa \varepsilon_0 A / d$, where κ is the dielectric constant, A is the area of one plate, and d is the distance between the plates.

From the definition of capacitance (Equation 19.8), $q = CV$. Thus, using Equation 19.10, we see that the charge q on a parallel plate capacitor that contains a dielectric is given by $q = (\kappa \varepsilon_0 A / d)V$. Since each dielectric occupies one-half of the volume between the plates, the area of each plate in contact with each material is $A/2$. Thus,

$$q_1 = \frac{\kappa_1 \varepsilon_0 (A/2)}{d} V = \frac{\kappa_1 \varepsilon_0 A}{2d} V \qquad \text{and} \qquad q_2 = \frac{\kappa_2 \varepsilon_0 (A/2)}{d} V = \frac{\kappa_2 \varepsilon_0 A}{2d} V$$

According to the problem statement, the total charge stored by the capacitor is

$$q_1 + q_2 = CV \qquad (1)$$

where q_1 and q_2 are the charges on the plates in contact with dielectrics 1 and 2, respectively.

Using the expressions for q_1 and q_2 above, Equation (1) becomes

$$CV = \frac{\kappa_1 \varepsilon_0 A}{2d} V + \frac{\kappa_2 \varepsilon_0 A}{2d} V = \frac{\kappa_1 \varepsilon_0 A + \kappa_2 \varepsilon_0 A}{2d} V = \frac{(\kappa_1 + \kappa_2) \varepsilon_0 A}{2d} V$$

This expression can be solved for C.

SOLUTION Solving for C, we obtain $\boxed{C = \dfrac{\varepsilon_0 A (\kappa_1 + \kappa_2)}{2d}}$

---

57. **REASONING** The charge that resides on the outer surface of the cell membrane is $q = CV$, according to Equation 19.8. Before we can use this expression, however, we must first determine the capacitance of the membrane. If we assume that the cell membrane behaves like a parallel plate capacitor filled with a dielectric, Equation 19.10 ($C = \kappa \varepsilon_0 A/d$) applies as well.

SOLUTION The capacitance of the cell membrane is

$$C = \frac{\kappa \varepsilon_0 A}{d} = \frac{(5.0)(8.85 \times 10^{-12} \text{ F/m})(5.0 \times 10^{-9} \text{ m}^2)}{1.0 \times 10^{-8} \text{ m}} = 2.2 \times 10^{-11} \text{ F}$$

a. The charge on the outer surface of the membrane is, therefore,

$$q = CV = (2.2 \times 10^{-11} \text{ F})(60.0 \times 10^{-3} \text{ V}) = \boxed{1.3 \times 10^{-12} \text{ C}}$$

b. If the charge in part (a) is due to K^+ ions with charge $+e$ ($e = 1.6 \times 10^{-19}$ C), the number of ions present on the outer surface of the membrane is

$$\frac{\text{Number of}}{K^+ \text{ ions}} = \frac{1.3 \times 10^{-12} \text{C}}{1.6 \times 10^{-19} \text{C}} = \boxed{8.1 \times 10^6}$$

61. **REASONING AND SOLUTION** Combining Equations 19.1 and 19.3, we have

$$W_{AB} = \text{EPE}_A - \text{EPE}_B = q_0(V_A - V_B) = (+1.6 \times 10^{-19}\,\text{C})(0.070\,\text{V}) = \boxed{1.1 \times 10^{-20}\,\text{J}}$$

---

67. **REASONING** If we assume that the motion of the proton and the electron is horizontal in the $+x$ direction, the motion of the proton is determined by Equation 2.8, $x = v_0 t + \frac{1}{2}a_p t^2$, where x is the distance traveled by the proton, v_0 is its initial speed, and a_p is its acceleration. If the distance between the capacitor places is d, then this relation becomes $\frac{1}{2}d = v_0 t + \frac{1}{2}a_p t^2$, or

$$d = 2v_0 t + a_p t^2 \tag{1}$$

We can solve Equation (1) for the initial speed v_0 of the proton, but, first, we must determine the time t and the acceleration a_p of the proton . Since the proton strikes the negative plate at the same instant the electron strikes the positive plate, we can use the motion of the electron to determine the time t.

For the electron, $\frac{1}{2}d = \frac{1}{2}a_e t^2$, where we have taken into account the fact that the electron is released from rest. Solving this expression for t we have $t = \sqrt{d/a_e}$. Substituting this expression into Equation (1), we have

$$d = 2v_0 \sqrt{\frac{d}{a_e}} + \left(\frac{a_p}{a_e}\right) d \tag{2}$$

The accelerations can be found by noting that the magnitudes of the forces on the electron and proton are equal, since these particles have the same magnitude of charge. The force on the electron is $F = eE = eV/d$, and the acceleration of the electron is, therefore,

$$a_e = \frac{F}{m_e} = \frac{eV}{m_e d} \tag{3}$$

Newton's second law requires that $m_e a_e = m_p a_p$, so that

$$\frac{a_p}{a_e} = \frac{m_e}{m_p} \tag{4}$$

Combining Equations (2), (3) and (4) leads to the following expression for v_0, the initial speed of the proton:

$$v_0 = \frac{1}{2}\left(1 - \frac{m_e}{m_p}\right)\sqrt{\frac{eV}{m_e}}$$

SOLUTION Substituting values into the expression above, we find

$$v_0 = \frac{1}{2}\left(1 - \frac{9.11\times10^{-31}\,\text{kg}}{1.67\times10^{-27}\,\text{kg}}\right)\sqrt{\frac{(1.60\times10^{-19}\,\text{C})(175\ \text{V})}{9.11\times10^{-31}\,\text{kg}}} = \boxed{2.77\times10^{6}\,\text{m/s}}$$

3. **REASONING** Electric current is the amount of charge flowing per unit time (see Equation 20.1). Thus, the amount of charge is the current times the time. Furthermore, the potential difference is the difference in electric potential energy per unit charge (see Equation 19.4), so that, once the amount of charge has been calculated, we can determine the energy by multiplying the potential difference by the charge.

SOLUTION

a. According to Equation 20.1, the current I is the amount of charge Δq divided by the time Δt, or $I = \Delta q/\Delta t$. Therefore, the additional charge that passes through the machine in normal mode versus standby mode is

$$q_{additional} = \underbrace{I_{normal}\Delta t}_{\Delta q \text{ in normal mode}} - \underbrace{I_{standby}\Delta t}_{\Delta q \text{ in standby mode}} = \left(I_{normal} - I_{standby}\right)\Delta t$$

$$= (0.110 \text{ A} - 0.067 \text{ A})(60.0 \text{ s}) = \boxed{2.6 \text{ C}}$$

b. According to Equation 19.4, the potential difference ΔV is the difference $\Delta(\text{EPE})$ in the electric potential energy divided by the charge $q_{additional}$, or $\Delta V = \Delta(\text{EPE})/q_{additional}$. As a result, the additional energy used in the normal mode as compared to the standby mode is

$$\Delta(\text{EPE}) = q_{additional}\Delta V = (2.6 \text{ C})(120 \text{ V}) = \boxed{310 \text{ J}}$$

9. **REASONING** The number N of protons that strike the target is equal to the amount of electric charge Δq striking the target divided by the charge e of a proton, $N = (\Delta q)/e$. From Equation 20.1, the amount of charge is equal to the product of the current I and the time Δt. We can combine these two relations to find the number of protons that strike the target in 15 seconds.

The heat Q that must be supplied to change the temperature of the aluminum sample of mass m by an amount ΔT is given by Equation 12.4 as $Q = cm\Delta T$, where c is the specific heat capacity of aluminum. The heat is provided by the kinetic energy of the protons and is equal to the number of protons that strike the target times the kinetic energy per proton. Using this reasoning, we can find the change in temperature of the block for the 15 second-time interval.

SOLUTION

a. The number N of protons that strike the target is

$$N = \frac{\Delta q}{e} = \frac{I\,\Delta t}{e} = \frac{(0.50 \times 10^{-6}\ \text{A})(15\ \text{s})}{1.6 \times 10^{-19}\ \text{C}} = \boxed{4.7 \times 10^{13}}$$

b. The amount of heat Q provided by the kinetic energy of the protons is

$$Q = (4.7 \times 10^{13}\ \text{protons})(4.9 \times 10^{-12}\ \text{J/proton}) = 230\ \text{J}$$

Since $Q = cm\Delta T$ and since Table 12.2 gives the specific heat of aluminum as $c = 9.00 \times 10^2$ J/(kg·C°), the change in temperature of the block is

$$\Delta T = \frac{Q}{cm} = \frac{230\ \text{J}}{\left(9.00 \times 10^2\ \dfrac{\text{J}}{\text{kg} \cdot \text{C}^\circ}\right)(15 \times 10^{-3}\ \text{kg})} = \boxed{17\ \text{C}^\circ}$$

---

15. **REASONING** The resistance of a metal wire of length L, cross-sectional area A and resistivity ρ is given by Equation 20.3: $R = \rho L / A$. Solving for A, we have $A = \rho L / R$. We can use this expression to find the ratio of the cross-sectional area of the aluminum wire to that of the copper wire.

SOLUTION Forming the ratio of the areas and using resistivity values from Table 20.1, we have

$$\frac{A_{\text{aluminum}}}{A_{\text{copper}}} = \frac{\rho_{\text{aluminum}} L / R}{\rho_{\text{copper}} L / R} = \frac{\rho_{\text{aluminum}}}{\rho_{\text{copper}}} = \frac{2.82 \times 10^{-8}\ \Omega \cdot \text{m}}{1.72 \times 10^{-8}\ \Omega \cdot \text{m}} = \boxed{1.64}$$

---

19. **REASONING** The resistance of a metal wire of length L, cross-sectional area A and resistivity ρ is given by Equation 20.3: $R = \rho L / A$. The volume V_2 of the new wire will be the same as the original volume V_1 of the wire, where volume is the product of length and cross-sectional area. Thus, $V_1 = V_2$ or $A_1 L_1 = A_2 L_2$. Since the new wire is three times longer than the first wire, we can write

$$A_1 L_1 = A_2 L_2 = A_2 (3L_1) \quad \text{or} \quad A_2 = A_1 / 3$$

We can form the ratio of the resistances, use this expression for the area A_2, and find the new resistance.

SOLUTION The resistance of the new wire is determined as follows:

$$\frac{R_2}{R_1} = \frac{\rho L_2 / A_2}{\rho L_1 / A_1} = \frac{L_2 A_1}{L_1 A_2} = \frac{(3L_1) A_1}{L_1 (A_1 / 3)} = 9$$

Solving for R_2, we find that

$$R_2 = 9 R_1 = 9(21.0 \ \Omega) = \boxed{189 \ \Omega}$$

---

27. **REASONING** According to Equation 6.10b, the energy used is Energy $= Pt$, where P is the power and t is the time. According to Equation 20.6a, the power is $P = IV$, where I is the current and V is the voltage. Thus, Energy $= IVt$, and we apply this result first to the dryer and then to the computer.

SOLUTION The energy used by the dryer is

$$\text{Energy} = Pt = IVt = (16 \text{ A})(240 \text{ V})(45 \text{ min}) \underbrace{\left(\frac{60 \text{ s}}{1.00 \text{ min}} \right)}_{\substack{\text{Converts minutes} \\ \text{to seconds}}} = 1.04 \times 10^7 \text{ J}$$

For the computer, we have

$$\text{Energy} = 1.04 \times 10^7 \text{ J} = IVt = (2.7 \text{ A})(120 \text{ V})t$$

Solving for t we find

$$t = \frac{1.04 \times 10^7 \text{ J}}{(2.7 \text{ A})(120 \text{ V})} = 3.21 \times 10^4 \text{ s} = \left(3.21 \times 10^4 \text{ s} \right) \left(\frac{1.00 \text{ h}}{3600 \text{ s}} \right) = \boxed{8.9 \text{ h}}$$

---

31. **REASONING AND SOLUTION** As a function of temperature, the resistance of the wire is given by Equation 20.5: $R = R_0 \left[1 + \alpha (T - T_0) \right]$, where α is the temperature coefficient of resistivity. From Equation 20.6c, we have $P = V^2 / R$. Combining these two equations, we have

$$P = \frac{V^2}{R_0 \left[1 + \alpha (T - T_0) \right]} = \frac{P_0}{1 + \alpha (T - T_0)}$$

where $P_0 = V^2 / R_0$, since the voltage is constant. But $P = \frac{1}{2} P_0$, so we find

$$\frac{P_0}{2} = \frac{P_0}{1 + \alpha\left(T - T_0\right)} \qquad \text{or} \qquad 2 = 1 + \alpha\left(T - T_0\right)$$

Solving for T, we find

$$T = \frac{1}{\alpha} + T_0 = \frac{1}{0.0045 \ (\text{C}°)^{-1}} + 28° = \boxed{250 \ °\text{C}}$$

---

35. **REASONING** The average power is given by Equation 20.15c as $\overline{P} = V_{rms}^2 / R$. In this expression the rms voltage V_{rms} appears. However, we seek the peak voltage V_0. The relation between the two types of voltage is given by Equation 20.13 as $V_{rms} = V_0 / \sqrt{2}$, so we can obtain the peak voltage by using Equation 20.13 to substitute into Equation 20.15c.

SOLUTION Substituting V_{rms} from Equation 20.13 into Equation 20.15c gives

$$\overline{P} = \frac{V_{rms}^2}{R} = \frac{\left(V_0 / \sqrt{2}\right)^2}{R} = \frac{V_0^2}{2R}$$

Solving for the peak voltage V_0 gives

$$V_0 = \sqrt{2R\overline{P}} = \sqrt{2\left(4.0 \ \Omega\right)\left(55 \ \text{W}\right)} = \boxed{21 \ \text{V}}$$

---

39. **REASONING**
a. We can obtain the frequency of the alternating current by comparing this specific expression for the current with the more general one in Equation 20.8.

b. The resistance of the light bulb is, according to Equation 20.14, equal to the rms-voltage divided by the rms-current. The rms-voltage is given, and we can obtain the rms-current by dividing the peak current by $\sqrt{2}$, as expressed by Equation 20.12.

c. The average power is given by Equation 20.15a as the product of the rms-current and the rms-voltage.

SOLUTION
a. By comparing $I = \left(0.707 \ \text{A}\right)\sin\left[\left(314 \ \text{Hz}\right) t\right]$ with the general expression (see Equation 20.8) for the current in an ac circuit, $I = I_0 \sin 2\pi f t$, we see that

$$2\pi f t = \left(314 \ \text{Hz}\right) t \qquad \text{or} \qquad f = \frac{314 \ \text{Hz}}{2\pi} = \boxed{50.0 \ \text{Hz}}$$

b. The resistance is equal to V_{rms}/I_{rms}, where the rms-current is related to the peak current I_0 by $I_{rms} = I_0 / \sqrt{2}$. Thus, the resistance of the light bulb is

$$R = \frac{V_{rms}}{I_{rms}} = \frac{V_{rms}}{\dfrac{I_0}{\sqrt{2}}} = \frac{\sqrt{2}(120.0 \text{ V})}{0.707 \text{ A}} = \boxed{2.40 \times 10^2 \ \Omega} \tag{20.14}$$

c. The average power is the product of the rms-current and rms-voltage:

$$\bar{P} = I_{rms}V_{rms} = \left(\frac{I_0}{\sqrt{2}}\right)V_{rms} = \left(\frac{0.707 \text{ A}}{\sqrt{2}}\right)(120.0 \text{ V}) = \boxed{60.0 \text{ W}} \tag{20.15a}$$

---

41. ***REASONING*** Using Ohm's law (Equation 20.2) we can write an expression for the voltage across the original circuit as $V = I_0 R_0$. When the additional resistor R is inserted in series, assuming that the battery remains the same, the voltage across the new combination is given by $V = I(R + R_0)$. Since V is the same in both cases, we can write $I_0 R_0 = I(R + R_0)$. This expression can be solved for R_0.

SOLUTION Solving for R_0, we have

$$I_0 R_0 - I R_0 = IR \quad \text{or} \quad R_0(I_0 - I) = IR$$

Therefore, we find that

$$R_0 = \frac{IR}{I_0 - I} = \frac{(12.0 \text{ A})(8.00 \ \Omega)}{15.0 \text{ A} - 12.0 \text{ A}} = \boxed{32 \ \Omega}$$

---

43. ***REASONING*** The equivalent series resistance R_s is the sum of the resistances of the three resistors. The potential difference V can be determined from Ohm's law according to $V = IR_s$.

SOLUTION
a. The equivalent resistance is

$$R_s = 25 \ \Omega + 45 \ \Omega + 75 \ \Omega = \boxed{145 \ \Omega}$$

b. The potential difference across the three resistors is

$$V = IR_s = (0.51 \text{ A})(145 \ \Omega) = \boxed{74 \text{ V}}$$

47. ***REASONING*** The resistance of one of the wires in the extension cord is given by

Equation 20.3: $R = \rho L / A$, where the resistivity of copper is $\rho = 1.72 \times 10^{-8}\ \Omega \cdot m$, according to Table 20.1. Since the two wires in the cord are in series with each other, their total resistance is $R_{cord} = R_{wire\ 1} + R_{wire\ 2} = 2\rho L / A$. Once we find the equivalent resistance of the entire circuit (extension cord + trimmer), Ohm's law can be used to find the voltage applied to the trimmer.

SOLUTION
a. The resistance of the extension cord is

$$R_{cord} = \frac{2\rho L}{A} = \frac{2(1.72 \times 10^{-8}\ \Omega \cdot m)(46\ m)}{1.3 \times 10^{-6}\ m^2} = \boxed{1.2\ \Omega}$$

b. The total resistance of the circuit (cord + trimmer) is, since the two are in series,

$$R_s = 1.2\ \Omega + 15.0\ \Omega = 16.2\ \Omega$$

Therefore from Ohm's law (Equation 20.2: $V = IR$), the current in the circuit is

$$I = \frac{V}{R_s} = \frac{120\ V}{16.2\ \Omega} = 7.4\ A$$

Finally, the voltage applied to the trimmer alone is (again using Ohm's law),

$$V_{trimmer} = (7.4\ A)(15.0\ \Omega) = \boxed{110\ V}$$

---

51. ***REASONING AND SOLUTION*** Since the circuit elements are in parallel, the equivalent resistance can be obtained directly from Equation 20.17:

$$\frac{1}{R_p} = \frac{1}{R_1} + \frac{1}{R_2} = \frac{1}{16\ \Omega} + \frac{1}{8.0\ \Omega} \qquad \text{or} \qquad \boxed{R_p = 5.3\ \Omega}$$

---

55. ***REASONING*** Since the resistors are connected in parallel, the voltage across each one is the same and can be calculated from Ohm's Law (Equation 20.2: $V = IR$). Once the voltage across each resistor is known, Ohm's law can again be used to find the current in the second resistor. The total power consumed by the parallel combination can be found calculating the power consumed by each resistor from Equation 20.6b: $P = I^2 R$. Then, the total power consumed is the sum of the power consumed by each resistor.

SOLUTION Using data for the second resistor, the voltage across the resistors is equal to

$$V = IR = (3.00 \text{ A})(64.0 \text{ } \Omega) = 192 \text{ V}$$

a. The current through the 42.0-Ω resistor is

$$I = \frac{V}{R} = \frac{192 \text{ V}}{42.0 \text{ } \Omega} = \boxed{4.57 \text{ A}}$$

b. The power consumed by the 42.0-Ω resistor is

$$P = I^2 R = (4.57 \text{ A})^2 (42.0 \text{ } \Omega) = 877 \text{ W}$$

while the power consumed by the 64.0-Ω resistor is

$$P = I^2 R = (3.00 \text{ A})^2 (64.0 \text{ } \Omega) = 576 \text{ W}$$

Therefore the total power consumed by the two resistors is 877 W + 576 W = $\boxed{1450 \text{ W}}$.

---

59. **REASONING** The equivalent resistance of the three devices in parallel is R_p, and we can find the value of R_p by using our knowledge of the total power consumption of the circuit; the value of R_p can be found from Equation 20.6c, $P = V^2 / R_p$. Ohm's law (Equation 20.2, $V = IR$) can then be used to find the current through the circuit.

SOLUTION
a. The total power used by the circuit is $P = 1650 \text{ W} + 1090 \text{ W} + 1250 \text{ W} = 3990 \text{ W}$. The equivalent resistance of the circuit is

$$R_p = \frac{V^2}{P} = \frac{(120 \text{ V})^2}{3990 \text{ W}} = \boxed{3.6 \text{ } \Omega}$$

b. The total current through the circuit is

$$I = \frac{V}{R_p} = \frac{120 \text{ V}}{3.6 \text{ } \Omega} = \boxed{33 \text{ A}}$$

This current is larger than the rating of the circuit breaker; therefore, the $\boxed{\text{breaker will open}}$.

63. **REASONING** To find the current, we will use Ohm's law, together with the proper equivalent resistance. The coffee maker and frying pan are in series, so their equivalent resistance is given by Equation 20.16 as $R_{coffee} + R_{pan}$. This total resistance is in parallel with the resistance of the bread maker, so the equivalent resistance of the parallel combination can be obtained from Equation 20.17 as $R_p^{-1} = (R_{coffee} + R_{pan})^{-1} + R_{bread}^{-1}$.

SOLUTION Using Ohm's law and the expression developed above for R_p^{-1}, we find

$$I = \frac{V}{R_p} = V \left(\frac{1}{R_{coffee} + R_{pan}} + \frac{1}{R_{bread}} \right) = (120 \text{ V}) \left(\frac{1}{14 \,\Omega + 16 \,\Omega} + \frac{1}{23 \,\Omega} \right) = \boxed{9.2 \text{ A}}$$

---

65. **REASONING** When two or more resistors are in series, the equivalent resistance is given by Equation 20.16: $R_s = R_1 + R_2 + R_3 + \ldots$. Likewise, when resistors are in parallel, the expression to be solved to find the equivalent resistance is given by Equation 20.17: $\frac{1}{R_p} = \frac{1}{R_1} + \frac{1}{R_2} + \frac{1}{R_3} + \ldots$. We will successively apply these to the individual resistors in the figure in the text beginning with the resistors on the right side of the figure.

SOLUTION Since the 4.0-Ω and the 6.0-Ω resistors are in series, the equivalent resistance of the combination of those two resistors is 10.0 Ω. The 9.0-Ω and 8.0-Ω resistors are in parallel; their equivalent resistance is 4.24 Ω. The equivalent resistances of the parallel combination (9.0 Ω and 8.0 Ω) and the series combination (4.0 Ω and the 6.0 Ω) are in parallel; therefore, their equivalent resistance is 2.98 Ω. The 2.98-Ω combination is in series with the 3.0-Ω resistor, so that equivalent resistance is 5.98 Ω. Finally, the 5.98-Ω combination and the 20.0-Ω resistor are in parallel, so the equivalent resistance between the points A and B is $\boxed{4.6 \,\Omega}$.

---

71. **REASONING** Since we know that the current in the 8.00-Ω resistor is 0.500 A, we can use Ohm's law ($V = IR$) to find the voltage across the 8.00-Ω resistor. The 8.00-Ω resistor and the 16.0-Ω resistor are in parallel; therefore, the voltages across them are equal. Thus, we can also use Ohm's law to find the current through the 16.0-Ω resistor. The currents that flow through the 8.00-Ω and the 16.0-Ω resistors combine to give the total current that flows through the 20.0-Ω resistor. Similar reasoning can be used to find the current through the 9.00-Ω resistor.

SOLUTION

a. The voltage across the 8.00-Ω resistor is $V_8 = (0.500 \text{ A})(8.00 \,\Omega) = 4.00 \text{ V}$. Since this is also the voltage that is across the 16.0-Ω resistor, we find that the current through the

16.0-Ω resistor is $I_{16} = (4.00 \text{ V})/(16.0\Omega) = 0.250 \text{ A}$. Therefore, the total current that flows through the 20.0-Ω resistor is

$$I_{20} = 0.500 \text{ A} + 0.250 \text{ A} = \boxed{0.750 \text{ A}}$$

b. The 8.00-Ω and the 16.0-Ω resistors are in parallel, so their equivalent resistance can be obtained from Equation 20.17, $\dfrac{1}{R_p} = \dfrac{1}{R_1} + \dfrac{1}{R_2} + \dfrac{1}{R_3} + \dots$, and is equal to 5.33 Ω. Therefore, the equivalent resistance of the upper branch of the circuit is $R_{upper} = 5.33 \ \Omega + 20.0 \ \Omega = 25.3 \ \Omega$, since the 5.33-$\Omega$ resistance is in series with the 20.0-Ω resistance. Using Ohm's law, we find that the voltage across the upper branch must be $V = (0.750 \text{ A})(25.3 \ \Omega) = 19.0 \text{ V}$. Since the lower branch is in parallel with the upper branch, the voltage across both branches must be the same. Therefore, the current through the 9.00-Ω resistor is, from Ohm's law,

$$I_9 = \frac{V_{lower}}{R_9} = \frac{19.0 \text{ V}}{9.00 \ \Omega} = \boxed{2.11 \text{ A}}$$

---

73. **REASONING** The terminal voltage of the battery is given by $V_{terminal} = \text{Emf} - Ir$, where r is the internal resistance of the battery. Since the terminal voltage is observed to be one-half of the emf of the battery, we have $V_{terminal} = \text{Emf}/2$ and $I = \text{Emf}/(2r)$. From Ohm's law, the equivalent resistance of the circuit is $R = \text{emf}/I = 2r$. We can also find the equivalent resistance of the circuit by considering that the identical bulbs are in parallel across the battery terminals, so that the equivalent resistance of the N bulbs is found from

$$\frac{1}{R_p} = \frac{N}{R_{bulb}} \qquad \text{or} \qquad R_p = \frac{R_{bulb}}{N}$$

This equivalent resistance is in series with the battery, so we find that the equivalent resistance of the circuit is

$$R = 2r = \frac{R_{bulb}}{N} + r$$

This expression can be solved for N.

SOLUTION Solving the above expression for N, we have

$$N = \frac{R_{bulb}}{2r - r} = \frac{R_{bulb}}{r} = \frac{15 \ \Omega}{0.50 \ \Omega} = \boxed{30}$$

79. **REASONING** The current I can be found by using Kirchhoff's loop rule. Once the current is known, the voltage between points A and B can be determined.

SOLUTION
a. We assume that the current is directed clockwise around the circuit. Starting at the upper-left corner and going clockwise around the circuit, we set the potential drops equal to the potential rises:

$$\underbrace{(5.0\ \Omega)I + (27\ \Omega)I + 10.0\ V + (12\ \Omega)I + (8.0\ \Omega)I}_{\text{Potential drops}} = \underbrace{30.0\ V}_{\text{Potential rises}}$$

Solving for the current gives $\boxed{I = 0.38\ A}$.

b. The voltage between points A and B is

$$V_{AB} = 30.0\ V - (0.38\ A)(27\ \Omega) = \boxed{2.0 \times 10^1\ V}$$

c. $\boxed{\text{Point } B}$ is at the higher potential.

---

85. **REASONING** We begin by labeling the currents in the three resistors. The drawing below shows the directions chosen for these currents. The directions are arbitrary, and if any of them is incorrect, then the analysis will show that the corresponding value for the current is negative.

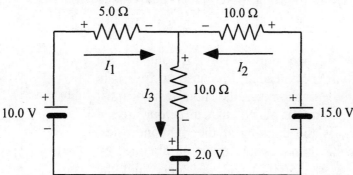

We then mark the resistors with the plus and minus signs that serve as an aid in identifying the potential drops and rises for the loop rule, recalling that conventional current is always directed from a higher potential (+) toward a lower potential (−). Thus, given the directions chosen for I_1, I_2, and I_3, the plus and minus signs *must* be those shown in the drawing. We can now use Kirchhoff's rules to find the voltage across the 5.0-Ω resistor.

SOLUTION   Applying the loop rule to the left loop (and suppressing units for convenience) gives

$$5.0 I_1 + 10.0 I_3 + 2.0 = 10.0 \tag{1}$$

Similarly, for the right loop,

$$10.0 I_2 + 10.0 I_3 + 2.0 = 15.0 \qquad (2)$$

If we apply the junction rule to the upper junction, we obtain

$$I_1 + I_2 = I_3 \qquad (3)$$

Subtracting Equation (2) from Equation (1) gives

$$5.0 I_1 - 10.0 I_2 = -5.0 \qquad (4)$$

We now multiply Equation (3) by 10 and add the result to Equation (2); the result is

$$10.0 I_1 + 20.0 I_2 = 13.0 \qquad (5)$$

If we then multiply Equation (4) by 2 and add the result to Equation (5), we obtain $20.0 I_1 = 3.0$, or solving for I_1, we obtain $I_1 = 0.15 \text{ A}$. The fact that I_1 is positive means that the current in the drawing has the correct direction. The voltage across the 5.0-Ω resistor can be found from Ohm's law:

$$V = (0.15 \text{ A})(5.0 \text{ } \Omega) = \boxed{0.75 \text{ V}}$$

Current flows from the higher potential to the lower potential, and the current through the 5.0-Ω flows from left to right, so the $\boxed{\text{left end of the resistor}}$ is at the higher potential.

---

87. **REASONING**   Since only 0.100 mA out of the available 60.0 mA is needed to cause a full-scale deflection of the galvanometer, the shunt resistor must allow the excess current of 59.9 mA to detour around the meter coil, as the drawing at the right indicates. The value for the shunt resistance can be obtained by recognizing that the 50.0-Ω coil resistance and the shunt resistance are in parallel, both being connected between points A and B in the drawing. Thus, the voltage across each resistance is the same.

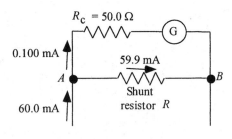

SOLUTION Expressing voltage as the product of current and resistance, we find that

$$\underbrace{\left(59.9\times10^{-3}\text{ A}\right)(R)}_{\text{Voltage across shunt resistance}}=\underbrace{\left(0.100\times10^{-3}\text{ A}\right)(50.0\,\Omega)}_{\text{Voltage across coil resistance}}$$

$$R=\frac{\left(0.100\times10^{-3}\text{ A}\right)(50.0\,\Omega)}{59.9\times10^{-3}\text{ A}}=\boxed{0.0835\,\Omega}$$

---

91. **REASONING AND SOLUTION** For the 20.0 V scale

$$V_1=I(R_1+R_c)$$

For the 30.0 V scale

$$V_2=I(R_2+R_c)$$

Subtracting and rearranging yields

$$I=\frac{V_2-V_1}{R_2-R_1}=\frac{30.0\text{ V}-20.0\text{ V}}{2930\,\Omega-1680\,\Omega}=\boxed{8.00\times10^{-3}\text{ A}}$$

Substituting this value into either of the equations for V_1 or V_2 gives $R_c=\boxed{820\,\Omega}$.

---

95. **REASONING AND SOLUTION** Let C_0 be the capacitance of an empty capacitor. Then the capacitances are as follows, according to the discussion following Equation 19.10:

$$C_1=3.00C_0\quad\text{and}\quad C_2=5.00C_0$$

The series capacitance of the two is

$$\frac{1}{C_s}=\frac{1}{2.50C_0}+\frac{1}{4.00C_0}\qquad\text{or}\qquad C_s=1.54C_0$$

Now

$$\kappa C_0=1.54C_0\qquad\text{or}\qquad \kappa=\boxed{1.54}$$

---

101. **REASONING** The foil effectively converts the capacitor into two capacitors in series. Equation 19.10 gives the expression for the capacitance of a capacitor of plate area A and plate separation d (no dielectric): $C_0=\varepsilon_0 A/d$. We can use this expression to determine the capacitance of the individual capacitors created by the presence of the foil. Then using the fact that the "two capacitors" are in series, we can use Equation 20.19 to find the equivalent capacitance of the system.

SOLUTION   Since the foil is placed one-third of the way from one plate of the original capacitor to the other, we have $d_1 = (2/3)d$, and $d_2 = (1/3)d$. Then

$$C_1 = \frac{\varepsilon_0 A}{(2/3)d} = \frac{3\varepsilon_0 A}{2d}$$

and

$$C_2 = \frac{\varepsilon_0 A}{(1/3)d} = \frac{3\varepsilon_0 A}{d}$$

Since these two capacitors are effectively in series, it follows that

$$\frac{1}{C_s} = \frac{1}{C_1} + \frac{1}{C_2} = \frac{1}{3\varepsilon_0 A/(2d)} + \frac{1}{3\varepsilon_0 A/d} = \frac{3d}{3\varepsilon_0 A} = \frac{d}{\varepsilon_0 A}$$

But $C_0 = \varepsilon_0 A/d$, so that $d/(\varepsilon_0 A) = 1/C_0$, and we have

$$\frac{1}{C_s} = \frac{1}{C_0} \qquad \text{or} \qquad \boxed{C_s = C_0}$$

---

103. **REASONING**   The charge q on a discharging capacitor in a RC circuit is given by Equation 20.22: $q = q_0 e^{-t/RC}$, where q_0 is the original charge at time $t = 0$ s. Once t (time for one pulse) and the ratio q/q_0 are known, this expression can be solved for C.

SOLUTION   Since the pacemaker delivers 81 pulses per minute, the time for one pulse is

$$\frac{1 \text{ min}}{81 \text{ pulses}} \times \frac{60.0 \text{ s}}{1.00 \text{ min})} = 0.74 \text{ s/pulse}$$

Since one pulse is delivered every time the fully-charged capacitor loses 63.2% of its original charge, the charge remaining is 36.8% of the original charge. Thus, we have $q = (0.368)q_0$, or $q/q_0 = 0.368$.

From Equation 20.22, we have

$$\frac{q}{q_0} = e^{-t/RC}$$

Taking the natural logarithm of both sides, we have,

$$\ln\left(\frac{q}{q_0}\right) = -\frac{t}{RC}$$

Solving for C, we find

$$C = \frac{-t}{R\,\ln(q/q_0)} = \frac{-(0.74\text{ s})}{(1.8\times10^6\ \Omega)\,\ln(0.368)} = \boxed{4.1\times10^{-7}\text{ F}}$$

---

109. **REASONING AND SOLUTION** The charges stored on capacitors in series are equal and equal to the charge separated by the battery. The total energy stored in the capacitors is

$$\text{Energy} = \frac{Q^2}{2C_1} + \frac{Q^2}{2C_2}$$

$$\text{Energy} = \frac{Q^2}{2}\left(\frac{1}{C_1} + \frac{1}{C_2}\right)$$

According to Equation 20.19, the quantity in the parentheses is just the reciprocal of the equivalent capacitance C of the circuit, so

$$\boxed{\text{Energy} = \frac{Q^2}{2C}}$$

---

111. **REASONING AND SOLUTION** Ohm's law (Equation 20.2), $V = IR$, gives the result directly:

$$R = \frac{V}{I} = \frac{9.0\text{ V}}{0.11\text{ A}} = \boxed{82\ \Omega}$$

---

117. **REASONING** The power P delivered to the circuit is, according to Equation 20.6c, $P = V^2/R_{12345}$, where V is the voltage of the battery and R_{12345} is the equivalent resistance of the five-resistor circuit. The voltage and power are known, so that the equivalent resistance can be calculated. We will use our knowledge of resistors wired in series and parallel to evaluate R_{12345} in terms of the resistance R of each resistor. In this manner we will find the value for R.

SOLUTION First we note that all the resistors are equal, so $R_1 = R_2 = R_3 = R_4 = R_5 = R$. We can find the equivalent resistance R_{12345} as follows. The resistors R_3 and R_4 are in series, so the equivalent resistance R_{34} of these two is $R_{34} = R_3 + R_4 = 2R$. The resistors R_2, R_{34}, and R_5 are in parallel, and the reciprocal of the equivalent resistance R_{2345} is

$$\frac{1}{R_{2345}} = \frac{1}{R_2} + \frac{1}{R_{34}} + \frac{1}{R_5} = \frac{1}{R} + \frac{1}{2R} + \frac{1}{R} = \frac{5}{2R}$$

so $R_{2345} = 2R/5$. The resistor R_1 is in series with R_{2345}, and the equivalent resistance of this combination is the equivalent resistance of the circuit. Thus, we have

$$R_{12345} = R_1 + R_{2345} = R + \frac{2R}{5} = \frac{7R}{5}$$

The power delivered to the circuit is

$$P = \frac{V^2}{R_{12345}} = \frac{V^2}{\left(\dfrac{7R}{5}\right)}$$

Solving for the resistance R, we find that

$$R = \frac{5V^2}{7P} = \frac{5(45 \text{ V})^2}{7(58 \text{ W})} = \boxed{25 \text{ }\Omega}$$

---

119. **REASONING**

a. The greatest voltage for the battery is the voltage that generates the maximum current I that the circuit can tolerate. Once this maximum current is known, the voltage can be calculated according to Ohm's law, as the current times the equivalent circuit resistance for the three resistors in series. To determine the maximum current we note that the power P dissipated in each resistance R is $P = I^2 R$ according to Equation 20.6b. Since the power rating and resistance are known for each resistor, the maximum current that can be tolerated by a resistor is $I = \sqrt{P/R}$. By examining this maximum current for each resistor, we will be able to identify the maximum current that the circuit can tolerate.

b. The battery delivers power to the circuit that is given by the battery voltage times the current, according to Equation 20.6a.

SOLUTION

a. Solving Equation 20.6b for the current, we find that the maximum current for each resistor is as follows:

$$I = \sqrt{\frac{P}{R}} = \sqrt{\frac{4.0 \text{ W}}{2.0 \text{ }\Omega}} = 1.4 \text{ A} \qquad I = \sqrt{\frac{10.0 \text{ W}}{12.0 \text{ }\Omega}} = 0.913 \text{ A} \qquad I = \sqrt{\frac{5.0 \text{ W}}{3.0 \text{ }\Omega}} = 1.3 \text{ A}$$
$$\underbrace{\qquad\qquad\qquad}_{2.0\text{-}\Omega \text{ resistor}} \qquad \underbrace{\qquad\qquad}_{12.0\text{-}\Omega \text{ resistor}} \qquad \underbrace{\qquad\qquad}_{3.0\text{-}\Omega \text{ resistor}}$$

The smallest of these three values is 0.913 A and is the maximum current that the circuit can tolerate. Since the resistors are connected in series, the equivalent resistance of the circuit is

$$R_S = 2.0 \text{ }\Omega + 12.0 \text{ }\Omega + 3.0 \text{ }\Omega = 17.0 \text{ }\Omega$$

Using Ohm's law with this equivalent resistance and the maximum current of 0.913 A reveals that the maximum battery voltage is

$$V = IR_S = (0.913 \text{ A})(17.0 \text{ }\Omega) = \boxed{15.5 \text{ V}}$$

b. The power delivered by the battery in part (a) is given by Equation 20.6a as

$$P = IV = (0.913 \text{ A})(15.5 \text{ V}) = \boxed{14.2 \text{ W}}$$

---

121. **REASONING** We will ignore any changes in length due to thermal expansion. Although the resistance of each section changes with temperature, the total resistance of the composite does not change with temperature. Therefore,

$$\underbrace{\left(R_{\text{tungsten}}\right)_0 + \left(R_{\text{carbon}}\right)_0}_{\text{At room temperature}} = \underbrace{R_{\text{tungsten}} + R_{\text{carbon}}}_{\text{At temperature } T}$$

From Equation 20.5, we know that the temperature dependence of the resistance for a wire of resistance R_0 at temperature T_0 is given by $R = R_0[1 + \alpha(T - T_0)]$, where α is the temperature coefficient of resistivity. Thus,

$$\left(R_{\text{tungsten}}\right)_0 + \left(R_{\text{carbon}}\right)_0 = \left(R_{\text{tungsten}}\right)_0 (1 + \alpha_{\text{tungsten}} \Delta T) + \left(R_{\text{carbon}}\right)_0 (1 + \alpha_{\text{carbon}} \Delta T)$$

Since ΔT is the same for each wire, this simplifies to

$$\left(R_{\text{tungsten}}\right)_0 \alpha_{\text{tungsten}} = -\left(R_{\text{carbon}}\right)_0 \alpha_{\text{carbon}} \tag{1}$$

This expression can be used to find the ratio of the resistances. Once this ratio is known, we can find the ratio of the lengths of the sections with the aid of Equation 20.3 ($L = RA/\rho$).

SOLUTION From Equation (1), the ratio of the resistances of the two sections of the wire is

$$\frac{\left(R_{\text{tungsten}}\right)_0}{\left(R_{\text{carbon}}\right)_0} = -\frac{\alpha_{\text{carbon}}}{\alpha_{\text{tungsten}}} = -\frac{-0.0005\,[(C°)^{-1}]}{0.0045\,[(C°)^{-1}]} = \frac{1}{9}$$

Thus, using Equation 20.3, we find the ratio of the tungsten and carbon lengths to be

$$\frac{L_{\text{tungsten}}}{L_{\text{carbon}}} = \frac{\left(R_0 A / \rho\right)_{\text{tungsten}}}{\left(R_0 A / \rho\right)_{\text{carbon}}} = \frac{\left(R_{\text{tungsten}}\right)_0}{\left(R_{\text{carbon}}\right)_0}\left(\frac{\rho_{\text{carbon}}}{\rho_{\text{tungsten}}}\right) = \left(\frac{1}{9}\right)\left(\frac{3.5\times10^{-5}\,\Omega\cdot\text{m}}{5.6\times10^{-8}\,\Omega\cdot\text{m}}\right) = \boxed{70}$$

where we have used resistivity values from Table 20.1 and the fact that the two sections have the same cross-sectional areas.

---

123. **REASONING** When two or more capacitors are in series, the equivalent capacitance of the combination can be obtained from Equation 20.19, $\frac{1}{C_s} = \frac{1}{C_1} + \frac{1}{C_2} + \frac{1}{C_3}\ldots$. Equation 20.18 gives the equivalent capacitance for two or more capacitors in parallel: $C_p = C_1 + C_2 + C_3 + \ldots$. The energy stored in a capacitor is given by $\frac{1}{2}CV^2$, according to Equation 19.11. Thus, the energy stored in the series combination is $\frac{1}{2}C_s V_s^2$, where

$$\frac{1}{C_s} = \frac{1}{7.0\,\mu\text{F}} + \frac{1}{3.0\,\mu\text{F}} = 0.476\,(\mu\text{F})^{-1} \quad \text{or} \quad C_s = \frac{1}{0.476\,(\mu\text{F})^{-1}} = 2.10\,\mu\text{F}$$

Similarly, the energy stored in the parallel combination is $\frac{1}{2}C_p V_p^2$ where

$$C_p = 7.0\,\mu\text{F} + 3.0\,\mu\text{F} = 10.0\,\mu\text{F}$$

The voltage required to charge the parallel combination of the two capacitors to the same total energy as the series combination can be found by equating the two energy expressions and solving for V_p.

SOLUTION Equating the two expressions for the energy, we have

$$\tfrac{1}{2}C_s V_s^2 = \tfrac{1}{2}C_p V_p^2$$

Solving for V_p, we obtain the result

$$V_p = V_s \sqrt{\frac{C_s}{C_p}} = (24\text{ V})\sqrt{\frac{2.10\,\mu\text{F}}{10.0\,\mu\text{F}}} = \boxed{11\text{V}}$$

CHAPTER 21 | *MAGNETIC FORCES AND MAGNETIC FIELDS*

1. **REASONING AND SOLUTION** The magnitude of the force can be determined using Equation 21.1, $F = |q| vB \sin \theta$, where θ is the angle between the velocity and the magnetic field. The direction of the force is determined by using Right-Hand Rule No. 1.

a. $F = |q| vB \sin 30.0° = (8.4 \times 10^{-6} \text{ C})(45 \text{ m/s})(0.30 \text{ T}) \sin 30.0° = \boxed{5.7 \times 10^{-5} \text{ N}}$, directed $\boxed{\text{into the paper}}$.

b. $F = |q| vB \sin 90.0° = (8.4 \times 10^{-6} \text{ C})(45 \text{ m/s})(0.30 \text{ T}) \sin 90.0° = \boxed{1.1 \times 10^{-4} \text{ N}}$, directed $\boxed{\text{into the paper}}$.

c. $F = |q| vB \sin 150° = (8.4 \times 10^{-6} \text{ C})(45 \text{ m/s})(0.30 \text{ T}) \sin 150° = \boxed{5.7 \times 10^{-5} \text{ N}}$, directed $\boxed{\text{into the paper}}$.

---

3. **REASONING** According to Equation 21.1, the magnitude of the magnetic force on a moving charge is $F = |q_0| vB \sin \theta$. Since the magnetic field points due north and the proton moves eastward, $\theta = 90.0°$. Furthermore, since the magnetic force on the moving proton balances its weight, we have $mg = |q_0| vB \sin \theta$, where m is the mass of the proton. This expression can be solved for the speed v.

SOLUTION Solving for the speed v, we have

$$v = \frac{mg}{|q_0| B \sin \theta} = \frac{(1.67 \times 10^{-27} \text{ kg})(9.80 \text{ m/s}^2)}{(1.6 \times 10^{-19} \text{ C})(2.5 \times 10^{-5} \text{ T}) \sin 90.0°} = \boxed{4.1 \times 10^{-3} \text{ m/s}}$$

---

11. **REASONING** The direction in which the electrons are deflected can be determined using Right-Hand Rule No. 1 and reversing the direction of the force (RHR-1 applies to positive charges, and electrons are negatively charged).

Each electron experiences an acceleration a given by Newton's second law of motion, $a = F/m$, where F is the net force and m is the mass of the electron. The only force acting on the electron is the magnetic force, $F = |q_0| vB \sin \theta$, so it is the net force. The speed v of the electron is related to its kinetic energy KE by the relation $KE = \frac{1}{2}mv^2$. Thus, we have enough information to find the acceleration.

SOLUTION

a. According to RHR-1, if you extend your right hand so that your fingers point along the direction of the magnetic field **B** and your thumb points in the direction of the velocity **v** of a *positive* charge, your palm will face in the direction of the force **F** on the positive charge.

For the electron in question, the fingers of the right hand should be oriented downward (direction of **B**) with the thumb pointing to the east (direction of **v**). The palm of the right hand points due north (the direction of **F** on a positive charge). Since the electron is negatively charged, it will be deflected $\boxed{\text{due south}}$.

b. The acceleration of an electron is given by Newton's second law, where the net force is the magnetic force. Thus,

$$a = \frac{F}{m} = \frac{|q_0| vB \sin \theta}{m}$$

Since the kinetic energy is $KE = \frac{1}{2}mv^2$, the speed of the electron is $v = \sqrt{2(KE)/m}$. Thus, the acceleration of the electron is

$$a = \frac{|q_0| vB \sin \theta}{m} = \frac{|q_0| \sqrt{\dfrac{2(KE)}{m}} \, B \sin \theta}{m}$$

$$= \frac{\left(1.60 \times 10^{-19} \text{ C}\right) \sqrt{\dfrac{2\left(2.40 \times 10^{-15} \text{ J}\right)}{9.11 \times 10^{-31} \text{ kg}}} \left(2.00 \times 10^{-5} \text{ T}\right) \sin 90.0°}{9.11 \times 10^{-31} \text{ kg}} = \boxed{2.55 \times 10^{14} \text{ m/s}^2}$$

---

13. REASONING AND SOLUTION

a. The speed of a proton can be found from Equation 21.2 $\left(r = \dfrac{mv}{|q|B} \right)$,

$$v = \frac{|q| Br}{m} = \frac{(1.6 \times 10^{-19} \text{ C})(0.30 \text{ T})(0.25 \text{ m})}{1.67 \times 10^{-27} \text{ kg}} = \boxed{7.2 \times 10^6 \text{ m/s}}$$

b. The magnitude F_c of the centripetal force is given by Equation 5.3,

$$F_c = \frac{mv^2}{r} = \frac{(1.67 \times 10^{-27} \text{ kg})(7.2 \times 10^6 \text{ m/s})^2}{0.25 \text{ m}} = \boxed{3.5 \times 10^{-13} \text{ N}}$$

---

17. **REASONING AND SOLUTION** The radius of the circular path of a charged particle in a magnetic field is given by Equation 21.2 $\left(r = \dfrac{mv}{|q| B} \right)$.

a. For an electron

$$r = \frac{mv}{|q| B} = \frac{\left(9.11 \times 10^{-31} \text{ kg}\right)\left(9.0 \times 10^6 \text{ m/s}\right)}{\left(1.6 \times 10^{-19} \text{ C}\right)\left(1.2 \times 10^{-7} \text{ T}\right)} = \boxed{4.3 \times 10^2 \text{ m}}$$

b. For a proton, only the mass changes in the calculation above. Using $m = 1.67 \times 10^{-27}$ kg, we obtain $r = \boxed{7.8 \times 10^5 \text{ m}}$.

---

23. **REASONING** When the proton moves in the magnetic field, its trajectory is a circular path. The proton will just miss the opposite plate if the distance between the plates is equal to the radius of the path. The radius is given by Equation 21.2 as $r = mv / \left(|q| B\right)$. This relation can be used to find the magnitude B of the magnetic field, since values for all the other variables are known.

SOLUTION Solving the relation $r = mv / \left(|q| B\right)$ for the magnitude of the magnetic field, and realizing that the radius is equal to the plate separation, we find that

$$B = \frac{mv}{|q| r} = \frac{\left(1.67 \times 10^{-27} \text{ kg}\right)\left(3.5 \times 10^6 \text{ m/s}\right)}{\left(1.60 \times 10^{-19} \text{ C}\right)(0.23 \text{ m})} = \boxed{0.16 \text{ T}}$$

The values for the mass and the magnitude of the charge (which is the same as that of the electron) have been taken from the inside of the front cover.

---

25. **REASONING** From the discussion in Section 21.3, we know that when a charged particle moves perpendicular to a magnetic field, the trajectory of the particle is a circle. The drawing at the right shows a particle moving in the plane of the paper (the magnetic field is perpendicular to the paper). If the particle is moving initially through the coordinate origin and to the right (along the +*x* axis), the subsequent circular path of the particle will intersect the y axis at the greatest possible value, which is equal to twice the radius *r* of the circle.

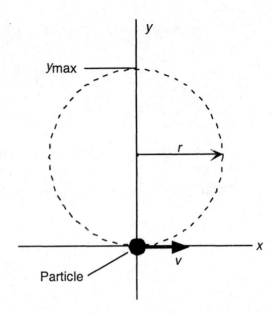

SOLUTION

a. From the drawing above, it can be seen that the largest value of y is equal to the diameter (2*r*) of the circle. When the particle passes through the coordinate origin its velocity must be parallel to the +*x* axis. Thus, the angle is $\boxed{\theta = 0°}$.

b. The maximum value of y is twice the radius *r* of the circle. According to Equation 21.2, the radius of the circular path is $r = mv/\left(\left|q\right|B\right)$. The maximum value y_{max} is, therefore,

$$y_{max} = 2r = 2\left(\frac{mv}{\left|q\right|B}\right) = 2\left[\frac{\left(3.8\times10^{-8}\ \text{kg}\right)\left(44\ \text{m/s}\right)}{\left(7.3\times10^{-6}\ \text{C}\right)\left(1.6\ \text{T}\right)}\right] = \boxed{0.29\ \text{m}}$$

---

31. **REASONING** The magnitude *F* of the magnetic force experienced by the wire is given by $F = ILB\sin\theta$ (Equation 21.3), where *I* is the current, *L* is the length of the wire, *B* is the magnitude of the earth's magnetic field, and θ is the angle between the direction of the current and the magnetic field. Since all the variables are known except *B*, we can use this relation to find its value.

SOLUTION Solving $F = ILB\sin\theta$ for the magnitude of the magnetic field, we have

$$B = \frac{F}{IL\sin\theta} = \frac{0.15\ \text{N}}{\left(75\ \text{A}\right)\left(45\ \text{m}\right)\sin 60.0°} = \boxed{5.1 \times 10^{-5}\ \text{T}}$$

35. **REASONING** According to Equation 21.3, the magnetic force has a magnitude of $F = ILB \sin \theta$, where I is the current, B is the magnitude of the magnetic field, L is the length of the wire, and $\theta = 90°$ is the angle of the wire with respect to the field.

SOLUTION Using Equation 21.3, we find that

$$L = \frac{F}{IB \sin\theta} = \frac{7.1 \times 10^{-5} \text{ N}}{(0.66 \text{ A})(4.7 \times 10^{-5} \text{ T}) \sin 58°} = \boxed{2.7 \text{ m}}$$

---

39. **REASONING** Since the rod does not rotate about the axis at P, the net torque relative to that axis must be zero; $\Sigma \tau = 0$ (Equation 9.2). There are two torques that must be considered, one due to the magnetic force and another due to the weight of the rod. We consider both of these to act at the rod's center of gravity, which is at the geometrical center of the rod (length $= L$), because the rod is uniform. According to Right-Hand Rule No. 1, the magnetic force acts perpendicular to the rod and is directed up and to the left in the drawing. Therefore, the magnetic torque is a counterclockwise (positive) torque. Equation 21.3 gives the magnitude F of the magnetic force as $F = ILB \sin 90.0°$, since the current is perpendicular to the magnetic field. The weight is mg and acts downward, producing a clockwise (negative) torque. The magnitude of each torque is the magnitude of the force times the lever arm (Equation 9.1). Thus, we have for the torques:

$$\tau_{\text{magnetic}} = +\underbrace{(ILB)}_{\text{force}} \underbrace{(L/2)}_{\text{lever arm}} \quad \text{and} \quad \tau_{\text{weight}} = -\underbrace{(mg)}_{\text{force}} \underbrace{\left[(L/2)\cos\theta\right]}_{\text{lever arm}}$$

Setting the sum of these torques equal to zero will enable us to find the angle θ that the rod makes with the ground.

SOLUTION Setting the sum of the torques equal to zero gives $\Sigma \tau = \tau_{\text{magnetic}} + \tau_{\text{weight}} = 0$, and we have

$$+(ILB)(L/2) - (mg)\left[(L/2)\cos\theta\right] = 0 \quad \text{or} \quad \cos\theta = \frac{ILB}{mg}$$

$$\theta = \cos^{-1}\left[\frac{(4.1 \text{ A})(0.45 \text{ m})(0.36 \text{ T})}{(0.094 \text{ kg})(9.80 \text{ m/s}^2)}\right] = \boxed{44°}$$

47. **REASONING** The torque on the loop is given by Equation 21.4, $\tau = NIAB\sin\phi$. From the drawing in the text, we see that the angle ϕ between the normal to the plane of the loop and the magnetic field is $90° - 35° = 55°$. The area of the loop is 0.70 m $\times$ 0.50 m $= 0.35$ m².

SOLUTION
a. The magnitude of the net torque exerted on the loop is

$$\tau = NIAB\sin\phi = (75)(4.4 \text{ A})(0.35 \text{ m}^2)(1.8 \text{ T})\sin 55° = \boxed{170 \text{ N}\cdot\text{m}}$$

b. As discussed in the text, when a current-carrying loop is placed in a magnetic field, the loop tends to rotate such that its normal becomes aligned with the magnetic field. The normal to the loop makes an angle of 55° with respect to the magnetic field. Since this angle decreases as the loop rotates, the $\boxed{35° \text{ angle increases}}$.

---

53. **REASONING AND SOLUTION**
a. In Figure 21.26a the magnetic field that exists at the location of each wire points upward. Since the current in each wire is the same, the fields at the locations of the wires also have the same magnitudes. Therefore, a single external field that points $\boxed{\text{downward}}$ will cancel the mutual repulsion of the wires, if this external field has a magnitude that equals that of the field produced by either wire.

b. Equation 21.5 gives the magnitude of the field produced by a long straight wire. The external field must have this magnitude:

$$B = \frac{\mu_0 I}{2\pi r} = \frac{\left(4\pi \times 10^{-7} \text{ T}\cdot\text{m/A}\right)(25 \text{ A})}{2\pi(0.016 \text{ m})} = \boxed{3.1 \times 10^{-4} \text{ T}}$$

---

55. **REASONING** The magnitude B of the magnetic field in the interior of a solenoid that has a length much greater than its diameter is given by $B = \mu_0 nI$ (Equation 21.7), where $\mu_0 = 4\pi \times 10^{-7}$ T·m/A is the permeability of free space, n is the number of turns per meter of the solenoid's length, and I is the current in the wire of the solenoid. Since B and I are given, we can solve Equation 21.7 for n.

SOLUTION Solving Equation 21.7 for n, we find that the number of turns per meter of length is

$$n = \frac{B}{\mu_0 I} = \frac{7.0 \text{ T}}{\left(4\pi \times 10^{-7} \text{ T}\cdot\text{m/A}\right)\left(2.0 \times 10^2 \text{ A}\right)} = \boxed{2.8 \times 10^4 \text{ turns/m}}$$

57. **REASONING** The magnitude of the magnetic field at the center of a circular loop of current is given by Equation 21.6 as $B = N\mu_0 I/(2R)$, where N is the number of turns, μ_0 is the permeability of free space, I is the current, and R is the radius of the loop. The field is perpendicular to the plane of the loop. Magnetic fields are vectors, and here we have two fields, each perpendicular to the plane of the loop producing it. Therefore, the two field vectors are perpendicular, and we must add them as vectors to get the net field. Since they are perpendicular, we can use the Pythagorean theorem to calculate the magnitude of the net field.

SOLUTION Using Equation 21.6 and the Pythagorean theorem, we find that the magnitude of the net magnetic field at the common center of the two loops is

$$B_{net} = \sqrt{\left(\frac{N\mu_0 I}{2R}\right)^2 + \left(\frac{N\mu_0 I}{2R}\right)^2} = \sqrt{2}\left(\frac{N\mu_0 I}{2R}\right)$$

$$= \frac{\sqrt{2}(1)(4\pi \times 10^{-7} \text{ T·m/A})(1.7 \text{ A})}{2(0.040 \text{ m})} = \boxed{3.8 \times 10^{-5} \text{ T}}$$

---

65. **REASONING** According to Equation 21.6 the magnetic field at the center of a circular, current-carrying loop of N turns and radius r is $B = N\mu_0 I/(2r)$. The number of turns N in the coil can be found by dividing the total length L of the wire by the circumference after it has been wound into a circle. The current in the wire can be found by using Ohm's law, $I = V/R$.

SOLUTION The number of turns in the wire is

$$N = \frac{L}{2\pi r}$$

The current in the wire is

$$I = \frac{V}{R} = \frac{12.0 \text{ V}}{(5.90 \times 10^{-3} \text{ }\Omega/\text{m})L} = \frac{2.03 \times 10^3}{L} \text{ A}$$

Therefore, the magnetic field at the center of the coil is

$$B = N\left(\frac{\mu_0 I}{2r}\right) = \left(\frac{L}{2\pi r}\right)\left(\frac{\mu_0 I}{2r}\right) = \frac{\mu_0 L I}{4\pi r^2}$$

$$= \frac{(4\pi \times 10^{-7} \text{ T·m/A})L\left(\frac{2.03 \times 10^3}{L} \text{ A}\right)}{4\pi(0.140 \text{ m})^2} = \boxed{1.04 \times 10^{-2} \text{ T}}$$

69. **REASONING AND SOLUTION** Ampère's law in the form of Equation 21.8 indicates that $\Sigma B_{\parallel} \Delta \ell = \mu_0 I$. Since the magnetic field is everywhere perpendicular to the plane of the paper, it is everywhere perpendicular to the circular path and has no component $B_{\parallel}$ that is parallel to the circular path. Therefore, Ampère's law reduces to $\Sigma B_{\parallel} \Delta \ell = 0 = \mu_0 I$, so that

$$\boxed{\text{the net current passing through the circular surface is zero}}.$$

75. **REASONING** The magnitude τ of the torque that acts on a current-carrying coil placed in a magnetic field is specified by $\tau = NIAB \sin \phi$ (Equation 21.4), where N is the number of loops in the coil, I is the current, A is the area of one loop, B is the magnitude of the magnetic field, and ϕ is the angle between the normal to the coil and the magnetic field. All the variables in this relation are known except for the current, which can, therefore, be obtained.

SOLUTION Solving the equation $\tau = NIAB \sin \phi$ for the current I and noting that $\phi = 90.0°$ since τ is specified to be the maximum torque, we have

$$I = \frac{\tau}{NAB \sin \phi} = \frac{5.8 \text{ N} \cdot \text{m}}{(1200)(1.1 \times 10^{-2} \text{ m}^2)(0.20 \text{ T}) \sin 90.0°} = \boxed{2.2 \text{ A}}$$

77. **REASONING AND SOLUTION** The current associated with the lightning bolt is

$$I = \frac{\Delta q}{\Delta t} = \frac{15 \text{ C}}{1.5 \times 10^{-3} \text{ s}} = 1.0 \times 10^4 \text{ A}$$

The magnetic field near this current is given by

$$B = \frac{\mu_0 I}{2 \pi r} = \frac{(4\pi \times 10^{-7} \text{ T} \cdot \text{m/A})(1.0 \times 10^4 \text{ A})}{2\pi (25 \text{ m})} = \boxed{8.0 \times 10^{-5} \text{ T}}$$

81. **REASONING** The particle travels in a semicircular path of radius r, where r is given by Equation 21.2 $\left(r = \frac{mv}{|q|B} \right)$. The time spent by the particle in the magnetic field is given by $t = s/v$, where s is the distance traveled by the particle and v is its speed. The distance s is equal to one-half the circumference of a circle ($s = \pi r$).

SOLUTION We find that

$$t = \frac{s}{v} = \frac{\pi r}{v} = \frac{\pi}{v}\left(\frac{mv}{|q|B}\right) = \frac{\pi m}{|q|B} = \frac{\pi(6.0\times10^{-8}\text{ kg})}{(7.2\times10^{-6}\text{ C})(3.0\text{ T})} = \boxed{8.7\times10^{-3}\text{ s}}$$

---

85. **REASONING**   The magnetic moment of the orbiting electron can be found from the expression *Magnetic moment = NIA*. For this situation, $N = 1$. Thus, we need to find the current and the area for the orbiting charge.

SOLUTION   The current for the orbiting charge is, by definition (see Equation 20.1), $I = \Delta q / \Delta t$, where Δq is the amount of charge that passes a given point during a time interval Δt. Since the charge ($\Delta q = e$) passes by a given point once per revolution, we can find the current by dividing the total orbiting charge by the period T of revolution.

$$I = \frac{\Delta q}{T} = \frac{\Delta q}{2\pi r / v} = \frac{(1.6\times10^{-19}\text{ C})(2.2\times10^{6}\text{ m/s})}{2\pi(5.3\times10^{-11}\text{ m})} = 1.06\times10^{-3}\text{ A}$$

The area of the orbiting charge is

$$A = \pi r^2 = \pi(5.3\times10^{-11}\text{ m})^2 = 8.82\times10^{-21}\text{ m}^2$$

Therefore, the magnetic moment is

$$\textit{Magnetic moment} = NIA = (1)(1.06\times10^{-3}\text{ A})(8.82\times10^{-21}\text{ m}^2) = \boxed{9.3\times10^{-24}\text{ A}\cdot\text{m}^2}$$

5. **REASONING AND SOLUTION** For the three rods in the drawing in the text, we have the following:

Rod A: The motional emf is $\boxed{\text{zero}}$, because the velocity of the rod is parallel to the direction of the magnetic field, and the charges do not experience a magnetic force.

Rod B: The motional emf ξ is, according to Equation 22.1,

$$\xi = vBL = (2.7 \text{ m/s})(0.45 \text{ T})(1.3 \text{ m}) = \boxed{1.6 \text{ V}}$$

The positive end of Rod B is $\boxed{\text{end 2}}$.

Rod C: The motional emf is $\boxed{\text{zero}}$, because the magnetic force F on each charge is directed perpendicular to the length of the rod. For the ends of the rod to become charged, the magnetic force must be directed parallel to the length of the rod.

___

9. **REASONING** The minimum length d of the rails is the speed v of the rod times the time t, or $d = vt$. We can obtain the speed from the expression for the motional emf given in Equation 22.1. Solving this equation for the speed gives $v = \dfrac{\xi}{BL}$, where ξ is the motional emf, B is the magnitude of the magnetic field, and L is the length of the rod. Thus, the length of the rails is $d = vt = \left(\dfrac{\xi}{BL}\right)t$. While we have no value for the motional emf, we do know that the bulb dissipates a power of $P = 60.0$ W, and has a resistance of $R = 240\ \Omega$. Power is related to the emf and the resistance according to $P = \dfrac{\xi^2}{R}$ (Equation 20.6c), which can be solved to show that $\xi = \sqrt{PR}$. Substituting this expression into the equation for d gives

$$d = \left(\frac{\xi}{BL}\right)t = \left(\frac{\sqrt{PR}}{BL}\right)t$$

SOLUTION Using the above expression for the minimum necessary length of the rails, we find that

$$d = \left(\frac{\sqrt{PR}}{BL} \right) t = \left[\frac{\sqrt{(60.0 \text{ W})(240 \text{ }\Omega)}}{(0.40 \text{ T})(0.60 \text{ m})} \right] (0.50 \text{ s}) = \boxed{250 \text{ m}}$$

---

13. **REASONING** The general expression for the magnetic flux through an area A is given by Equation 22.2: $\Phi = BA\cos\phi$, where B is the magnitude of the magnetic field and ϕ is the angle of inclination of the magnetic field **B** with respect to the normal to the area.

The magnetic flux through the door is a maximum when the magnetic field lines are perpendicular to the door and $\phi_1 = 0.0°$ so that $\Phi_1 = \Phi_{max} = BA(\cos 0.0°) = BA$.

SOLUTION When the door rotates through an angle ϕ_2, the magnetic flux that passes through the door decreases from its maximum value to one-third of its maximum value. Therefore, $\Phi_2 = \frac{1}{3}\Phi_{max}$, and we have

$$\Phi_2 = BA\cos\phi_2 = \frac{1}{3}BA \qquad \text{or} \qquad \cos\phi_2 = \frac{1}{3} \qquad \text{or} \qquad \phi_2 = \cos^{-1}\left(\frac{1}{3}\right) = \boxed{70.5°}$$

---

17. **REASONING** The general expression for the magnetic flux through an area A is given by Equation 22.2: $\Phi = BA\cos\phi$, where B is the magnitude of the magnetic field and ϕ is the angle of inclination of the magnetic field **B** with respect to the normal to the surface.

SOLUTION Since the magnetic field **B** is parallel to the surface for the triangular ends and the bottom surface, the flux through each of these three surfaces is $\boxed{0 \text{ Wb}}$.

The flux through the 1.2 m by 0.30 m face is

$$\Phi = BA\cos\phi = (0.25 \text{ T})(1.2 \text{ m})(0.30 \text{ m})\cos 0.0° = \boxed{0.090 \text{ Wb}}$$

For the 1.2 m by 0.50 m side, the area makes an angle ϕ with the magnetic field **B**, where

$$\phi = 90° - \tan^{-1}\left(\frac{0.30 \text{ m}}{0.40 \text{ m}}\right) = 53°$$

Therefore,

$$\Phi = BA\cos\phi = (0.25 \text{ T})(1.2 \text{ m})(0.50 \text{ m})\cos 53° = \boxed{0.090 \text{ Wb}}$$

21. **REASONING** According to Equation 22.3, the average emf induced in a coil of N loops is $\xi = -N\Delta\Phi/\Delta t$.

SOLUTION For the circular coil in question, the flux through a single turn changes by

$$\Delta\Phi = BA\cos 45° - BA\cos 90° = BA\cos 45°$$

during the interval of $\Delta t = 0.010$ s. Therefore, for N turns, Faraday's law gives the magnitude of the emf as

$$|\xi| = \left| -N \frac{BA\cos 45°}{\Delta t} \right|$$

Since the loops are circular, the area A of each loop is equal to πr^2. Solving for B, we have

$$B = \frac{|\xi|\Delta t}{N\pi r^2 \cos 45°} = \frac{(0.065 \text{ V})(0.010 \text{ s})}{(950)\pi(0.060 \text{ m})^2 \cos 45°} = \boxed{8.6\times 10^{-5} \text{ T}}$$

---

27. **REASONING** The energy dissipated in the resistance is given by Equation 6.10b as the power P dissipated multiplied by the time t, Energy $= Pt$. The power, according to Equation 20.6c, is the square of the induced emf ξ divided by the resistance R, $P = \xi^2/R$. The induced emf can be determined from Faraday's law of electromagnetic induction, Equation 22.3.

SOLUTION Expressing the energy consumed as Energy $= Pt$, and substituting in $P = \xi^2/R$, we find

$$\text{Energy} = Pt = \frac{\xi^2 t}{R}$$

The induced emf is given by Faraday's law as $\xi = -N\left(\Delta\Phi/\Delta t\right)$, and the resistance R is equal to the resistance per unit length $(3.3 \times 10^{-2} \ \Omega/\text{m})$ times the length of the circumference of the loop, $2\pi r$. Thus, the energy dissipated is

$$\text{Energy} = \frac{\left(-N\frac{\Delta\Phi}{\Delta t}\right)^2 t}{\left(3.3\times10^{-2}\ \Omega/\text{m}\right)2\pi r} = \frac{\left[-N\left(\frac{\Phi-\Phi_0}{t-t_0}\right)\right]^2 t}{\left(3.3\times10^{-2}\ \Omega/\text{m}\right)2\pi r}$$

$$= \frac{\left[-N\left(\frac{BA\cos\phi - B_0 A\cos\phi}{t-t_0}\right)\right]^2 t}{\left(3.3\times10^{-2}\ \Omega/\text{m}\right)2\pi r} = \frac{\left[-NA\cos\phi\left(\frac{B-B_0}{t-t_0}\right)\right]^2 t}{\left(3.3\times10^{-2}\ \Omega/\text{m}\right)2\pi r}$$

$$= \frac{\left[-(1)\pi(0.12\ \text{m})^2(\cos 0°)\left(\frac{0.60\ \text{T}-0\ \text{T}}{0.45\ \text{s}-0\ \text{s}}\right)\right]^2 (0.45\ \text{s})}{\left(3.3\times10^{-2}\ \Omega/\text{m}\right)2\pi(0.12\ \text{m})} = \boxed{6.6\times10^{-2}\ \text{J}}$$

---

31. ***REASONING*** According to Ohm's law, the average current I induced in the coil is given by $I = |\xi|/R$, where $|\xi|$ is the magnitude of the induced emf and R is the resistance of the coil. To find the induced emf, we use Faraday's law of electromagnetic induction

SOLUTION The magnitude of the induced emf can be found from Faraday's law of electromagnetic induction and is given by Equation 22.3:

$$|\xi| = \left|-N\frac{\Delta\Phi}{\Delta t}\right| = \left|-N\frac{\Delta(BA\cos 0°)}{\Delta t}\right| = NA\frac{\Delta B}{\Delta t}$$

We have used the fact that the field within a long solenoid is perpendicular to the cross-sectional area A of the solenoid and makes an angle of $0°$ with respect to the normal to the area. The field is given by Equation 21.7 as $B = \mu_0 nI$, so the change ΔB in the field is $\Delta B = \mu_0 n\Delta I$, where ΔI is the change in the current. The induced current is, then,

$$I = \frac{NA\frac{\Delta B}{\Delta t}}{R} = \frac{NA\frac{\mu_0 n\Delta I}{\Delta t}}{R}$$

$$= \frac{(10)(6.0\times10^{-4}\ \text{m}^2)\dfrac{(4\pi\times10^{-7}\ \text{T}\cdot\text{m/A})(400\ \text{turns/m})(0.40\ \text{A})}{(0.050\ \text{s})}}{1.5\ \Omega} = \boxed{1.6\times10^{-5}\ \text{A}}$$

33. ***REASONING AND SOLUTION***   If the north and south poles of the magnet are interchanged, the currents in the ammeter would simply be reversed in direction. This follows from Lenz's law. As the south pole approaches the coil in Figure 22.1*b*, the field that the coil experiences is becoming stronger and points toward the south pole. To oppose the increasing flux from this field (as specified by Lenz's law) the current induced in the coil must produce a field that points away from the approaching south pole. RHR-2 indicates that to do this, the induced current must flow into the ammeter on the right and out of the ammeter on the left. Another way to look at things is to realize that, with this direction for the induced current, the coil becomes an electromagnet with a south pole located at its left side. This induced south pole opposes the motion of the approaching south pole of the moving magnet.

Therefore, in Figure 22.1*b*, the current will flow ┃ right to left ┃ through the ammeter.

Similar reasoning (with the poles of the magnet interchanged in Figure 22.1) leads to the conclusion that in Figure 22.1*c* the induced current must flow into the ammeter on the left and out of the ammeter on the right.

Thus, in Figure 22.1*c*, the current will flow ┃ left to right ┃ through the ammeter.

---

35. ***REASONING*** In solving this problem, we apply Lenz's law, which essentially says that the change in magnetic flux must be opposed by the induced magnetic field.

SOLUTION
a. The magnetic field due to the wire in the vicinity of position 1 is directed out of the paper. The coil is moving closer to the wire into a region of higher magnetic field, so the flux through the coil is increasing. Lenz's law demands that the induced field counteract this increase. The direction of the induced field, therefore, must be into the paper. The current in the coil must be ┃ clockwise ┃.

b. At position 2 the magnetic field is directed into the paper and is decreasing as the coil moves away from the wire. The induced magnetic field, therefore, must be directed into the paper, so the current in the coil must be ┃ clockwise ┃.

---

41. ***REASONING AND SOLUTION***
a. From the drawing, we see that the period of the generator (the time for one full cycle) is 0.42 s; therefore, the frequency of the generator is

$$f = \frac{1}{T} = \frac{1}{0.42 \text{ s}} = \boxed{2.4 \text{ Hz}}$$

b. The angular speed of the generator is related to its frequency by $\omega = 2\pi f$, so the angular speed is

$$\omega = 2\pi(2.4 \text{ Hz}) = \boxed{15 \text{ rad/s}}$$

c. The maximum emf ξ_0 induced in a rotating planar coil is given by $\xi_0 = NAB\omega$ (see Equation 22.4). The magnitude of the magnetic field can be found by solving this expression for B and noting from the drawing that $\xi_0 = 28$ V:

$$B = \frac{\xi_0}{NA\omega} = \frac{28 \text{ V}}{(150)(0.020 \text{ m}^2)(15 \text{ rad/s})} = \boxed{0.62 \text{ T}}$$

---

47. **REASONING** The peak emf ξ_0 produced by a generator is related to the number N of turns in the coil, the area A of the coil, the magnitude B of the magnetic field, and the angular speed ω_{coil} of the coil by $\xi_0 = NAB\omega_{\text{coil}}$ (see Equation 22.4). We are given that the angular speed ω_{coil} of the coil is 38 times as great as the angular speed ω_{tire} of the tire. Since the tires roll without slipping, the angular speed of a tire is related to the linear speed v of the bike by $\omega_{\text{tire}} = v/r$ (see Section 8.6), where r is the radius of a tire. The speed of the bike after 5.1 s can be found from its acceleration and the fact that the bike starts from rest.

SOLUTION The peak emf produced by the generator is

$$\xi_0 = NAB\omega_{\text{coil}} \tag{22.4}$$

Since the angular speed of the coil is 38 times as great as the angular speed of the tire, $\omega_{\text{coil}} = 38\,\omega_{\text{tire}}$. Substituting this expression into Equation 22.4 gives $\xi_0 = NAB\omega_{\text{coil}} = NAB(38\omega_{\text{tire}})$. Since the tire rolls without slipping, the angular speed of the tire is related to the linear speed v (the speed at which its axle is moving forward) by $\omega_{\text{tire}} = v/r$ (Equation 8.12), where r is the radius of the tire. Substituting this result into the expression for ξ_0 yields

$$\xi_0 = NAB(38\omega_{\text{tire}}) = NAB\left(38\frac{v}{r}\right) \tag{1}$$

The velocity of the car is given by $v = v_0 + at$ (Equation 2.4), where v_0 is the initial velocity, a is the acceleration and t is the time. Substituting this relation into Equation (1), and noting that $v_0 = 0$ m/s since the bike starts from rest, we find that

$$\xi_0 = NAB\left(38\frac{v}{r}\right) = NAB\left[38\frac{(v_0 + at)}{r}\right]$$

$$= (125)(3.86\times10^{-3} \text{ m}^2)(0.0900 \text{ T})(38)\left[\frac{0 \text{ m/s}+(0.550 \text{ m/s}^2)(5.10 \text{ s})}{0.300 \text{ m}}\right] = \boxed{15.4 \text{ V}}$$

---

49. **REASONING** The energy density is given by Equation 22.11 as

$$\text{Energy density} = \frac{\text{Energy}}{\text{Volume}} = \frac{1}{2\mu_0}B^2$$

The energy stored is the energy density times the volume.

SOLUTION The volume is the area A times the height h. Therefore, the energy stored is

$$\text{Energy} = \frac{B^2 Ah}{2\mu_0} = \frac{(7.0\times10^{-5} \text{ T})^2(5.0\times10^8 \text{ m}^2)(1500 \text{ m})}{2(4\pi\times10^{-7} \text{ T}\cdot\text{m/A})} = \boxed{1.5\times10^9 \text{ J}}$$

---

55. **REASONING AND SOLUTION** From the results of Example 13, the self-inductance L of a long solenoid is given by $L = \mu_0 n^2 A\ell$. Solving for the number of turns n per unit length gives

$$n = \sqrt{\frac{L}{\mu_0 A\ell}} = \sqrt{\frac{1.4\times10^{-3} \text{ H}}{(4\pi\times10^{-7}\text{T}\cdot\text{m/A})(1.2\times10^{-3} \text{ m}^2)(0.052 \text{ m})}} = 4.2\times10^3 \text{ turns/m}$$

Therefore, the total number of turns N is the product of n and the length ℓ of the solenoid:

$$N = n\ell = (4.2\times10^3 \text{ turns/m})(0.052 \text{ m}) = \boxed{220 \text{ turns}}$$

---

63. **REASONING** The ratio $\left(I_s/I_p\right)$ of the current in the secondary coil to that in the primary coil is equal to the ratio $\left(N_p/N_s\right)$ of the number of turns in the primary coil to that in the secondary coil. This relation can be used directly to find the current in the primary coil.

SOLUTION Solving the relation $\left(I_s/I_p\right) = \left(N_p/N_s\right)$ (Equation 22.13) for I_p gives

$$I_p = I_s\left(\frac{N_s}{N_p}\right) = (1.6 \text{ A})\left(\frac{1}{8}\right) = \boxed{0.20 \text{ A}}$$

---

67. ***REASONING*** The power used to heat the wires is given by Equation 20.6b: $P = I^2 R$. Before we can use this equation, however, we must determine the total resistance R of the wire and the current that flows through the wire.

SOLUTION

a. The total resistance of one of the wires is equal to the resistance per unit length multiplied by the length of the wire. Thus, we have

$$(5.0 \times 10^{-2} \ \Omega/\text{km})(7.0 \ \text{km}) = 0.35 \ \Omega$$

and the total resistance of the transmission line is twice this value or $0.70 \ \Omega$. According to Equation 20.6a ($P = IV$), the current flowing into the town is

$$I = \frac{P}{V} = \frac{1.2 \times 10^6 \ \text{W}}{1200 \ \text{V}} = 1.0 \times 10^3 \ \text{A}$$

Thus, the power used to heat the wire is

$$P = I^2 R = \left(1.0 \times 10^3 \ \text{A}\right)^2 (0.70 \ \Omega) = \boxed{7.0 \times 10^5 \ \text{W}}$$

b. According to the transformer equation (Equation 22.12), the stepped-up voltage is

$$V_s = V_p \left(\frac{N_s}{N_p} \right) = (1200 \ \text{V}) \left(\frac{100}{1} \right) = 1.2 \times 10^5 \ \text{V}$$

According to Equation 20.6a ($P = IV$), the current in the wires is

$$I = \frac{P}{V} = \frac{1.2 \times 10^6 \ \text{W}}{1.2 \times 10^5 \ \text{V}} = 1.0 \times 10^1 \ \text{A}$$

The power used to heat the wires is now

$$P = I^2 R = \left(1.0 \times 10^1 \ \text{A}\right)^2 (0.70 \ \Omega) = \boxed{7.0 \times 10^1 \ \text{W}}$$

71. **_REASONING_** We can use the information given in the problem statement to determine the area of the coil A. Since it is square, the length of one side is $\ell = \sqrt{A}$.

SOLUTION According to Equation 22.4, the maximum emf ξ_0 induced in the coil is $\xi_0 = NAB\omega$. Therefore, the length of one side of the coil is

$$\ell = \sqrt{A} = \sqrt{\frac{\xi_0}{NB\omega}} = \sqrt{\frac{75.0 \text{ V}}{(248)(0.170 \text{ T})(79.1 \text{ rad/s})}} = \boxed{0.150 \text{ m}}$$

---

73. **_REASONING_** The current I produces a magnetic field, and hence a magnetic flux, that passes through the loops A and B. Since the current decreases to zero when the switch is opened, the magnetic flux also decreases to zero. According to Lenz's law, the current induced in each coil will have a direction such that the induced magnetic field will oppose the original flux change.

SOLUTION
a. The drawing in the text shows that the magnetic field at coil A is perpendicular to the plane of the coil and points down (when viewed from above the table top). When the switch is opened, the magnetic flux through coil A decreases to zero. According to Lenz's law, the induced magnetic field produced by coil A must oppose this change in flux. Since the magnetic field points down and is decreasing, the induced magnetic field must also point down. According to Right-Hand Rule No. 2 (RHR-2), the induced current must be $\boxed{\text{clockwise}}$ around loop A.

b. The drawing in the text shows that the magnetic field at coil B is perpendicular to the plane of the coil and points up (when viewed from above the table top). When the switch is opened, the magnetic flux through coil B decreases to zero. According to Lenz's law, the induced magnetic field produced by coil B must oppose this change in flux. Since the magnetic field points up and is decreasing, the induced magnetic field must also point up. According to RHR-2, the induced current must be $\boxed{\text{counterclockwise}}$ around loop B.

---

75. **_REASONING AND SOLUTION_** The emf ξ induced in the coil of an ac generator is given by Equation 22.4 as

$$\xi = NAB\omega \sin\omega t = (500)(1.2\times10^{-2} \text{ m}^2)(0.13 \text{ T})(34 \text{ rad/s}) \sin 27° = \boxed{12 \text{ V}}$$

79. ***REASONING*** According to Equation 22.3, the average emf ξ induced in a single loop ($N = 1$) is $\xi = -\Delta\Phi / \Delta t$. Since the magnitude of the magnetic field is changing, the area of the loop remains constant, and the direction of the field is parallel to the normal to the loop, the change in flux through the loop is given by $\Delta\Phi = (\Delta B)A$. Thus the magnitude $|\xi|$ of the induced emf in the loop is given by $|\xi| = |-(\Delta B) A / \Delta t|$.

Similarly, when the area of the loop is changed and the field B has a given value, we find the magnitude of the induced emf to be $|\xi| = |-B(\Delta A)/\Delta t|$.

SOLUTION
a. The magnitude of the induced emf when the field changes in magnitude is

$$|\xi| = \left|-A\left(\frac{\Delta B}{\Delta t}\right)\right| = (0.018 \text{ m}^2)(0.20 \text{ T/s}) = \boxed{3.6 \times 10^{-3} \text{ V}}$$

b. At a particular value of B (when B is changing), the rate at which the area must change can be obtained from

$$|\xi| = \left|-\frac{B\Delta A}{\Delta t}\right| \quad \text{or} \quad \frac{\Delta A}{\Delta t} = \frac{|\xi|}{B} = \frac{3.6 \times 10^{-3} \text{ V}}{1.8 \text{ T}} = \boxed{2.0 \times 10^{-3} \text{ m}^2 / \text{s}}$$

In order for the induced emf to be zero, the magnitude of the magnetic field and the area of the loop must change in such a way that the flux remains constant. Since the magnitude of the magnetic field is increasing, the area of the loop must decrease, if the flux is to remain constant. Therefore, $\boxed{\text{the area of the loop must be shrunk}}$.

CHAPTER 23 | ALTERNATING CURRENT CIRCUITS

1. **REASONING** The voltage across the capacitor reaches its maximum instantaneous value when the generator voltage reaches its maximum instantaneous value. The maximum value of the capacitor voltage first occurs one-fourth of the way, or one-quarter of a period, through a complete cycle (see the voltage curve in Figure 23.4).

   SOLUTION The period of the generator is $T = 1/f = 1/(5.00 \text{ Hz}) = 0.200 \text{ s}$. Therefore, the least amount of time that passes before the instantaneous voltage across the capacitor reaches its maximum value is $\frac{1}{4}T = \frac{1}{4}(0.200 \text{ s}) = \boxed{5.00 \times 10^{-2} \text{ s}}$.

---

3. **REASONING** The rms current in a capacitor is $I_{rms} = V_{rms}/X_C$, according to Equation 23.1. The capacitive reactance is $X_C = 1/(2\pi f C)$, according to Equation 23.2. For the first capacitor, we use $C = C_1$ in these expressions. For the two capacitors in parallel, we use $C = C_P$, where C_P is the equivalent capacitance from Equation 20.18 ($C_P = C_1 + C_2$). Taking the difference between the currents and using the given data, we can obtain the desired value for C_2. The capacitance C_1 is unknown, but it will be eliminated algebraically from the calculation.

   SOLUTION Using Equations 23.1 and 23.2, we find that the current in a capacitor is

   $$ I_{rms} = \frac{V_{rms}}{X_C} = \frac{V_{rms}}{1/(2\pi f C)} = V_{rms} \, 2\pi f C $$

   Applying this result to the first capacitor and the parallel combination of the two capacitors, we obtain

   $$ \underbrace{I_1 = V_{rms} \, 2\pi f C_1}_{\text{Single capacitor}} \quad \text{and} \quad \underbrace{I_{Combination} = V_{rms} \, 2\pi f \left(C_1 + C_2\right)}_{\text{Parallel combination}} $$

   Subtracting I_1 from $I_{Combination}$, reveals that

   $$ I_{Combination} - I_1 = V_{rms} \, 2\pi f \left(C_1 + C_2\right) - V_{rms} \, 2\pi f C_1 = V_{rms} \, 2\pi f C_2 $$

   Solving for C_2 gives

$$C_2 = \frac{I_{\text{Combination}} - I_1}{V_{\text{rms}} \, 2\pi f} = \frac{0.18 \text{ A}}{(24 \text{ V})2\pi(440 \text{ Hz})} = \boxed{2.7 \times 10^{-6} \text{ F}}$$

---

9. **REASONING** The individual reactances are given by Equations 23.2 and 23.4, respectively,

Capacitive reactance $\qquad X_C = \dfrac{1}{2\pi f C}$

Inductive reactance $\qquad X_L = 2\pi f L$

When the reactances are equal, we have $X_C = X_L$, from which we find

$$\frac{1}{2\pi f C} = 2\pi f L \qquad \text{or} \qquad 4\pi^2 f^2 LC = 1$$

The last expression may be solved for the frequency f.

SOLUTION Solving for f with $L = 52 \times 10^{-3}$ H and $C = 76 \times 10^{-6}$ F, we obtain

$$f = \frac{1}{2\pi\sqrt{LC}} = \frac{1}{2\pi\sqrt{(52 \times 10^{-3} \text{ H})(76 \times 10^{-6} \text{ F})}} = \boxed{8.0 \times 10^1 \text{ Hz}}$$

---

13. **REASONING** Since the capacitor and the inductor are connected in parallel, the voltage across each of these elements is the same or $V_L = V_C$. Using Equations 23.3 and 23.1, respectively, this becomes $I_{\text{rms}} X_L = I_{\text{rms}} X_C$. Since the currents in the inductor and capacitor are equal, this relation simplifies to $X_L = X_C$. Therefore, we can find the value of the inductance by equating the expressions (Equations 23.4 and 23.2) for the inductive reactance and the capacitive reactance, and solving for L.

SOLUTION Since $X_L = X_C$, we have

$$2\pi f L = \frac{1}{2\pi f C}$$

Therefore, the value of the inductance is

$$L = \frac{1}{4\pi^2 f^2 C} = \frac{1}{4\pi^2 (60.0 \text{ Hz})^2 (40.0 \times 10^{-6} \text{ F})} = 0.176 \text{ H} = \boxed{176 \text{ mH}}$$

19. *REASONING* The voltage supplied by the generator can be found from Equation 23.6, $V_{rms} = I_{rms}Z$. The value of I_{rms} is given in the problem statement, so we must obtain the impedance of the circuit.

SOLUTION The impedance of the circuit is, according to Equation 23.7,

$$Z = \sqrt{R^2 + (X_L - X_C)^2} = \sqrt{(275\ \Omega)^2 + (648\ \Omega - 415\ \Omega)^2} = 3.60 \times 10^2\ \Omega$$

The rms voltage of the generator is

$$V_{rms} = I_{rms}Z = (0.233\ \text{A})(3.60 \times 10^2\ \Omega) = \boxed{83.9\ \text{V}}$$

---

21. *REASONING* We can use the equations for a series RCL circuit to solve this problem provided that we set $X_C = 0\ \Omega$ since there is no capacitor in the circuit. The current in the circuit can be found from Equation 23.6, $V_{rms} = I_{rms}Z$, once the impedance of the circuit has been obtained. Equation 23.8, $\tan\phi = (X_L - X_C)/R$, can then be used (with $X_C = 0\ \Omega$) to find the phase angle between the current and the voltage.

SOLUTION The inductive reactance is (Equation 23.4)

$$X_L = 2\pi f L = 2\pi(106\ \text{Hz})(0.200\ \text{H}) = 133\ \Omega$$

The impedance of the circuit is

$$Z = \sqrt{R^2 + (X_L - X_C)^2} = \sqrt{R^2 + X_L^2} = \sqrt{(215\ \Omega)^2 + (133\ \Omega)^2} = 253\ \Omega$$

a. The current through each circuit element is, using Equation 23.6,

$$I_{rms} = \frac{V_{rms}}{Z} = \frac{234\ \text{V}}{253\ \Omega} = \boxed{0.925\ \text{A}}$$

b. The phase angle between the current and the voltage is, according to Equation 23.8 (with $X_C = 0\ \Omega$),

$$\tan\phi = \frac{X_L - X_C}{R} = \frac{X_L}{R} = \frac{133\ \Omega}{215\ \Omega} = 0.619 \qquad \text{or} \qquad \phi = \tan^{-1}(0.619) = \boxed{31.8°}$$

25. ***REASONING*** We can use the equations for a series RCL circuit to solve this problem, provided that we set $X_L = 0$ since there is no inductance in the circuit. Thus, according to Equations 23.6 and 23.7, the current in the circuit is $I_{rms} = V_{rms} / \sqrt{R^2 + X_C^2}$. When the frequency f is very large, the capacitive reactance is zero, or $X_C = 0$, in which case the current becomes I_{rms} (large f) $= V_{rms} / R$. When the current I_{rms} in the circuit is one-half the value of I_{rms} (large f) that exists when the frequency is very large, we have

$$\frac{I_{rms}}{I_{rms}(\text{large } f)} = \frac{1}{2}$$

We can use these expressions to write the ratio above in terms of the resistance and the capacitive reactance. Once the capacitive reactance is known, the frequency can be determined.

SOLUTION The ratio of the currents is

$$\frac{I_{rms}}{I_{rms}(\text{large } f)} = \frac{V_{rms} / \sqrt{R^2 + X_C^2}}{V_{rms} / R} = \frac{R}{\sqrt{R^2 + X_C^2}} = \frac{1}{2} \qquad \text{or} \qquad \frac{R^2}{R^2 + X_C^2} = \frac{1}{4}$$

Taking the reciprocal of this result gives

$$\frac{R^2 + X_C^2}{R^2} = 4 \qquad \text{or} \qquad 1 + \frac{X_C^2}{R^2} = 4$$

Therefore,

$$\frac{X_C}{R} = \sqrt{3}$$

According to Equation 23.2, $X_C = 1/(2\pi f C)$, so it follows that

$$\frac{X_C}{R} = \frac{1/(2\pi f C)}{R} = \sqrt{3}$$

Thus,

$$f = \frac{1}{2\pi R C \sqrt{3}} = \frac{1}{2\pi (85\,\Omega)(4.0 \times 10^{-6}\ \text{F})\sqrt{3}} = \boxed{270\ \text{Hz}}$$

31. **_REASONING_** The resonant frequency f_0 of a series RCL circuit depends on the inductance L and capacitance C through the relation $f_0 = \dfrac{1}{2\pi\sqrt{LC}}$ (Equation 23.10). Since all the variables are known except L, we can use this relation to find the inductance.

SOLUTION Solving Equation 23.10 for the inductance gives

$$L = \frac{1}{4\pi^2 f_0^2 C} = \frac{1}{4\pi^2 \left(690\times10^3 \text{ Hz}\right)^2 \left(2.0\times10^{-9} \text{ F}\right)} = \boxed{2.7\times10^{-5} \text{ H}}$$

---

33. **_REASONING_** The current in an RCL circuit is given by Equation 23.6, $I_{rms} = V_{rms}/Z$, where the impedance Z of the circuit is given by Equation 23.7 as $Z = \sqrt{R^2 + (X_L - X_C)^2}$. The current is a maximum when the impedance is a minimum for a given generator voltage. The minimum impedance occurs when the frequency is f_0, corresponding to the condition that $X_L = X_C$, or $2\pi f_0 L = 1/(2\pi f_0 C)$. Solving for the frequency f_0, called the resonant frequency, we find that

$$f_0 = \frac{1}{2\pi\sqrt{LC}}$$

Note that the resonant frequency depends on the inductance and the capacitance, but does not depend on the resistance.

SOLUTION
a. The frequency at which the current is a maximum is

$$f_0 = \frac{1}{2\pi\sqrt{LC}} = \frac{1}{2\pi\sqrt{(17.0\times10^{-3} \text{ H})(12.0\times10^{-6} \text{ F})}} = \boxed{352 \text{ Hz}}$$

b. The maximum value of the current occurs when $f = f_0$. This occurs when $X_L = X_C$, so that $Z = R$. Therefore, according to Equation 23.6, we have

$$I_{rms} = \frac{V_{rms}}{Z} = \frac{V_{rms}}{R} = \frac{155 \text{ V}}{10.0 \ \Omega} = \boxed{15.5 \text{ A}}$$

39. **REASONING**   Since we know the values of the resonant frequency of the circuit, the capacitance, and the generator voltage, we can find the value of the inductance from Equation 23.10, the expression for the resonant frequency. The resistance can be found from energy considerations at resonance; the power factor is given by $\cos\phi$, where the phase angle ϕ is given by Equation 23.8, $\tan\phi = (X_L - X_C)/R$.

SOLUTION

a. Solving Equation 23.10 for the inductance L, we find that

$$L = \frac{1}{4\pi^2 f_0^2 C} = \frac{1}{4\pi^2 (1.30\times10^3 \text{ Hz})^2 (5.10\times10^{-6} \text{ F})} = \boxed{2.94\times10^{-3} \text{ H}}$$

b. At resonance, $f = f_0$, and the current is a maximum. This occurs when $X_L = X_C$, so that $Z = R$. Thus, the average power $\overline{P}$ provided by the generator is $\overline{P} = V_{rms}^2/R$, and solving for R we find

$$R = \frac{V_{rms}^2}{\overline{P}} = \frac{(11.0 \text{ V})^2}{25.0 \text{ W}} = \boxed{4.84 \text{ }\Omega}$$

c. When the generator frequency is 2.31 kHz, the individual reactances are

$$X_C = \frac{1}{2\pi f C} = \frac{1}{2\pi(2.31\times10^3 \text{ Hz})(5.10\times10^{-6} \text{ F})} = 13.5 \text{ }\Omega$$

$$X_L = 2\pi f L = 2\pi(2.31\times10^3 \text{ Hz})(2.94\times10^{-3} \text{ H}) = 42.7 \text{ }\Omega$$

The phase angle ϕ is, from Equation 23.8,

$$\phi = \tan^{-1}\left(\frac{X_L - X_C}{R}\right) = \tan^{-1}\left(\frac{42.7 \text{ }\Omega - 13.5 \text{ }\Omega}{4.84 \text{ }\Omega}\right) = 80.6°$$

The power factor is then given by

$$\cos\phi = \cos 80.6° = \boxed{0.163}$$

43. ***REASONING AND SOLUTION*** The rms voltage can be calculated using $V = IX_C$, where the capacitive reactance X_C can be found using

$$X_C = \frac{1}{2\pi f C} = \frac{1}{2\pi (3.4 \times 10^3 \text{ Hz})(0.86 \times 10^{-6} \text{F})} = 54 \ \Omega$$

The voltage is, therefore,

$$V = IX_C = (35 \times 10^{-3} \text{ A})(54 \ \Omega) = \boxed{1.9 \text{ V}}$$

47. ***REASONING*** The rms current can be calculated from Equation 23.3, $I_{rms} = V_{rms} / X_L$, provided that the inductive reactance is obtained first. Then the peak value of the current I_0 supplied by the generator can be calculated from the rms current I_{rms} by using Equation 20.12, $I_0 = \sqrt{2} \ I_{rms}$.

SOLUTION At the frequency of $f = 620 \text{ Hz}$, we find, using Equations 23.4 and 23.3, that

$$X_L = 2\pi f L = 2\pi (620 \text{ Hz})(8.2 \times 10^{-3} \text{ H}) = 32 \ \Omega$$

$$I_{rms} = \frac{V_{rms}}{X_L} = \frac{10.0 \text{ V}}{32 \ \Omega} = 0.31 \text{ A}$$

Therefore, from Equation 20.12, we find that the *peak value* I_0 of the current supplied by the generator must be

$$I_0 = \sqrt{2} \ I_{rms} = \sqrt{2} \ (0.31 \text{ A}) = \boxed{0.44 \text{ A}}$$

51. ***REASONING*** At the resonant frequency f_0, we have $C = 1/(4\pi^2 f_0^2 L)$. We want to determine some series combination of capacitors whose equivalent capacitance C_s' is such that $f_0' = 3 f_0$. Thus,

$$C_s' = \frac{1}{4\pi^2 f_0'^2 L} = \frac{1}{4\pi^2 (3 f_0)^2 L} = \frac{1}{9} \left(\frac{1}{4\pi^2 f_0^2 L} \right) = \frac{1}{9} C$$

The equivalent capacitance of a series combination of capacitors is $1/C_s' = 1/C_1 + 1/C_2 + ...$ If we require that all the capacitors have the same capacitance C, the equivalent capacitance is

$$\frac{1}{C_s'} = \frac{1}{C} + \frac{1}{C} + \cdots = \frac{n}{C}$$

where n is the total number of identical capacitors. Using the result above, we find that

$$\frac{1}{C'_s} = \frac{1}{\frac{1}{9}C} = \frac{n}{C} \qquad \text{or} \qquad n = 9$$

Therefore, the number of *additional* capacitors that must be inserted in series in the circuit so that the frequency triples is $n' = n - 1 = \boxed{8}$.

3. **REASONING AND SOLUTION** This is a standard exercise in units conversion. We first determine the number of meters in one light-year. The distance that light travels in one year is

$$d = ct = (3.00 \times 10^8 \text{ m/s})(1.00 \text{ year}) \left(\frac{365.25 \text{ days}}{1 \text{ year}} \right) \left(\frac{24 \text{ hours}}{1.0 \text{ day}} \right) \left(\frac{3600 \text{ s}}{1 \text{ hour}} \right) = 9.47 \times 10^{15} \text{ m}$$

Thus, 1 light year = 9.47×10^{15} m. Then, the distance to Alpha Centauri is

$$(4.3 \text{ light-years}) \left(\frac{9.47 \times 10^{15} \text{ m}}{1 \text{ light-year}} \right) = \boxed{4.1 \times 10^{16} \text{ m}}$$

5. **REASONING** The equation that represents the wave mathematically is $y = A \sin(2\pi f t - 2\pi x / \lambda)$. In this expression the amplitude is $A = 156$ N/C. The wavelength λ can be calculated using Equation 16.1, and we obtain

$$\lambda = \frac{c}{f} = \frac{3.00 \times 10^8 \text{ m/s}}{1.50 \times 10^8 \text{ Hz}} = 2.00 \text{ m}$$

SOLUTION
a. For $t = 0$ s, the wave expression becomes

$$y = A \sin(2\pi f t - 2\pi x / \lambda) = 156 \sin \left[2\pi f(0) - \frac{2\pi x}{2.00} \right] = -156 \sin \left(\frac{2\pi x}{2.00} \right) = -156 \sin(\pi x)$$

In this result, the units are suppressed for convenience. The following table gives the values of the electric field obtained using this version of the wave expression with the given values of the position x. The term πx is in radians when x is in meters, and conversion from radians to degrees is accomplished using the fact that $2\pi \text{ rad} = 360°$.

x	$y = -156 \sin(\pi x)$
0 m	$-156 \sin(0) = -156 \sin(0°) = 0$
0.50 m	$-156 \sin(0.50\pi) = -156 \sin(90°) = -156$
1.00 m	$-156 \sin(1.00\pi) = -156 \sin(180°) = 0$
1.50 m	$-156 \sin(1.50\pi) = -156 \sin(270°) = +156$
2.00 m	$-156 \sin(2.00\pi) = -156 \sin(360°) = 0$

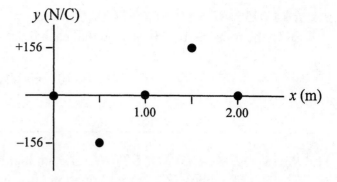

These values for the electric field are plotted in the graph shown at the right.

b. For $t = T/4$, we use the fact that $f = 1/T$, and the wave expression becomes

$$y = A\sin\left(2\pi ft - 2\pi x/\lambda\right) = 156\sin\left[2\pi\left(\frac{1}{T}\right)\left(\frac{T}{4}\right) - \frac{2\pi x}{2.00}\right] = 156\sin\left(\frac{\pi}{2} - \pi x\right) = 156\cos\left(\pi x\right)$$

In this result, the units are suppressed for convenience. The following table gives the values of the electric field obtained using this version of the wave expression with the given values of the position x. The term πx is in radians when x is in meters, and conversion from radians to degrees is accomplished using the fact that $2\pi\,\text{rad} = 360°$.

x	$y = 156\cos\left(\pi x\right)$
0 m	$156\cos(0) = 156\cos(0°) = +156$
0.50 m	$156\cos(0.50\,\pi) = 156\cos(90°) = 0$
1.00 m	$156\cos(1.00\,\pi) = 156\cos(180°) = -156$
1.50 m	$156\cos(1.50\,\pi) = 156\cos(270°) = 0$
2.00 m	$156\cos(2.00\,\pi) = 156\cos(360°) = +156$

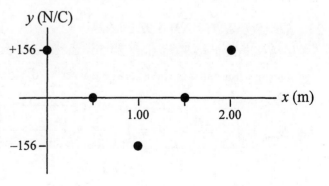

These values for the electric field are plotted in the graph shown at the right.

9. *REASONING AND SOLUTION* According to Equation 16.1, the wavelength is $\lambda = c/f$. Since the rods can be adjusted so that each one has a length of $\lambda/4$, the length of each rod is

$$L = \frac{\lambda}{4} = \frac{c/f}{4} = \frac{\left(3.00 \times 10^8 \text{ m/s}\right)/\left(60 \times 10^6 \text{ Hz}\right)}{4} = \boxed{1.25 \text{ m}}$$

---

11. *REASONING AND SOLUTION* The number of wavelengths that can fit across the width W of your thumb is W/λ. From Equation 16.1, we know that $\lambda = c/f$, so

$$\text{No. of wavelengths} = \frac{W}{\lambda} = \frac{Wf}{c} = \frac{(2.0 \times 10^{-2} \text{ m})(5.5 \times 10^{14} \text{ Hz})}{3.0 \times 10^8 \text{ m/s}} = \boxed{3.7 \times 10^4}$$

---

15. *REASONING* We proceed by first finding the time t for sound waves to travel between the astronauts. Since this is the same time it takes for the electromagnetic waves to travel to earth, the distance between earth and the spaceship is $d_{\text{earth-ship}} = ct$.

SOLUTION The time it takes for sound waves to travel at 343 m/s through the air between the astronauts is

$$t = \frac{d_{\text{astronaut}}}{v_{\text{sound}}} = \frac{1.5 \text{ m}}{343 \text{ m/s}} = 4.4 \times 10^{-3} \text{ s}$$

Therefore, the distance between the earth and the spaceship is

$$d_{\text{earth-ship}} = ct = (3.0 \times 10^8 \text{ m/s})(4.4 \times 10^{-3} \text{ s}) = \boxed{1.3 \times 10^6 \text{ m}}$$

---

25. *REASONING AND SOLUTION*

a. According to Equation 24.5b, the average intensity is $\overline{S} = c\varepsilon_0 E_{\text{rms}}^2$. In addition, the average intensity is the average power $\overline{P}$ divided by the area A. Therefore,

$$E_{\text{rms}} = \sqrt{\frac{\overline{S}}{c\varepsilon_0}} = \sqrt{\frac{\overline{P}}{c\varepsilon_0 A}} = \sqrt{\frac{1.20 \times 10^4 \text{ W}}{(3.00 \times 10^8 \text{ m/s})[8.85 \times 10^{-12} \text{C}^2/(\text{N} \cdot \text{m}^2)](135 \text{ m}^2)}} = \boxed{183 \text{ N/C}}$$

b. Then, from $E_{\text{rms}} = cB_{\text{rms}}$ (Equation 24.3), we have

$$B_{\text{rms}} = \frac{E_{\text{rms}}}{c} = \frac{183 \text{ N/C}}{3.00 \times 10^8 \text{ m/s}} = \boxed{6.10 \times 10^{-7} \text{ T}}$$

27. **REASONING AND SOLUTION** The energy is equal to the power P multiplied by the time t. The power, on the other hand, is equal to product of the intensity S of the wave and the area A through which the wave passes.

$$\text{Energy} = P\,t = (SA)t = (1390 \text{ W/m}^2)(25 \text{ m} \times 45 \text{ m})(3600 \text{ s}) = \boxed{5.6 \times 10^9 \text{ J}}$$

37. **REASONING** The Doppler effect for electromagnetic radiation is given by Equation 24.6, $f_o = f_s\left(1 \pm v_{rel}/c\right)$, where f_o is the observed frequency, f_s is the frequency emitted by the source, and v_{rel} is the speed of the source relative to the observer. As discussed in the text, the plus sign applies when the source and the observer are moving toward one another, while the minus sign applies when they are moving apart.

SOLUTION
a. At location A, the galaxy is moving away from the earth with a relative speed of

$$v_{rel} = (1.6 \times 10^6 \text{ m/s}) - (0.4 \times 10^6 \text{ m/s}) = 1.2 \times 10^6 \text{ m/s}$$

Therefore, the minus sign in Equation 24.6 applies and the observed frequency for the light from region A is

$$f_o = f_s\left(1 - \frac{v_{rel}}{c}\right) = (6.200 \times 10^{14} \text{ Hz})\left(1 - \frac{1.2 \times 10^6 \text{ m/s}}{3.0 \times 10^8 \text{ m/s}}\right) = \boxed{6.175 \times 10^{14} \text{ Hz}}$$

b. Similarly, at location B, the galaxy is moving away from the earth with a relative speed of

$$v_{rel} = (1.6 \times 10^6 \text{ m/s}) + (0.4 \times 10^6 \text{ m/s}) = 2.0 \times 10^6 \text{ m/s}$$

The observed frequency for the light from region B is

$$f_o = f_s\left(1 - \frac{v_{rel}}{c}\right) = (6.200 \times 10^{14} \text{ Hz})\left(1 - \frac{2.0 \times 10^6 \text{ m/s}}{3.0 \times 10^8 \text{ m/s}}\right) = \boxed{6.159 \times 10^{14} \text{ Hz}}$$

41. **REASONING AND SOLUTION** The average intensity of light leaving each polarizer is given by Malus' Law : $\bar{S} = \bar{S}_0 \cos^2 \theta$ (Equation 24.7). Solving for the angle θ and noting that $\bar{S} = 0.100\,\bar{S}_0$ (since 90.0% of the intensity is absorbed), we have

$$\theta = \cos^{-1}\sqrt{\frac{0.100\,\bar{S}_0}{\bar{S}_0}} = \boxed{71.6°}$$

47. **REASONING** The average intensity of light leaving each analyzer is given by Malus' Law (Equation 24.7). Thus, intensity of the light transmitted through the first analyzer is

$$\overline{S}_1 = \overline{S}_0 \cos^2 27°$$

Similarly, the intensity of the light transmitted through the second analyzer is

$$\overline{S}_2 = \overline{S}_1 \cos^2 27° = \overline{S}_0 \cos^4 27°$$

And the intensity of the light transmitted through the third analyzer is

$$\overline{S}_3 = \overline{S}_2 \cos^2 27° = \overline{S}_0 \cos^6 27°$$

If we generalize for the Nth analyzer, we deduce that

$$\overline{S}_N = \overline{S}_{N-1} \cos^2 27° = \overline{S}_0 \cos^{2N} 27°$$

Since we want the light reaching the photocell to have an intensity that is reduced by a least a factor of one hundred relative to the first analyzer, we want $\overline{S}_N / \overline{S}_0 = 0.010$. Therefore, we need to find N such that $\cos^{2N} 27° = 0.010$. This expression can be solved for N.

SOLUTION Taking the common logarithm of both sides of the last expression gives

$$2N \log(\cos 27°) = \log 0.010 \quad \text{or} \quad N = \frac{\log 0.010}{2 \log (\cos 27°)} = \boxed{20}$$

---

51. **REASONING** The electromagnetic wave will be picked up by the radio when the resonant frequency f_0 of the circuit in Figure 24.4 is equal to the frequency of the broadcast wave, or $f_0 = 1400$ kHz. This frequency, in turn, is related to the capacitance C and inductance L of the circuit via $f_0 = 1/(2\pi\sqrt{LC})$ (Equation 23.10). Since C is known, we can use this relation to find the inductance.

SOLUTION Solving the relation $f_0 = 1/(2\pi\sqrt{LC})$ for the inductance L, we find that

$$L = \frac{1}{4\pi^2 f_0^2 C} = \frac{1}{4\pi^2 (1400 \times 10^3 \text{ Hz})^2 (8.4 \times 10^{-11} \text{ F})} = \boxed{1.5 \times 10^{-4} \text{ H}}$$

53. **REASONING** The rms value E_{rms} of the electric field is related to the average energy density $\bar{u}$ of the microwave radiation according to $\bar{u} = \varepsilon_0 E_{rms}^2$ (Equation 24.2b).

SOLUTION Solving for E_{rms} gives

$$E_{rms} = \sqrt{\frac{\bar{u}}{\varepsilon_0}} = \sqrt{\frac{4 \times 10^{-14} \text{ J/m}^3}{8.85 \times 10^{-12} \text{ C}^2/(\text{N} \cdot \text{m}^2)}} = \boxed{0.07 \text{ N/C}}$$

---

55. **REASONING** Since the incident beam is unpolarized, the intensity of the light transmitted by the first sheet of polarizing material is one-half the intensity of the incident beam. The beams striking the second and third sheets of polarizing material are polarized, so the average intensity $\bar{S}$ of the light transmitted by each sheet is given by Malus' law, $\bar{S} = \bar{S}_0 \cos^2 \theta$, where $\bar{S}_0$ is the average intensity of the light incident on each sheet.

SOLUTION The average intensity $\bar{S}_1$ of the light leaving the first sheet is one-half the intensity of the incident beam, so $\bar{S}_1 = \frac{1}{2}\left(1260.0 \text{ W/m}^2\right) = 630.0 \text{ W/m}^2$. The intensity $\bar{S}_2$ of the light leaving the second sheet of polarizing material is given by Malus' law, Equation 24.7, $\bar{S}_2 = \bar{S}_1 \cos^2 \theta$, where θ is the angle between the polarization of the incident beam and the transmission axis of the second sheet:

$$\bar{S}_2 = \left(630.0 \text{ W/m}^2\right) \cos^2 \left(55.0° - 19.0°\right) = 412 \text{ W/m}^2$$

The intensity $\bar{S}_3$ of the light leaving the third sheet of polarizing material is $\bar{S}_3 = \bar{S}_2 \cos^2 \theta$, where θ is the angle between the polarization of the incident beam and the transmission axis of the third sheet:

$$\bar{S}_3 = \left(412 \text{ W/m}^2\right) \cos^2 \left(100.0° - 55.0°\right) = \boxed{206 \text{ W/m}^2}$$

59. ***REASONING AND SOLUTION*** The sun radiates sunlight (electromagnetic waves) uniformly in all directions, so the intensity at a distance r from the sun is given by Equation 16.9 as $S = P/(4\pi r^2)$, where P is the power radiated by the sun. The power that strikes an area $A_\perp$ oriented perpendicular to the direction in which the sunlight is radiated is $P' = SA_\perp$, according to Equation 16.8. The 0.75-m^2 patch of flat land on the equator at point Q is not perpendicular to the direction of the sunlight, however.

The figure at the right shows that

$$A_\perp = (0.75 \text{ m}^2) \cos 27°$$

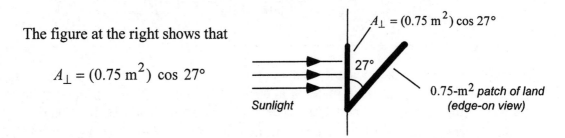

$A_\perp = (0.75 \text{ m}^2) \cos 27°$

27°

Sunlight

0.75-m^2 *patch of land* (*edge-on view*)

Therefore, the power striking the patch of land is

$$P' = SA_\perp = \left(\frac{P}{4\pi r^2}\right)(0.75 \text{ m}^2) \cos 27°$$

$$= \left[\frac{3.9 \times 10^{26} \text{ W}}{4\pi (1.5 \times 10^{11} \text{ m})^2}\right](0.75 \text{ m}^2) \cos 27° = \boxed{920 \text{ W}}$$

CHAPTER 25 | *THE REFLECTION OF LIGHT: MIRRORS*

1. ***REASONING AND SOLUTION***   The drawing at the right shows a ray diagram in which the reflected rays have been projected behind the mirror. We can see by inspection of this drawing that, after the rays reflect from the plane mirror, the angle α between them is still $\boxed{10°}$.

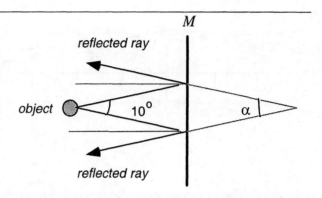

5. ***REASONING***   The geometry is shown below. According to the law of reflection, the incident ray, the reflected ray, and the normal to the surface all lie in the same plane, and the angle of reflection θ_r equals the angle of incidence θ_i. We can use the law of reflection and the properties of triangles to determine the angle θ at which the ray leaves M_2.

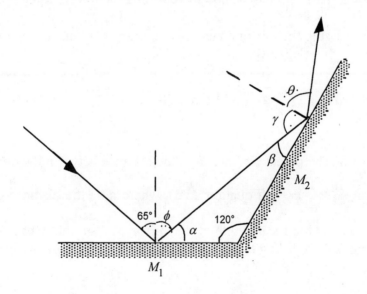

SOLUTION   From the law of reflection, we know that $\phi = 65°$. We see from the figure that $\phi + \alpha = 90°$, or $\alpha = 90° - \phi = 90° - 65° = 25°$. From the figure and the fact that the sum of the interior angles in any triangle is 180°, we have $\alpha + \beta + 120° = 180°$. Solving for β, we find that $\beta = 180° - (120° + 25°) = 35°$. Therefore, since $\beta + \gamma = 90°$, we find that the angle γ is given by $\gamma = 90° - \beta = 90° - 35° = 55°$. Since γ is the angle of incidence of the ray on mirror M_2, we know from the law of reflection that $\boxed{\theta = 55°}$.

15. **REASONING AND SOLUTION** The ray diagram is shown below. (Note: $f = -50.0$ cm and $d_o = 25.0$ cm)

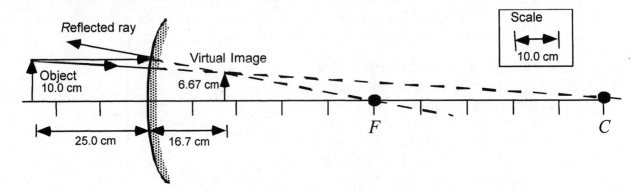

a. The ray diagram indicates that ⎢ the image is 16.7 cm behind the mirror ⎢.

b. The ray diagram indicates that ⎢ the image height is 6.67 cm ⎢.

---

19. **REASONING** This problem can be solved using the mirror equation, Equation 25.3.

SOLUTION Using the mirror equation with $d_i = +26$ cm and $f = 12$ cm, we find

$$\frac{1}{d_o} = \frac{1}{f} - \frac{1}{d_i} = \frac{1}{12 \text{ cm}} - \frac{1}{26 \text{ cm}} \qquad \text{or} \qquad \boxed{d_o = +22 \text{ cm}}$$

---

24. **REASONING** The object distance d_o is the distance between the object and the mirror. It is found from $\frac{1}{d_o} + \frac{1}{d_i} = \frac{1}{f}$ (Equation 25.3), where d_i is the distance between the mirror and the image, and f is the focal length of the mirror. We are told that the image appears in front of the mirror, so, according to the sign conventions for spherical mirrors, the image distance must be positive: $d_i = +97$ cm.

SOLUTION Solving $\frac{1}{d_o} + \frac{1}{d_i} = \frac{1}{f}$ (Equation 25.3) for d_o yields

$$\frac{1}{d_o} = \frac{1}{f} - \frac{1}{d_i} \qquad \text{or} \qquad d_o = \frac{1}{\dfrac{1}{f} - \dfrac{1}{d_i}} = \frac{1}{\dfrac{1}{(+42 \text{ cm})} - \dfrac{1}{(+97 \text{ cm})}} = \boxed{74 \text{ cm}}$$

---

27. **REASONING** When paraxial light rays that are parallel to the principal axis strike a convex mirror, the reflected rays diverge after being reflected, and appear to originate from the focal point F behind the mirror (see Figure 25.16). We can treat the sun as being infinitely far from the mirror, so it is reasonable to treat the incident rays as paraxial rays that are parallel to the principal axis.

SOLUTION
a. Since the sun is infinitely far from the mirror and its image is a virtual image that lies *behind* the mirror, we can conclude that the mirror is a $\boxed{\text{convex mirror}}$.

b. With $d_i = -12.0$ cm and $d_o = \infty$, the mirror equation (Equation 25.3) gives

$$\frac{1}{f} = \frac{1}{d_o} + \frac{1}{d_i} = \frac{1}{\infty} + \frac{1}{d_i} = \frac{1}{d_i}$$

Therefore, the focal length f lies 12.0 cm behind the mirror (this is consistent with the reasoning above that states that, after being reflected, the rays appear to originate from the focal point behind the mirror). In other words, $f = -12.0$ cm. Then, according to Equation 25.2, $f = -\frac{1}{2}R$, and the radius of curvature is

$$R = -2f = -2(-12.0 \text{ cm}) = \boxed{24.0 \text{ cm}}$$

---

29. **REASONING** We need to know the focal length of the mirror and can obtain it from the mirror equation, Equation 25.3, as applied to the first object:

$$\frac{1}{d_{o1}} + \frac{1}{d_{i1}} = \frac{1}{14.0 \text{ cm}} + \frac{1}{-7.00 \text{ cm}} = \frac{1}{f} \qquad \text{or} \qquad f = -14.0 \text{ cm}$$

According to the magnification equation, Equation 25.4, the image height h_i is related to the object height h_o as follows: $h_i = mh_o = \left(-d_i / d_o\right)h_o$.

SOLUTION Applying this result to each object, we find that $h_{i2} = h_{i1}$, or

$$\left(\frac{-d_{i2}}{d_{o2}}\right)h_{o2} = \left(\frac{-d_{i1}}{d_{o1}}\right)h_{o1}$$

Therefore,

$$d_{i2} = d_{o2}\left(\frac{d_{i1}}{d_{o1}}\right)\left(\frac{h_{o1}}{h_{o2}}\right)$$

Using the fact that $h_{o2} = 2h_{o1}$, we have

$$d_{i2} = d_{o2}\left(\frac{d_{i1}}{d_{o1}}\right)\left(\frac{h_{o1}}{h_{o2}}\right) = d_{o2}\left(\frac{-7.00 \text{ cm}}{14.0 \text{ cm}}\right)\left(\frac{h_{o1}}{2h_{o1}}\right) = -0.250\, d_{o2}$$

Using this result in the mirror equation, as applied to the second object, we find that

$$\frac{1}{d_{o2}} + \frac{1}{d_{i2}} = \frac{1}{f}$$

or

$$\frac{1}{d_{o2}} + \frac{1}{-0.250\, d_{o2}} = \frac{1}{-14.0 \text{ cm}}$$

Therefore,

$$\boxed{d_{o2} = +42.0 \text{ cm}}$$

---

37. **REASONING**   We have seen that a convex mirror always forms a *virtual image* as shown in Figure 25.21a of the text, where the image is *upright* and *smaller* than the object. These characteristics should bear out in the results of our calculations.

SOLUTION   The radius of curvature of the convex mirror is 68 cm. Therefore, the focal length is, from Equation 25.2, $f = -(1/2)R = -34$ cm. Since the image is virtual, we know that $d_i = -22$ cm.

a. With $d_i = -22$ cm and $f = -34$ cm, the mirror equation gives

$$\frac{1}{d_o} = \frac{1}{f} - \frac{1}{d_i} = \frac{1}{-34 \text{ cm}} - \frac{1}{-22 \text{ cm}} \qquad \text{or} \qquad \boxed{d_o = +62 \text{ cm}}$$

b. According to the magnification equation, the magnification is

$$m = -\frac{d_i}{d_o} = -\frac{-22 \text{ cm}}{62 \text{ cm}} = \boxed{+0.35}$$

c. Since the magnification m is positive, the image is $\boxed{\text{upright}}$.

d. Since the magnification m is less than one, the image is $\boxed{\text{smaller}}$ than the object.

---

41. **REASONING** This problem can be solved by using the mirror equation, Equation 25.3, and the magnification equation, Equation 25.4.

SOLUTION

a. Using the mirror equation with $d_i = d_o$ and $f = R/2$, we have

$$\frac{1}{d_o} = \frac{1}{f} - \frac{1}{d_i} = \frac{1}{R/2} - \frac{1}{d_o} \qquad \text{or} \qquad \frac{2}{d_o} = \frac{2}{R}$$

Therefore, we find that $\boxed{d_o = R}$.

b. According to the magnification equation, the magnification is

$$m = -\frac{d_i}{d_o} = -\frac{d_o}{d_o} = \boxed{-1}$$

c. Since the magnification m is negative, the image is $\boxed{\text{inverted}}$.

---

43. **REASONING** The mirror equation relates the object and image distances to the focal length. The magnification equation relates the magnification to the object and image distances. The problem neither gives nor asks for information about the image distance. Therefore, we can solve the magnification equation for the image distance and substitute the result into the mirror equation to obtain an expression relating the object distance, the magnification, and the focal length. This expression can be applied to both mirrors A and B to obtain the ratio of the focal lengths.

SOLUTION The magnification equation gives the magnification as $m = -d_i/d_o$. Solving for d_i, we obtain $d_i = -md_o$. Substituting this result into the mirror equation, we obtain

$$\frac{1}{d_o} + \frac{1}{d_i} = \frac{1}{d_o} + \frac{1}{(-md_o)} = \frac{1}{f} \qquad \text{or} \qquad \frac{1}{d_o}\left(1 - \frac{1}{m}\right) = \frac{1}{f} \qquad \text{or} \qquad f = \frac{d_o m}{m-1}$$

Applying this result for the focal length f to each mirror gives

$$f_A = \frac{d_o m_A}{m_A - 1} \qquad \text{and} \qquad f_B = \frac{d_o m_B}{m_B - 1}$$

Dividing the expression for f_A by the expression for f_B, we find

$$\frac{f_A}{f_B} = \frac{d_0 m_A / (m_A - 1)}{d_0 m_B / (m_B - 1)} = \frac{m_A (m_B - 1)}{m_B (m_A - 1)} = \frac{4.0(2.0 - 1)}{2.0(4.0 - 1)} = \boxed{0.67}$$

---

47. **REASONING AND SOLUTION**   We can see from the upper triangle in the drawing that

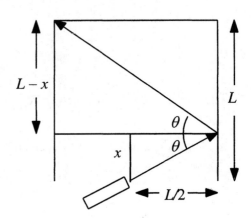

$$\tan \theta = (L - x)/L$$

We also see from the lower triangle that

$$\tan \theta = x/(L/2)$$

Equating these two expressions gives

$$\frac{L - x}{L} = \frac{x}{\frac{1}{2}L} \qquad \text{or} \qquad x = \frac{1}{3}L$$

Therefore,

$$\theta = \tan^{-1}\left(\frac{x}{\frac{1}{2}L}\right) = \tan^{-1}\left(\frac{\frac{1}{3}L}{\frac{1}{2}L}\right) = \boxed{33.7°}$$

CHAPTER 26 THE REFRACTION OF LIGHT: LENSES AND OPTICAL INSTRUMENTS

1. **REASONING** The substance can be identified from Table 26.1 if its index of refraction is known. The index of refraction n is defined as the speed of light c in a vacuum divided by the speed of light v in the substance (Equation 26.1), both of which are known.

SOLUTION Using Equation 26.1, we find that

$$n = \frac{c}{v} = \frac{2.998 \times 10^8 \text{ m/s}}{2.201 \times 10^8 \text{ m/s}} = 1.362$$

Referring to Table 26.1, we see that the substance is $\boxed{\text{ethyl alcohol}}$.

---

5. **REASONING** Since the light will travel in glass at a constant speed v, the time it takes to pass perpendicularly through the glass is given by $t = d/v$, where d is the thickness of the glass. The speed v is related to the vacuum value c by Equation 26.1: $n = c/v$.

SOLUTION Substituting for v from Equation 26.1 and substituting values, we obtain

$$t = \frac{d}{v} = \frac{nd}{c} = \frac{(1.5)(4.0 \times 10^{-3} \text{ m})}{3.00 \times 10^8 \text{ m/s}} = \boxed{2.0 \times 10^{-11} \text{ s}}$$

---

9. **REASONING AND SOLUTION**

a. We know from the law of reflection (Section 25.2), that the angle of reflection is equal to the angle of incidence, so the reflected ray is reflected at $\boxed{43°}$.

b. Snell's law of refraction (Equation 26.2: $n_1 \sin \theta_1 = n_2 \sin \theta_2$ can be used to find the angle of refraction. Table 26.1 indicates that the index of refraction of water is 1.333. Solving for θ_2 and substituting values, we find that

$$\sin \theta_2 = \frac{n_1 \sin \theta_1}{n_2} = \frac{(1.000)(\sin 43°)}{1.333} = 0.51 \cdot \qquad \text{or} \qquad \theta_2 = \sin^{-1} 0.51 = \boxed{31°}$$

---

13. **REASONING** We will use the geometry of the situation to determine the angle of incidence. Once the angle of incidence is known, we can use Snell's law to find the index of refraction of the unknown liquid. The speed of light v in the liquid can then be determined.

SOLUTION From the drawing in the text, we see that the angle of incidence at the liquid-air interface is

$$\theta_1 = \tan^{-1}\left(\frac{5.00 \text{ cm}}{6.00 \text{ cm}}\right) = 39.8°$$

The drawing also shows that the angle of refraction is 90.0°. Thus, according to Snell's law (Equation 26.2: $n_1 \sin \theta_1 = n_2 \sin \theta_2$), the index of refraction of the unknown liquid is

$$n_1 = \frac{n_2 \sin\theta_2}{\sin\theta_1} = \frac{(1.000)(\sin 90.0°)}{\sin 39.8°} = 1.56$$

From Equation 26.1 ($n = c/v$), we find that the speed of light in the unknown liquid is

$$v = \frac{c}{n_1} = \frac{3.00 \times 10^8 \text{ m/s}}{1.56} = \boxed{1.92 \times 10^8 \text{ m/s}}$$

---

21. The drawing at the right shows the geometry of the situation using the same notation as that in Figure 26.6. In addition to the text's notation, let t represent the thickness of the pane, let L represent the length of the ray in the pane, let x (shown twice in the figure) equal the displacement of the ray, and let the difference in angles $\theta_1 - \theta_2$ be given by ϕ.

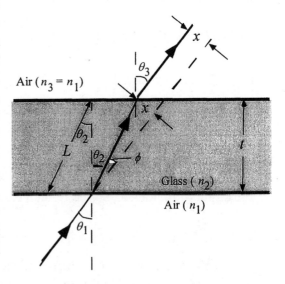

   We wish to find the amount x by which the emergent ray is displaced relative to the incident ray. This can be done by applying Snell's law at each interface, and then making use of the geometric and trigonometric relations in the drawing.

SOLUTION If we apply Snell's law (see Equation 26.2) to the bottom interface we obtain $n_1 \sin \theta_1 = n_2 \sin \theta_2$. Similarly, if we apply Snell's law at the top interface where the ray emerges, we have $n_2 \sin \theta_2 = n_3 \sin \theta_3 = n_1 \sin \theta_3$. Comparing this with Snell's law at the bottom face, we see that $n_1 \sin \theta_1 = n_1 \sin \theta_3$, from which we can conclude that $\theta_3 = \theta_1$. Therefore, the emerging ray is parallel to the incident ray.

From the geometry of the ray and thickness of the pane, we see that $L \cos \theta_2 = t$, from which it follows that $L = t / \cos \theta_2$. Furthermore, we see that $x = L \sin \phi = L \sin\left(\theta_1 - \theta_2\right)$. Substituting for L, we find

$$x = L \sin(\theta_1 - \theta_2) = \frac{t \, \sin(\theta_1 - \theta_2)}{\cos \theta_2}$$

Before we can use this expression to determine a numerical value for x, we must find the value of θ_2. Solving the expression for Snell's law at the bottom interface for θ_2, we have

$$\sin \theta_2 = \frac{n_1 \sin \theta_1}{n_2} = \frac{(1.000) \, (\sin 30.0°)}{1.52} = 0.329 \qquad \text{or} \qquad \theta_2 = \sin^{-1} 0.329 = 19.2°$$

Therefore, the amount by which the emergent ray is displaced relative to the incident ray is

$$x = \frac{t \, \sin\left(\theta_1 - \theta_2\right)}{\cos \theta_2} = \frac{(6.00 \text{ mm}) \sin\left(30.0° - 19.2°\right)}{\cos 19.2°} = \boxed{1.19 \text{ mm}}$$

---

23. **REASONING** Following the discussion in Conceptual Example 4, we have the drawing at the right to use as a guide. In this drawing the symbol d refers to depths in the water, while the symbol h refers to heights in the air above the water. Moreover, symbols with a prime denote apparent distances, and unprimed symbols denote actual distances. We will use Equation 26.3 to relate apparent distances to actual distances. In so doing, we will use the fact that the refractive index of air is essentially $n_{air} = 1$ and denote the refractive index of water by $n_w = 1.333$ (see Table 26.1).

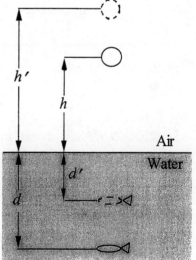

SOLUTION To the fish, the man appears to be a distance above the air-water interface that is given by Equation 26.3 as $h' = h\left(n_w / 1\right)$. Thus, measured above the eyes of the fish, the man appears to be located at a distance of

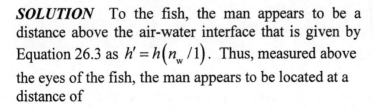

$$h' + d = h\left(\frac{n_w}{1}\right) + d \qquad\qquad (1)$$

To the man, the fish appears to be a distance below the air-water interface that is given by Equation 26.3 as $d' = d(1/n_w)$. Thus, measured below the man's eyes, the fish appears to be located at a distance of

$$h + d' = h + d\left(\frac{1}{n_w}\right) \tag{2}$$

Dividing Equation (1) by Equation (2) and using the fact that $h = d$, we find

$$\frac{h' + d}{h + d'} = \frac{h\left(\dfrac{n_w}{1}\right) + d}{h + d\left(\dfrac{1}{n_w}\right)} = \frac{n_w + 1}{1 + \dfrac{1}{n_w}} = n_w \tag{3}$$

In Equation (3), $h' + d$ is the distance we seek, and $h + d'$ is given as 2.0 m. Thus, we find

$$h' + d = n_w (h + d') = (1.333)(2.0 \text{ m}) = \boxed{2.7 \text{ m}}$$

---

27. ***REASONING AND SOLUTION*** According to Equation 26.4, the critical angle is related to the refractive indices n_1 and n_2 by $\sin \theta_c = n_2 / n_1$, where $n_1 > n_2$. Solving for n_1, we find

$$n_1 = \frac{n_2}{\sin \theta_c} = \frac{1.000}{\sin 40.5°} = \boxed{1.54}$$

---

35. ***REASONING*** Total internal reflection will occur at point P provided that the angle α in the drawing at the right exceeds the critical angle. This angle is determined by the angle θ_2 at which the light rays enter the quartz slab. We can determine θ_2 by using Snell's law of refraction and the incident angle, which is given as $\theta_1 = 34°$.

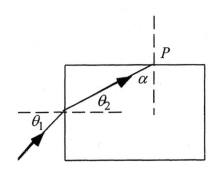

SOLUTION Using n for the refractive index of the fluid that surrounds the crystalline quartz slab and n_q for the refractive index of quartz and applying Snell's law give

$$n \sin \theta_1 = n_q \sin \theta_2 \qquad \text{or} \qquad \sin \theta_2 = \frac{n}{n_q} \sin \theta_1 \tag{1}$$

But when α equals the critical angle, we have from Equation 26.4 that

$$\sin \alpha = \sin \theta_c = \frac{n}{n_q} \tag{2}$$

According to the geometry in the drawing above, $\alpha = 90° - \theta_2$. As a result, Equation (2) becomes

$$\sin\left(90° - \theta_2\right) = \cos \theta_2 = \frac{n}{n_q} \tag{3}$$

Squaring Equation (3), using the fact that $\sin^2 \theta_2 + \cos^2 \theta_2 = 1$, and substituting from Equation (1), we obtain

$$\cos^2 \theta_2 = 1 - \sin^2 \theta_2 = 1 - \frac{n^2}{n_q^2}\sin^2 \theta_1 = \frac{n^2}{n_q^2} \tag{4}$$

Solving Equation (4) for n and using the value given in Table 26.1 for the refractive index of crystalline quartz, we find

$$n = \frac{n_q}{\sqrt{1+\sin^2 \theta_1}} = \frac{1.544}{\sqrt{1+\sin^2 34°}} = \boxed{1.35}$$

---

39. **REASONING** Brewster's law (Equation 26.5: $\tan \theta_B = n_2 / n_1$) relates the angle of incidence θ_B at which the reflected ray is completely polarized parallel to the surface to the indices of refraction n_1 and n_2 of the two media forming the interface. We can use Brewster's law for light incident from above to find the ratio of the refractive indices n_2/n_1. This ratio can then be used to find the Brewster angle for light incident from below on the same interface.

SOLUTION The index of refraction for the medium in which the incident ray occurs is designated by n_1. For the light striking from above $n_2 / n_1 = \tan \theta_B = \tan 65.0° = 2.14$. The same equation can be used when the light strikes from below if the indices of refraction are interchanged

$$\theta_B = \tan^{-1}\left(\frac{n_1}{n_2}\right) = \tan^{-1}\left(\frac{1}{n_2/n_1}\right) = \tan^{-1}\left(\frac{1}{2.14}\right) = \boxed{25.0°}$$

45. **REASONING** Because the refractive index of the glass depends on the wavelength (i.e., the color) of the light, the rays corresponding to different colors are bent by different amounts in the glass. We can use Snell's law (Equation 26.2: $n_1 \sin \theta_1 = n_2 \sin \theta_2$) to find the angle of refraction for the violet ray and the red ray. The angle between these rays can be found by the subtraction of the two angles of refraction.

SOLUTION In Table 26.2 the index of refraction for violet light in crown glass is 1.538, while that for red light is 1.520. According to Snell's law, then, the sine of the angle of refraction for the violet ray in the glass is $\sin \theta_2 = (1.000/1.538) \sin 45.00° = 0.4598$, so that

$$\theta_2 = \sin^{-1}(0.4598) = 27.37°$$

Similarly, for the red ray, $\sin \theta_2 = (1.000/1.520) \sin 45.00° = 0.4652$, from which it follows that

$$\theta_2 = \sin^{-1}(0.4652) = 27.72°$$

Therefore, the angle between the violet ray and the red ray in the glass is

$$27.72° - 27.37° = \boxed{0.35°}$$

---

47. **REASONING** We can use Snell's law (Equation 26.2: $n_1 \sin \theta_1 = n_2 \sin \theta_2$) at each face of the prism. At the first interface where the ray enters the prism, $n_1 = 1.000$ for air and $n_2 = n_g$ for glass. Thus, Snell's law gives

$$(1)\sin 60.0° = n_g \sin \theta_2 \qquad \text{or} \qquad \sin \theta_2 = \frac{\sin 60.0°}{n_g} \qquad (1)$$

We will represent the angles of incidence and refraction at the second interface as θ_1' and θ_2', respectively. Since the triangle is an equilateral triangle, the angle of incidence at the second interface, where the ray emerges back into air, is $\theta_1' = 60.0° - \theta_2$. Therefore, at the second interface, where $n_1 = n_g$ and $n_2 = 1.000$, Snell's law becomes

$$n_g \sin (60.0° - \theta_2) = (1)\sin \theta_2' \qquad (2)$$

We can now use Equations (1) and (2) to determine the angles of refraction θ_2' at which the red and violet rays emerge into the air from the prism.

SOLUTION

Red Ray   The index of refraction of flint glass at the wavelength of red light is $n_g = 1.662$. Therefore, using Equation (1), we can find the angle of refraction for the red ray as it enters the prism:

$$\sin \theta_2 = \frac{\sin 60.0°}{1.662} = 0.521 \qquad \text{or} \qquad \theta_2 = \sin^{-1} 0.521 = 31.4°$$

Substituting this value for θ_2 into Equation (2), we can find the angle of refraction at which the red ray emerges from the prism:

$$\sin \theta_2' = 1.662 \sin \left(60.0° - 31.4°\right) = 0.796 \qquad \text{or} \qquad \theta_2' = \sin^{-1} 0.796 = \boxed{52.7°}$$

Violet Ray   For violet light, the index of refraction for glass is $n_g = 1.698$. Again using Equation (1), we find

$$\sin \theta_2 = \frac{\sin 60.0°}{1.698} = 0.510 \qquad \text{or} \qquad \theta_2 = \sin^{-1} 0.510 = 30.7°$$

Using Equation (2), we find

$$\sin \theta_2' = 1.698 \sin \left(60.0° - 30.7°\right) = 0.831 \qquad \text{or} \qquad \theta_2' = \sin^{-1} 0.831 = \boxed{56.2°}$$

---

49. **REASONING** The ray diagram is constructed by drawing the paths of two rays from a point on the object. For convenience, we will choose the top of the object. The ray that is parallel to the principal axis will be refracted by the lens so that it passes through the focal point on the right of the lens. The ray that passes through the center of the lens passes through undeflected. The image is formed at the intersection of these two rays. In this case, the rays do not intersect on the right of the lens. However, if they are extended backwards they intersect on the left of the lens, locating a virtual, upright, and enlarged image.

SOLUTION

a. The ray-diagram, drawn to scale, is shown as follows:

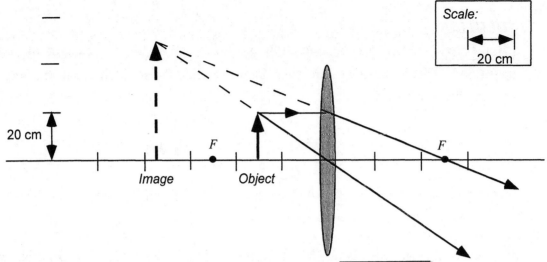

From the diagram, we see that the image distance is $\boxed{d_i = -75 \text{ cm}}$ and the magnification is $\boxed{+2.5}$. The negative image distance indicates that the image is virtual. The positive magnification indicates that the image is larger than the object.

b. From the thin-lens equation [Equation 26.6: $(1/d_o) + (1/d_i) = (1/f)$], we obtain

$$\frac{1}{d_i} = \frac{1}{f} - \frac{1}{d_o} = \frac{1}{50.0 \text{ m}} - \frac{1}{30.0 \text{ cm}} \quad \text{or} \quad \boxed{d_i = -75.0 \text{ cm}}$$

The magnification equation (Equation 26.7) gives the magnification to be

$$m = -\frac{d_i}{d_o} = -\frac{-75.0 \text{ cm}}{30.0 \text{ cm}} = \boxed{+2.50}$$

---

59. **REASONING AND SOLUTION**
    a. A real image must be projected on the drum; therefore, the lens in the copier must be a $\boxed{\text{converging lens}}$.

    b. If the document and its copy have the same size, but are inverted with respect to one another, the magnification equation (Equation 26.7) indicates that $m = -d_i/d_o = -1$. Therefore, $d_i/d_o = 1$ or $d_i = d_o$. Then, the thin-lens equation (Equation 26.6) gives

$$\frac{1}{d_i} + \frac{1}{d_o} = \frac{1}{f} = \frac{2}{d_o} \quad \text{or} \quad d_o = d_i = 2f$$

Therefore the document is located at a distance $\boxed{2f}$ from the lens.

c. Furthermore, the image is located at a distance of $\boxed{2f}$ from the lens.

61. **REASONING** The magnification equation (Equation 26.7) relates the object and image distances d_o and d_i, respectively, to the relative size of the of the image and object: $m = -(d_i / d_o)$. We consider two cases: in case 1, the object is placed 18 cm in front of a diverging lens. The magnification for this case is given by m_1. In case 2, the object is moved so that the magnification m_2 is reduced by a factor of 2 compared to that in case 1. In other words, we have $m_2 = \frac{1}{2} m_1$. Using Equation 26.7, we can write this as

$$-\frac{d_{i2}}{d_{o2}} = -\frac{1}{2}\left(\frac{d_{i1}}{d_{o1}}\right) \tag{1}$$

This expression can be solved for d_{o2}. First, however, we must find a numerical value for d_{i1}, and we must eliminate the variable d_{i2}.

SOLUTION
The image distance for case 1 can be found from the thin-lens equation [Equation 26.6: $(1/d_o) + (1/d_i) = (1/f)$]. The problem statement gives the focal length as $f = -12$ cm. Since the object is 18 cm in front of the diverging lens, $d_{o1} = 18$ cm. Solving for d_{i1}, we find

$$\frac{1}{d_{i1}} = \frac{1}{f} - \frac{1}{d_{o1}} = \frac{1}{-12 \text{ cm}} - \frac{1}{18 \text{ cm}} \quad \text{or} \quad d_{i1} = -7.2 \text{ cm}$$

where the minus sign indicates that the image is virtual. Solving Equation (1) for d_{o2}, we have

$$d_{o2} = 2 d_{i2}\left(\frac{d_{o1}}{d_{i1}}\right) \tag{2}$$

To eliminate d_{i2} from this result, we note that the thin-lens equation applied to case 2 gives

$$\frac{1}{d_{i2}} = \frac{1}{f} - \frac{1}{d_{o2}} = \frac{d_{o2} - f}{f d_{o2}} \quad \text{or} \quad d_{i2} = \frac{f d_{o2}}{d_{o2} - f}$$

Substituting this expression for d_{i2} into Equation (2), we have

$$d_{o2} = \left(\frac{2 f d_{o2}}{d_{o2} - f}\right)\left(\frac{d_{o1}}{d_{i1}}\right) \quad \text{or} \quad d_{o2} - f = 2 f\left(\frac{d_{o1}}{d_{i1}}\right)$$

Solving for d_{o2}, we find

$$d_{o2} = f\left[2\left(\frac{d_{o1}}{d_{i1}}\right)+1\right] = (-12\text{cm})\left[2\left(\frac{18\text{ cm}}{-7.2\text{ cm}}\right)+1\right] = \boxed{48\text{ cm}}$$

---

69. **REASONING** The problem can be solved using the thin-lens equation [Equation 26.6: $(1/d_o)+(1/d_i)=(1/f)$] twice in succession. We begin by using the thin lens-equation to find the location of the image produced by the converging lens; this image becomes the object for the diverging lens.

SOLUTION
a. The image distance for the converging lens is determined as follows:

$$\frac{1}{d_{i1}} = \frac{1}{f} - \frac{1}{d_{o1}} = \frac{1}{12.0\text{ cm}} - \frac{1}{36.0\text{ cm}} \quad \text{or} \quad d_{i1} = 18.0\text{ cm}$$

This image acts as the object for the diverging lens. Therefore,

$$\frac{1}{d_{i2}} = \frac{1}{f} - \frac{1}{d_{o2}} = \frac{1}{-6.00\text{ cm}} - \frac{1}{(30.0\text{ cm}-18.0\text{ cm})} \quad \text{or} \quad d_{i2} = -4.00\text{ cm}$$

Thus, the final image is located $\boxed{4.00\text{ cm to the left of the diverging lens}}$.

b. The magnification equation (Equation 26.7: $h_i/h_o = -d_i/d_o$) gives

$$\underbrace{m_c = -\frac{d_{i1}}{d_{o1}} = -\frac{18.0\text{ cm}}{36.0\text{ cm}} = -0.500}_{\text{Converging lens}} \qquad \underbrace{m_d = -\frac{d_{i2}}{d_{o2}} = -\frac{-4.00\text{ cm}}{12.0\text{ cm}} = 0.333}_{\text{Diverging lens}}$$

Therefore, the overall magnification is given by the product $m_c m_d = \boxed{-0.167}$.

c. Since the final image distance is negative, we can conclude that the image is $\boxed{\text{virtual}}$.

d. Since the overall magnification of the image is negative, the image is $\boxed{\text{inverted}}$.

e. The magnitude of the overall magnification is less than one; therefore, the final image is $\boxed{\text{smaller}}$.

71. **_REASONING_**   We begin by using the thin-lens equation [Equation 26.6: $(1/d_o) + (1/d_i) = (1/f)$] to locate the image produced by the lens. This image is then treated as the object for the mirror.

SOLUTION

a. The image distance from the diverging lens can be determined as follows:

$$\frac{1}{d_i} = \frac{1}{f} - \frac{1}{d_o} = \frac{1}{-8.00 \text{ cm}} - \frac{1}{20.0 \text{ cm}} \qquad \text{or} \qquad d_i = -5.71 \text{ cm}$$

The image produced by the lens is 5.71 cm to the left of the lens. The distance between this image and the concave mirror is 5.71 cm + 30.0 cm = 35.7 cm. The mirror equation [Equation 25.3: $(1/d_o) + (1/d_i) = (1/f)$] gives the image distance from the mirror:

$$\frac{1}{d_i} = \frac{1}{f} - \frac{1}{d_o} = \frac{1}{12.0 \text{ cm}} - \frac{1}{35.7 \text{ cm}} \qquad \text{or} \qquad \boxed{d_i = 18.1 \text{ cm}}$$

b. The image is $\boxed{\text{real}}$, because d_i is a positive number, indicating that the final image lies to the left of the concave mirror.

c. The image is $\boxed{\text{inverted}}$, because a diverging lens always produces an upright image, and the concave mirror produces an inverted image when the object distance is greater than the focal length of the mirror.

---

77. **_REASONING_** The closest she can read the magazine is when the image formed by the contact lens is at the near point of her eye, or $d_i = -138$ cm. The image distance is negative because the image is a virtual image (see Section 26.10). Since the focal length is also known, the object distance can be found from the thin-lens equation.

SOLUTION The object distance d_o is related to the focal length f and the image distance d_i by the thin-lens equation:

$$\frac{1}{d_o} = \frac{1}{f} - \frac{1}{d_i} = \frac{1}{35.1 \text{ cm}} - \frac{1}{-138 \text{ cm}} \qquad \text{or} \qquad d_o = \boxed{28.0 \text{ cm}} \qquad (26.6)$$

---

79. **_REASONING AND SOLUTION_** An optometrist prescribes contact lenses that have a focal length of 55.0 cm.

a. The focal length is positive (+55.0 cm); therefore, we can conclude that the lenses are $\boxed{\text{converging}}$.

b. As discussed in the text (see Section 26.10), farsightedness is corrected by converging lenses. Therefore, the person who wears these lens is $\boxed{\text{farsighted}}$.

c. If the lenses are designed so that objects no closer than 35.0 cm can be seen clearly, we have $d_o = 35.0$ cm. The thin-lens equation (Equation 26.6) gives the image distance:

$$\frac{1}{d_i} = \frac{1}{f} - \frac{1}{d_o} = \frac{1}{55.0 \text{ cm}} - \frac{1}{35.0 \text{ cm}} \qquad \text{or} \qquad d_i = -96.3 \text{ cm}$$

Thus, the near point is located $\boxed{96.3 \text{ cm}}$ from the eyes.

---

83. **REASONING** The angular size of a distant object in radians is approximately equal to the diameter of the object divided by the distance from the eye. We will use this definition to calculate the angular size of the quarter, and then, calculate the angular size of the sun; we can then form the ratio $\theta_{quarter} / \theta_{sun}$.

SOLUTION The angular sizes are

$$\theta_{quarter} \approx \frac{2.4 \text{ cm}}{70.0 \text{ cm}} = 0.034 \text{ rad} \qquad \text{and} \qquad \theta_{sun} \approx \frac{1.39 \times 10^9 \text{ m}}{1.50 \times 10^{11} \text{ m}} = 0.0093 \text{ rad}$$

Therefore, the ratio of the angular sizes is

$$\frac{\theta_{quarter}}{\theta_{sun}} = \frac{0.034 \text{ rad}}{0.0093 \text{ rad}} = \boxed{3.7}$$

---

89. **REASONING AND SOLUTION** The information given allows us to determine the near point for this farsighted person. With $f = 45.4$ cm and $d_o = 25.0$ cm, we find from the thin-lens equation that

$$\frac{1}{d_i} = \frac{1}{f} - \frac{1}{d_o} = \frac{1}{45.4 \text{ cm}} - \frac{1}{25.0 \text{ cm}} \qquad \text{or} \qquad d_i = -55.6 \text{ cm}$$

Therefore, this person's near point, N, is 55.6 cm. We now need to find the focal length of the magnifying glass based on the near point for a normal eye, i.e., $M = N/f + 1$ (Equation 26.10, with $d_i = -N$), where $N = 25.0$ cm. Solving for the focal length gives

$$f = \frac{N}{M-1} = \frac{25.0 \text{ cm}}{7.50 - 1} = 3.85 \text{ cm}$$

We can now determine the maximum angular magnification for the farsighted person

$$M = \frac{N}{f} + 1 = \frac{55.6 \text{ cm}}{3.85 \text{ cm}} + 1 = \boxed{15.4}$$

91. **REASONING** The angular magnification of a compound microscope is given by Equation 26.11:

$$M \approx -\frac{(L - f_e)N}{f_o f_e}$$

where f_o is the focal length of the objective, f_e is the focal length of the eyepiece, and L is the separation between the two lenses. This expression can be solved for f_o, the focal length of the objective.

SOLUTION Solving for f_o, we find that the focal length of the objective is

$$f_o = -\frac{(L - f_e)N}{f_e M} = -\frac{(16.0 \text{ cm} - 1.4 \text{ cm})(25 \text{ cm})}{(1.4 \text{ cm})(-320)} = \boxed{0.81 \text{ cm}}$$

97. **REASONING** Knowing the angles subtended at the unaided eye and with the telescope will allow us to determine the angular magnification of the telescope. Then, since the angular magnification is related to the focal lengths of the eyepiece and the objective, we will use the known focal length of the eyepiece to determine the focal length of the objective.

SOLUTION From Equation 26.12, we have

$$M = \frac{\theta'}{\theta} = -\frac{f_o}{f_e}$$

where θ is the angle subtended by the unaided eye and θ' is the angle subtended when the telescope is used. We note that θ' is negative, since the telescope produces an inverted image. Thus, using Equation 26.12, we find

$$f_o = -\frac{f_e \theta'}{\theta} = -\frac{(0.032 \text{ m})\left(-2.8 \times 10^{-3} \text{ rad}\right)}{8.0 \times 10^{-5} \text{ rad}} = \boxed{1.1 \text{ m}}$$

99. **REASONING AND SOLUTION** The angular magnification of an astronomical telescope, is given by Equation 26.12 as $M \approx -f_o / f_e$. Solving for the focal length of the eyepiece, we find

$$f_e \approx -\frac{f_o}{M} = -\frac{48.0 \text{ cm}}{(-184)} = \boxed{0.261 \text{ cm}}$$

---

105. **REASONING** The ray diagram is constructed by drawing the paths of two rays from a point on the object. For convenience, we choose the top of the object. The ray that is parallel to the principal axis will be refracted by the lens and pass through the focal point on the right side. The ray that passes through the center of the lens passes through undeflected. The image is formed at the intersection of these two rays on the right side of the lens.

SOLUTION The following ray diagram (to scale) shows that $\boxed{d_i = 18 \text{ cm}}$ and reveals a real, inverted, and enlarged image.

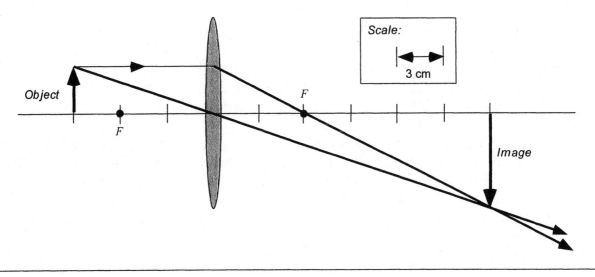

---

107. **REASONING AND SOLUTION**

a. The index of refraction n_2 of the liquid must match that of the glass, or $n_2 = \boxed{1.50}$.

b. When none of the light is transmitted into the liquid, the angle of incidence must be equal to or greater than the critical angle. According to Equation 26.4, the critical angle θ_c is given by $\sin \theta_c = n_2/n_1$, where n_2 is the index of refraction of the liquid and n_1 is that of the glass. Therefore,

$$n_2 = n_1 \sin \theta_c = (1.50) \sin 58.0° = \boxed{1.27}$$

If n_2 were larger than 1.27, the critical angle would also be larger, and light would be transmitted from the glass into the liquid. Thus, $n_2 = 1.27$ represents the largest index of refraction of the liquid such that none of the light is transmitted into the liquid.

---

109. **REASONING** The far point is 5.0 m from the right eye, and 6.5 m from the left eye. For an object infinitely far away ($d_o = \infty$), the image distances for the corrective lenses are then -5.0 m for the right eye and -6.5 m for the left eye, the negative sign indicating that the images are virtual images formed to the left of the lenses. The thin-lens equation [Equation 26.6: $(1/d_o) + (1/d_i) = (1/f)$] can be used to find the focal length. Then, Equation 26.8 can be used to find the refractive power for the lens for each eye.

SOLUTION Since the object distance d_o is essentially infinite, $1/d_o = 1/\infty = 0$, and the thin-lens equation becomes $1/d_i = 1/f$, or $d_i = f$. Therefore, for the right eye, $f = -5.0$ m, and the refractive power is (see Equation 26.8)

[**Right eye**] $\quad \dfrac{\text{Refractive power}}{\text{(in diopters)}} = \dfrac{1}{f} = \dfrac{1}{(-5.0 \text{ m})} = \boxed{-0.20 \text{ diopters}}$

Similarly, for the left eye, $f = -6.5$ m, and the refractive power is

[**Left eye**] $\quad \dfrac{\text{Refractive power}}{\text{(in diopters)}} = \dfrac{1}{f} = \dfrac{1}{(-6.5 \text{ m})} = \boxed{-0.15 \text{ diopters}}$

---

111. **REASONING AND SOLUTION** Equation 26.6 gives the thin-lens equation which relates the object and image distances d_o and d_i, respectively, to the focal length f of the lens: $(1/d_o) + (1/d_i) = (1/f)$.

The optical arrangement is similar to that in Figure 26.26. The problem statement gives values for the focal length ($f = 50.0$ mm) and the maximum lens-to-image-sensor distance ($d_i = 275$ mm). Therefore, the maximum distance that the object can be located in front of the lens at a distance that is as small as d_o, where d_o is found as follows:

$$\frac{1}{d_o} = \frac{1}{f} - \frac{1}{d_i} = \frac{1}{50.0 \text{ mm}} - \frac{1}{275 \text{ mm}} \quad \text{or} \quad \boxed{d_o = 61.1 \text{ mm}}$$

---

113. **REASONING** We begin by using Snell's law (Equation 26.2: $n_1 \sin \theta_1 = n_2 \sin \theta_2$) to find the index of refraction of the material. Then we will use Equation 26.1, the definition of the index of refraction ($n = c/v$) to find the speed of light in the material.

SOLUTION From Snell's law, the index of refraction of the material is

$$n_2 = \frac{n_1 \sin \theta_1}{\sin \theta_2} = \frac{(1.000) \sin 63.0°}{\sin 47.0°} = 1.22$$

Then, from Equation 26.1, we find that the speed of light v in the material is

$$v = \frac{c}{n_2} = \frac{3.00 \times 10^8 \text{ m/s}}{1.22} = \boxed{2.46 \times 10^8 \text{ m/s}}$$

---

119. **REASONING** The optical arrangement is similar to that in Figure 26.26. We begin with the thin-lens equation, [Equation 26.6: $(1/d_o) + (1/d_i) = (1/f)$]. Since the distance between the moon and the camera is so large, the object distance d_o is essentially infinite, and $1/d_o = 1/\infty = 0$. Therefore the thin-lens equation becomes $1/d_i = 1/f$ or $d_i = f$. The diameter of the moon's imagine on the slide film is equal to the image height h_i, as given by the magnification equation (Equation 26.7: $h_i / h_o = -d_i / d_o$).

When the slide is projected onto a screen, the situation is similar to that in Figure 26.27. In this case, the thin-lens and magnification equations can be used in their usual forms.

SOLUTION
a. Solving the magnification equation for h_i gives

$$h_i = -h_o \frac{d_i}{d_o} = (-3.48 \times 10^6 \text{ m}) \left(\frac{50.0 \times 10^{-3} \text{ m}}{3.85 \times 10^8 \text{ m}} \right) = -4.52 \times 10^{-4} \text{ m}$$

The diameter of the moon's image on the slide film is, therefore, $\boxed{4.52 \times 10^{-4} \text{ m}}$.

b. From the magnification equation, $h_i = -h_o (d_i / d_o)$. We need to find the ratio d_i / d_o. Beginning with the thin-lens equation, we have

$$\frac{1}{d_o} + \frac{1}{d_i} = \frac{1}{f} \quad \text{or} \quad \frac{1}{d_o} = \frac{1}{f} - \frac{1}{d_i} \quad \text{which leads to} \quad \frac{d_i}{d_o} = \frac{d_i}{f} - \frac{d_i}{d_i} = \frac{d_i}{f} - 1$$

Therefore,

$$h_i = -h_o \left(\frac{d_i}{f} - 1 \right) = -\left(4.52 \times 10^{-4} \text{ m} \right) \left(\frac{15.0 \text{ m}}{110.0 \times 10^{-3} \text{ m}} - 1 \right) = -6.12 \times 10^{-2} \text{ m}$$

The diameter of the image on the screen is $\boxed{6.12 \times 10^{-2} \text{ m}}$.

121. **REASONING** A contact lens is placed directly on the eye. Therefore, the object distance, which is the distance from the book to the lens, is 25.0 cm. The near point can be determined from the thin-lens equation [Equation 26.6: $(1/d_o) + (1/d_i) = (1/f)$].

SOLUTION
a. Using the thin-lens equation, we have

$$\frac{1}{d_i} = \frac{1}{f} - \frac{1}{d_o} = \frac{1}{65.0\text{ cm}} - \frac{1}{25.0\text{ cm}} \qquad \text{or} \qquad d_i = -40.6\text{ cm}$$

In other words, at age 40, the man's near point is 40.6 cm. Similarly, when the man is 45, we have

$$\frac{1}{d_i} = \frac{1}{f} - \frac{1}{d_o} = \frac{1}{65.0\text{ cm}} - \frac{1}{29.0\text{ cm}} \qquad \text{or} \qquad d_i = -52.4\text{ cm}$$

and his near point is 52.4 cm. Thus, the man's near point has changed by $52.4\text{ cm} - 40.6\text{ cm} = \boxed{11.8\text{ cm}}$.

b. With $d_o = 25.0$ cm and $d_i = -52.4$ cm, the focal length of the lens is found as follows:

$$\frac{1}{f} = \frac{1}{d_o} + \frac{1}{d_i} = \frac{1}{25.0\text{ cm}} + \frac{1}{(-52.4\text{ cm})} \qquad \text{or} \qquad \boxed{f = 47.8\text{ cm}}$$

---

123. **REASONING AND SOLUTION** Using the thin-lens equation, we can find the first image distance d_i with respect to the objective:

$$\frac{1}{d_i} = \frac{1}{f_o} - \frac{1}{d_o} = \frac{1}{1.500\text{ m}} - \frac{1}{114.00\text{ m}} \qquad \text{or} \qquad d_i = 1.520\text{ m} \qquad (1)$$

Using the magnification equation, we can find the linear magnification m achieved by the objective:

$$m = -\frac{d_i}{d_o} = -\frac{1.520\text{ m}}{114.00\text{ m}} = -0.01333 \qquad (2)$$

We then use this "first image" as the object for the second lens (the eyepiece). With the aid of the thin-lens equation, we can determine the distance d_i' of the final image with respect to the eyepiece:

$$\frac{1}{d_i'} = \frac{1}{f_e} - \frac{1}{d_o'} = \frac{1}{0.070 \text{ m}} - \frac{1}{0.050 \text{ m}} \qquad \text{or} \qquad d_i' = -0.18 \text{ m} \qquad (3)$$

In Equation (3) we have used $d_o' = 0.050 \text{ m}$. This follows because we know that the separation between the objective and the eyepiece is $L \approx f_o + f_e$ (see Example 17 in the text). Therefore, we know that $L \approx 1.500 \text{ m} + 0.070 \text{ m} = 1.570 \text{ m}$, and we can determine that the distance of the "first image" from the second lens (the eyepiece) is $L - 1.520 \text{ m} \approx 1.570 \text{ m} - 1.520 \text{ m} = 0.050 \text{ m}$.

Using the magnification equation, we can find the linear magnification m' achieved by the eyepiece:

$$m' = -\frac{d_i'}{d_o'} = -\frac{-0.18 \text{ m}}{0.050 \text{ m}} = +3.6 \qquad (4)$$

Using Equations (2) and (4), we find that the total linear magnification achieved by both the objective and the eyepiece is

$$m_{\text{Total}} = \frac{h_i'}{h_o} = m \times m' = (-0.01333)(+3.6) = -0.048 \qquad (5)$$

where h_i' is the height of the final image and h_o is the height of the initial object.

However, we need the angular magnification, which, according to Equation 26.9, is

$$M = \frac{\theta'}{\theta} = \frac{-\dfrac{h_i'}{d_i'}}{\dfrac{h_o}{d_o + f_o + f_e}} = \left(-\frac{h_i'}{h_o}\right)\left(\frac{d_o + f_o + f_e}{d_i'}\right) \qquad (6)$$

In Equation (6) $\theta' = -\dfrac{h_i'}{d_i'}$ is the angular size of the final image and $\theta = \dfrac{h_o}{d_o + f_o + f_e}$ is the angular size of the initial object.

Substituting Equations (3) and (5) into Equation (6) and using the given data for $d_o, f_o,$ and f_e, we find that

$$M = \left(-\frac{h_i'}{h_o}\right)\left(\frac{d_o + f_o + f_e}{d_i'}\right) = \left[-(-0.048)\right]\left(\frac{114.00 \text{ m} + 1.500 \text{ m} + 0.070 \text{ m}}{-0.18}\right) = \boxed{-31}$$

CHAPTER 27 | *INTERFERENCE AND THE WAVE NATURE OF LIGHT*

1. ***REASONING AND SOLUTION*** To decide whether constructive or destructive interference occurs, we need to determine the wavelength of the wave. For electromagnetic waves, Equation 16.1 can be written $f\lambda = c$, so that

$$\lambda = \frac{c}{f} = \frac{3.00 \times 10^8 \text{ m}}{536 \times 10^3 \text{ Hz}} = 5.60 \times 10^2 \text{ m}$$

Since the two wave sources are in phase, constructive interference occurs when the path difference is an integer number of wavelengths, and destructive interference occurs when the path difference is an odd number of half wavelengths. We find that the path difference is 8.12 km – 7.00 km = 1.12 km. The number of wavelengths in this path difference is $(1.12 \times 10^3 \text{ m})/(5.60 \times 10^2 \text{ m}) = 2.00$. Therefore, | constructive interference occurs |.

---

11. ***REASONING*** The light that travels through the plastic has a different path length than the light that passes through the unobstructed slit. Since the center of the screen now appears dark, rather than bright, destructive interference, rather than constructive interference occurs there. This means that the difference between the number of wavelengths in the plastic sheet and that in a comparable thickness of air is $\frac{1}{2}$.

SOLUTION The wavelength of the light in the plastic sheet is given by Equation 27.3 as

$$\lambda_{\text{plastic}} = \frac{\lambda_{\text{vacuum}}}{n} = \frac{586 \times 10^{-9} \text{ m}}{1.60} = 366 \times 10^{-9} \text{ m}$$

The number of wavelengths contained in a plastic sheet of thickness t is

$$N_{\text{plastic}} = \frac{t}{\lambda_{\text{plastic}}} = \frac{t}{366 \times 10^{-9} \text{ m}}$$

The number of wavelengths contained in an equal thickness of air is

$$N_{\text{air}} = \frac{t}{\lambda_{\text{air}}} = \frac{t}{586 \times 10^{-9} \text{ m}}$$

where we have used the fact that $\lambda_{air} \approx \lambda_{vacuum}$. Destructive interference occurs when the difference, $N_{plastic} - N_{air}$, in the number of wavelengths is $\frac{1}{2}$:

$$N_{plastic} - N_{air} = \frac{1}{2}$$

$$\frac{t}{366 \times 10^{-9} \text{ m}} - \frac{t}{586 \times 10^{-9} \text{ m}} = \frac{1}{2}$$

Solving this equation for t yields $t = 487 \times 10^{-9}$ m = $\boxed{487 \text{ nm}}$.

---

13. **REASONING** When the light strikes the film from above, the wave reflected from the top surface of the film undergoes a phase shift that is equivalent to one-half of a wavelength, since the light travels from a smaller refractive index (n_{air} = 1.00) toward a larger refractive index (n_{film} = 1.33). On the other hand, there is no phase shift when the light reflects from the bottom surface of the film, since the light travels from a larger refractive index (n_{film} = 1.33) toward a smaller refractive index (n_{air} = 1.00). Thus, the net phase change due to reflection from the two surfaces is equivalent to one-half of a wavelength in the film. This half-wavelength must be combined with the extra distance $2t$ traveled by the wave reflected from the bottom surface, where t is the film thickness. Thus, the condition for constructive interference is

$$\underbrace{2t}_{\substack{\text{Extra distance} \\ \text{traveled by wave} \\ \text{in the film}}} + \underbrace{\tfrac{1}{2}\lambda_{film}}_{\substack{\text{Half-wavelength} \\ \text{net phase change} \\ \text{due to reflection}}} = \underbrace{\lambda_{film}, \, 2\lambda_{film}, \, 3\lambda_{film}, \, \cdots}_{\substack{\text{Condition for} \\ \text{constructive interference}}}$$

We will use this relation to find the two smallest non-zero film thicknesses for which constructive interference occurs in the reflected light.

SOLUTION The smallest film thickness occurs when the condition for constructive interference is λ_{film}. Then, the relation above becomes

$$2t + \tfrac{1}{2}\lambda_{film} = \lambda_{film} \qquad \text{or} \qquad t = \tfrac{1}{4}\lambda_{film}$$

Since $\lambda_{film} = \dfrac{\lambda_{vacuum}}{n_{film}}$ (Equation 27.3), we have that

$$t = \tfrac{1}{4}\lambda_{film} = \tfrac{1}{4}\left(\frac{\lambda_{vacuum}}{n_{film}}\right) = \tfrac{1}{4}\left(\frac{691 \text{ nm}}{1.33}\right) = \boxed{1.30 \times 10^2 \text{ nm}}$$

The next smallest film thickness occurs when the condition for constructive interference is $2\lambda_{\text{film}}$. Then, we have

$$2t + \tfrac{1}{2}\lambda_{\text{film}} = 2\lambda_{\text{film}} \qquad \text{or} \qquad t = \tfrac{3}{4}\lambda_{\text{film}} = \tfrac{3}{4}\left(\frac{\lambda_{\text{vacuum}}}{n_{\text{film}}}\right) = \tfrac{3}{4}\left(\frac{691\text{ nm}}{1.33}\right) = \boxed{3.90 \times 10^2\text{ nm}}$$

---

19. **REASONING** To solve this problem, we must express the condition for constructive interference in terms of the film thickness t and the wavelength λ_{film} of the light in the soap film. We must also take into account any phase changes that occur upon reflection.

SOLUTION For the reflection at the top film surface, the light travels from air, where the refractive index is smaller ($n = 1.00$), toward the film, where the refractive index is larger ($n = 1.33$). Associated with this reflection there is a phase change that is equivalent to one-half of a wavelength. For the reflection at the bottom film surface, the light travels from the film, where the refractive index is larger ($n = 1.33$), toward air, where the refractive index is smaller ($n = 1.00$). Associated with this reflection, there is no phase change. As a result of these two reflections, there is a net phase change that is equivalent to one-half of a wavelength. To obtain the condition for constructive interference, this net phase change must be added to the phase change that arises because of the film thickness t, which is traversed twice by the light that penetrates it. For constructive interference we find that

$$2t + \tfrac{1}{2}\lambda_{\text{film}} = \lambda_{\text{film}}, 2\lambda_{\text{film}}, 3\lambda_{\text{film}}, \ldots$$

or

$$2t = \left(m + \tfrac{1}{2}\right)\lambda_{\text{film}}, \qquad \text{where } m = 0, 1, 2, \ldots$$

Equation 27.3 indicates that $\lambda_{\text{film}} = \lambda_{\text{vacuum}}/n$. Using this expression and the fact that $m = 0$ for the minimum thickness t, we find that the condition for constructive interference becomes

$$2t = \left(m + \tfrac{1}{2}\right)\lambda_{\text{film}} = \left(0 + \tfrac{1}{2}\right)\left(\frac{\lambda_{\text{vacuum}}}{n}\right)$$

or

$$t = \frac{\lambda_{\text{vacuum}}}{4n} = \frac{611\text{ nm}}{4(1.33)} = \boxed{115\text{ nm}}$$

---

25. **REASONING** This problem can be solved by using Equation 27.4 for the value of the angle θ when $m = 1$ (first dark fringe).

SOLUTION

a. When the slit width is $W = 1.8 \times 10^{-4}$ m and $\lambda = 675$ nm $= 675 \times 10^{-9}$ m, we find, according to Equation 27.4,

$$\theta = \sin^{-1}\left(m\frac{\lambda}{W}\right) = \sin^{-1}\left[(1)\frac{675 \times 10^{-9} \text{ m}}{1.8 \times 10^{-4} \text{ m}}\right] = \boxed{0.21°}$$

b. Similarly, when the slit width is $W = 1.8 \times 10^{-6}$ m and $\lambda = 675 \times 10^{-9}$ m, we find

$$\theta = \sin^{-1}\left[(1)\frac{675 \times 10^{-9} \text{ m}}{1.8 \times 10^{-6} \text{ m}}\right] = \boxed{22°}$$

---

27. **REASONING** The width W of the slit and the wavelength λ of the light are related to the angle θ defining the location of a dark fringe in the single-slit diffraction pattern, so we can determine the width from values for the wavelength and the angle. The wavelength is given. We can obtain the angle from the width given for the central bright fringe on the screen and the distance between the screen and the slit. To do this, we will use trigonometry and the fact that the width of the central bright fringe is defined by the first dark fringe on either side of the central bright fringe.

SOLUTION The angle that defines the location of a dark fringe in the diffraction pattern can be determined according to

$$\sin\theta = m\frac{\lambda}{W} \qquad m = 1, 2, 3,\dots \qquad (27.4)$$

Recognizing that we need the case for $m = 1$ (the first dark fringe on either side of the central bright fringe determines the width of the central bright fringe) and solving for W give

$$W = \frac{\lambda}{\sin\theta} \qquad (1)$$

Referring to Figure 27.22 in the text, we see that the width of the central bright fringe is $2y$, where trigonometry shows that

$$2y = 2L\tan\theta \quad \text{or} \quad \theta = \tan^{-1}\left(\frac{2y}{2L}\right) = \tan^{-1}\left[\frac{0.050 \text{ m}}{2(0.60 \text{ m})}\right] = 2.4°$$

Substituting this value for θ into Equation (1), we find that

$$W = \frac{\lambda}{\sin\theta} = \frac{510 \times 10^{-9} \text{ m}}{\sin 2.4°} = \boxed{1.2 \times 10^{-5} \text{ m}}$$

---

31. **REASONING** The width of the central bright fringe is defined by the location of the first dark fringe on either side of it. According to Equation 27.4, the angle θ locating the first dark fringe can be obtained from $\sin \theta = \lambda/W$, where λ is the wavelength of the light and W is the width of the slit. According to the drawing, $\tan \theta = y/L$, where y is half the width of the central bright fringe and L is the distance between the slit and the screen.

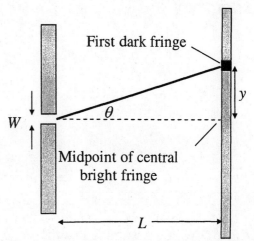

First dark fringe

W

θ

y

Midpoint of central bright fringe

L

SOLUTION Since the angle θ is small, we can use the fact that $\sin \theta \approx \tan \theta$. Since $\sin \theta = \lambda/W$ and $\tan \theta = y/L$, we have

$$\frac{\lambda}{W} = \frac{y}{L} \quad \text{or} \quad W = \frac{\lambda L}{y}$$

Applying this result to both slits gives

$$\frac{W_2}{W_1} = \frac{\lambda L / y_2}{\lambda L / y_1} = \frac{y_1}{y_2}$$

$$W_2 = W_1 \frac{y_1}{y_2} = \left(3.2 \times 10^{-5} \text{ m}\right) \frac{\frac{1}{2}(1.2 \text{ cm})}{\frac{1}{2}(1.9 \text{ cm})} = \boxed{2.0 \times 10^{-5} \text{ m}}$$

---

35. **REASONING** According to Rayleigh's criterion, the two taillights must be separated by a distance s sufficient to subtend an angle $\theta_{\min} \approx 1.22 \lambda / D$ at the pupil of the observer's eye. Recalling that this angle must be expressed in radians, we relate $\theta_{\min}$ to the distances s and L.

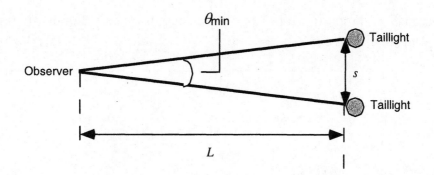

$\theta_{\min}$

Taillight

Observer

s

Taillight

L

SOLUTION The wavelength λ is 660 nm. Therefore, we have from Equation 27.6

$$\theta_{min} \approx 1.22\frac{\lambda}{D} = 1.22\left(\frac{660\times10^{-9}\ m}{7.0\times10^{-3}\ m}\right) = 1.2\times10^{-4}\ rad$$

According to Equation 8.1, the distance L between the observer and the taillights is

$$L = \frac{s}{\theta_{min}} = \frac{1.2\ m}{1.2\times10^{-4}\ rad} = \boxed{1.0\times10^{4}\ m}$$

41. **REASONING** Assuming that the angle θ_{min} is small, the distance y between the blood cells is given by

$$y = f\theta_{min} \qquad (8.1)$$

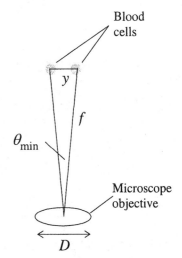

Blood cells

y

f

θ_{min}

Microscope objective

D

where f is the distance between the microscope objective and the cells (which is given as the focal length of the objective). However, the minimum angular separation θ_{min} of the cells is given by the Rayleigh criterion as $\theta_{min} = 1.22\ \lambda/D$ (Equation 27.6), where λ is the wavelength of the light and D is the diameter of the objective. These two relations can be used to find an expression for y in terms of λ.

SOLUTION
a. Substituting Equation 27.6 into Equation 8.1 yields

$$y = f\theta_{min} = f\left(\frac{1.22\lambda}{D}\right)$$

Since it is given that $f = D$, we see that $y = \boxed{1.22\ \lambda}$.

b. Because y is proportional to λ, the wavelength must be $\boxed{\text{shorter}}$ to resolve cells that are closer together.

43. **REASONING AND SOLUTION** According to Equation 27.7, the angles that correspond to the first-order ($m = 1$) maximum for the two wavelengths in question are:

a. For $\lambda = 660$ nm $= 660 \times 10^{-9}$ m,

$$\theta = \sin^{-1}\left(m\frac{\lambda}{d}\right) = \sin^{-1}\left[(1)\left(\frac{660 \times 10^{-9} \text{ m}}{1.1 \times 10^{-6} \text{ m}}\right)\right] = \boxed{37°}$$

b. For $\lambda = 410$ nm $= 410 \times 10^{-9}$ m,

$$\theta = \sin^{-1}\left(m\frac{\lambda}{d}\right) = \sin^{-1}\left[(1)\left(\frac{410 \times 10^{-9} \text{ m}}{1.1 \times 10^{-6} \text{ m}}\right)\right] = \boxed{22°}$$

---

45. **REASONING AND SOLUTION** The geometry of the situation is shown below.

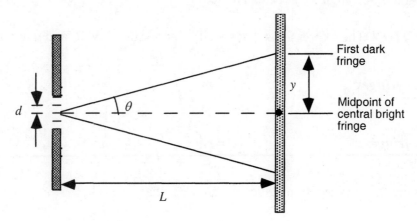

From the geometry, we have

$$\tan\theta = \frac{y}{L} = \frac{0.60 \text{ mm}}{3.0 \text{ mm}} = 0.20 \qquad \text{or} \qquad \theta = 11.3°$$

Then, solving Equation 27.7 with $m = 1$ for the separation d between the slits, we have

$$d = \frac{m\lambda}{\sin\theta} = \frac{(1)\left(780 \times 10^{-9} \text{ m}\right)}{\sin 11.3°} = \boxed{4.0 \times 10^{-6} \text{ m}}$$

---

51. **REASONING** For a diffraction grating, the angular position θ of a principal maximum on the screen is given by Equation 27.7 as $\sin\theta = m\lambda/d$ with $m = 0, 1, 2, 3, \ldots$

SOLUTION When the fourth-order principal maximum of light A exactly overlaps the third-order principal maximum of light B, we have

$$\sin\theta_A = \sin\theta_B$$

$$\frac{4\lambda_A}{d} = \frac{3\lambda_B}{d} \quad \text{or} \quad \frac{\lambda_A}{\lambda_B} = \boxed{\frac{3}{4}}$$

---

55. **REASONING** The angles θ that determine the locations of the dark and bright fringes in a Young's double-slit experiment are related to the integers m that identify the fringes, the wavelength λ of the light, and the separation d between the slits. Since values are given for m, λ, and d, the angles can be calculated.

SOLUTION The expressions that specify θ in terms of m, λ, and d are as follows:

Bright fringes $\qquad \sin\theta = m\dfrac{\lambda}{d} \qquad\qquad m = 0, 1, 2, 3, \ldots \qquad\qquad (27.1)$

Dark fringes $\qquad \sin\theta = \left(m + \dfrac{1}{2}\right)\dfrac{\lambda}{d} \qquad m = 0, 1, 2, 3, \ldots \qquad (27.2)$

Applying these expressions gives the answers that we seek.

a. $\qquad \sin\theta = \left(m + \dfrac{1}{2}\right)\dfrac{\lambda}{d} \quad \text{or} \quad \theta = \sin^{-1}\left[\left(0 + \dfrac{1}{2}\right)\dfrac{520\times10^{-9}\text{ m}}{1.4\times10^{-6}\text{ m}}\right] = \boxed{11°}$

b. $\qquad \sin\theta = m\dfrac{\lambda}{d} \qquad\quad \text{or} \quad \theta = \sin^{-1}\left[(1)\dfrac{520\times10^{-9}\text{ m}}{1.4\times10^{-6}\text{ m}}\right] = \boxed{22°}$

c. $\qquad \sin\theta = \left(m + \dfrac{1}{2}\right)\dfrac{\lambda}{d} \quad \text{or} \quad \theta = \sin^{-1}\left[\left(1 + \dfrac{1}{2}\right)\dfrac{520\times10^{-9}\text{ m}}{1.4\times10^{-6}\text{ m}}\right] = \boxed{34°}$

d. $\qquad \sin\theta = m\dfrac{\lambda}{d} \qquad\quad \text{or} \quad \theta = \sin^{-1}\left[(2)\dfrac{520\times10^{-9}\text{ m}}{1.4\times10^{-6}\text{ m}}\right] = \boxed{48°}$

57. **REASONING** The slit separation d is given by Equation 27.1 with $m = 1$; namely $d = \lambda / \sin\theta$. As shown in Example 1 in the text, the angle θ is given by $\theta = \tan^{-1}(y/L)$.

SOLUTION The angle θ is

$$\theta = \tan^{-1}\left(\frac{0.037 \text{ m}}{4.5 \text{ m}}\right) = 0.47°$$

Therefore, the slit separation d is

$$d = \frac{\lambda}{\sin\theta} = \frac{490 \times 10^{-9} \text{ m}}{\sin 0.47°} = \boxed{6.0 \times 10^{-5} \text{ m}}$$

---

61. **REASONING** The angle θ that locates the first-order maximum produced by a grating with 3300 lines/cm is given by Equation 27.7, $\sin\theta = m\lambda / d$, with the order of the fringes given by $m = 0, 1, 2, 3, \ldots$ Any two of the diffraction patterns will overlap when their angular positions are the same.

SOLUTION Since the grating has 3300 lines/cm, we have

$$d = \frac{1}{3300 \text{ lines/cm}} = 3.0 \times 10^{-4} \text{ cm} = 3.0 \times 10^{-6} \text{ m}$$

a. In first order, $m = 1$; therefore, for violet light,

$$\theta = \sin^{-1}\left(m\frac{\lambda}{d}\right) = \sin^{-1}\left[(1)\left(\frac{410 \times 10^{-9} \text{ m}}{3.0 \times 10^{-6} \text{ m}}\right)\right] = \boxed{7.9°}$$

Similarly for red light,

$$\theta = \sin^{-1}\left(m\frac{\lambda}{d}\right) = \sin^{-1}\left[(1)\left(\frac{660 \times 10^{-9} \text{ m}}{3.0 \times 10^{-6} \text{ m}}\right)\right] = \boxed{13°}$$

b. Repeating the calculation for the second order maximum ($m = 2$), we find that

($m = 2$)	
for violet	$\theta = 16°$
for red	$\theta = 26°$

c. Repeating the calculation for the third order maximum ($m = 3$), we find that

($m = 3$)	
for violet	$\theta = 24°$
for red	$\theta = 41°$

d. Comparisons of the values for θ calculated in parts (a), (b) and (c) show that the second and third orders overlap .

CHAPTER 28 | *SPECIAL RELATIVITY*

1. ***REASONING*** Since the "police car" is moving relative to the earth observer, the earth observer measures a greater time interval Δt between flashes. Since both the proper time Δt_0 (as observed by the officer) and the dilated time Δt (as observed by the person on earth) are known, the speed of the "police car" relative to the observer can be determined from the time dilation relation, Equation 28.1.

    SOLUTION According to Equation 28.1, the dilated time interval between flashes is $\Delta t = \Delta t_0 / \sqrt{1-(v^2/c^2)}$, where Δt_0 is the proper time. Solving for the speed v, we find

    $$v = c\sqrt{1-\left(\frac{\Delta t_0}{\Delta t}\right)^2} = (3.0\times10^8 \text{ m/s})\sqrt{1-\left(\frac{1.5 \text{ s}}{2.5 \text{ s}}\right)^2} = \boxed{2.4\times10^8 \text{ m/s}}$$

---

3. ***REASONING*** The expression for time dilation is, according to Equation 28.1,

    $$\Delta t = \frac{\Delta t_0}{\sqrt{1-v^2/c^2}}$$

    For a given event, it relates the proper time interval Δt_0 to the time interval Δt that would be measured by an observer moving at a speed v relative to the frame of reference in which the event takes place.

    We must consider two situations; in the first situation, the Klingon spacecraft has a speed of $0.75c$ with respect to the earth. In the second situation, the craft has a speed of $0.94c$ relative to the earth. We will refer to these two situations as A and B, respectively.

    Since the proper time interval always has the same value, $(\Delta t_0)_A = (\Delta t_0)_B$. We can express both sides of this expression using Equation 28.1. The result can be solved for Δt_B.

    SOLUTION Use of Equation 28.1 gives

    $$\Delta t_A \sqrt{1-v_A^2/c^2} = \Delta t_B \sqrt{1-v_B^2/c^2}$$

    $$\Delta t_B = \Delta t_A \frac{\sqrt{1-v_A^2/c^2}}{\sqrt{1-v_B^2/c^2}} = \Delta t_A \sqrt{\frac{1-(v_A/c)^2}{1-(v_B/c)^2}} = (37.0 \text{ h})\sqrt{\frac{1-(0.75c/c)^2}{1-(0.94c/c)^2}} = \boxed{72 \text{ h}}$$

---

9. ***REASONING*** All standard meter sticks at rest have a length of 1.00 m for observers who are at rest with respect to them. Thus, 1.00 m is the proper length L_0 of the meter stick. When the meter stick moves with speed v relative to an earth-observer, its length $L = 0.500$ m will be a contracted length. Since both L_0 and L are known, v can be found directly from Equation 28.2, $L = L_0\sqrt{1-\left(v^2/c^2\right)}$.

SOLUTION Solving Equation 28.2 for v, we find that

$$v = c\sqrt{1-\left(\frac{L}{L_0}\right)^2} = (3.00 \times 10^8 \text{ m/s})\sqrt{1-\left(\frac{0.500 \text{ m}}{1.00 \text{ m}}\right)^2} = \boxed{2.60 \times 10^8 \text{ m/s}}$$

---

11. ***REASONING AND SOLUTION*** The length L_0 that the person measures for the UFO when it lands is the proper length, since the UFO is at rest with respect to the person. Therefore, from Equation 28.2 we have

$$L_0 = \frac{L}{\sqrt{1-\dfrac{v^2}{c^2}}} = \frac{230 \text{ m}}{\sqrt{1-\dfrac{(0.90\,c)^2}{c^2}}} = \boxed{530 \text{ m}}$$

---

21. ***REASONING*** The height of the woman as measured by the observer is given by Equation 28.2 as $h = h_0\sqrt{1-(v/c)^2}$, where h_0 is her proper height. In order to use this equation, we must determine the speed v of the woman relative to the observer. We are given the magnitude of her relativistic momentum, so we can determine v from p.

SOLUTION According to Equation 28.3 $p = mv/\sqrt{1-v^2/c^2}$, so $mv = p\sqrt{1-v^2/c^2}$. Squaring both sides, we have

$$m^2v^2 = p^2(1-v^2/c^2) \qquad \text{or} \qquad m^2v^2 + p^2\frac{v^2}{c^2} = p^2$$

$$v^2\left(m^2 + \frac{p^2}{c^2}\right) = p^2 \qquad \text{or} \qquad v^2 = \frac{p^2}{m^2 + \dfrac{p^2}{c^2}}$$

Solving for v and substituting values, we have

$$v = \frac{p}{\sqrt{m^2 + \dfrac{p^2}{c^2}}} = \frac{2.0 \times 10^{10} \text{ kg} \cdot \text{m/s}}{\sqrt{(55 \text{ kg})^2 + \left(\dfrac{2.0 \times 10^{10} \text{ kg} \cdot \text{m/s}}{3.00 \times 10^8 \text{ m/s}}\right)^2}} = 2.3 \times 10^8 \text{ m/s}$$

Then, the height that the observer measures for the woman is

$$h = h_0 \sqrt{1 - \left(\frac{v}{c}\right)^2} = (1.6 \text{ m}) \sqrt{1 - \left(\frac{2.3 \times 10^8 \text{ m/s}}{3.0 \times 10^8 \text{ m/s}}\right)^2} = \boxed{1.0 \text{ m}}$$

---

23. **REASONING AND SOLUTION** The total momentum of the man/woman system is conserved, since no net external force acts on the system. Therefore, the final total momentum $p_m + p_w$ must equal the initial total momentum, which is zero. As a result, $p_m = -p_w$ where Equation 28.3 must be used for the momenta p_m and p_w. Thus, we find

$$\frac{m_m v_m}{\sqrt{1 - \left(v_m / c\right)^2}} = -\frac{m_w v_w}{\sqrt{1 - \left(v_w / c\right)^2}}$$

We know that m_m = 88 kg, m_w = 54 kg, and v_w = +2.5 m/s. Remember that c has the hypothetical value of 3.0 m/s. Solving for v_m, we find $\boxed{v_m = -2.0 \text{ m/s}}$.

---

25. **REASONING AND SOLUTION**
a. In Section 28.6 it is shown that when the speed of a particle is $0.01c$ (or less), the relativistic kinetic energy becomes nearly equal to the nonrelativistic kinetic energy. Since the speed of the particle here is $0.001c$, the ratio of the relativistic kinetic energy to the nonrelativistic kinetic energy is $\boxed{1.0}$.

b. Taking the ratio of the relativistic kinetic energy, Equation 28.6, to the nonrelativistic kinetic energy, $\frac{1}{2}mv^2$, we find that

$$\frac{mc^2 \left(\dfrac{1}{\sqrt{1 - (v^2 / c^2)}} - 1\right)}{\frac{1}{2}mv^2} = 2\left(\frac{c}{v}\right)^2 \left(\frac{1}{\sqrt{1 - (v^2 / c^2)}} - 1\right)$$

$$= 2\left(\frac{c}{0.970c}\right)^2 \left(\frac{1}{\sqrt{1 - (0.970c)^2 / c^2}} - 1\right) = \boxed{6.6}$$

29. **REASONING** According to the work-energy theorem, Equation 6.3, the work that must be done on the electron to accelerate it from rest to a speed of 0.990c is equal to the kinetic energy of the electron when it is moving at 0.990c.

SOLUTION Using Equation 28.6, we find that

$$KE = mc^2 \left(\frac{1}{\sqrt{1-(v^2/c^2)}} - 1 \right)$$

$$= (9.11 \times 10^{-31} \text{ kg})(3.00 \times 10^8 \text{ m/s})^2 \left(\frac{1}{\sqrt{1-(0.990c)^2/c^2}} - 1 \right) = \boxed{5.0 \times 10^{-13} \text{ J}}$$

---

35. **REASONING** Let's define the following relative velocities, assuming that the spaceship and exploration vehicle are moving in the positive direction.

v_{ES} = velocity of **E**xploration vehicle relative to the **S**paceship.
v_{EO} = velocity of **E**xploration vehicle relative to an **O**bserver on earth = +0.70c
v_{SO} = velocity of **S**paceship relative to an **O**bserver on earth = +0.50c

The velocity v_{ES} can be determined from the velocity-addition formula, Equation 28.8:

$$v_{ES} = \frac{v_{EO} + v_{OS}}{1 + \frac{v_{EO}v_{OS}}{c^2}}$$

The velocity v_{OS} of the observer on earth relative to the spaceship is not given. However, we know that v_{OS} is the negative of v_{SO}, so $v_{OS} = -v_{SO} = -(+0.50c) = -0.50c$.

SOLUTION The velocity of the exploration vehicle relative to the spaceship is

$$v_{ES} = \frac{v_{EO} + v_{OS}}{1 + \frac{v_{EO}v_{OS}}{c^2}} = \frac{+0.70c + (-0.50c)}{1 + \frac{(+0.70c)(-0.50c)}{c^2}} = +0.31c$$

The speed of the exploration vehicle relative to the spaceship is the magnitude of this result or $\boxed{0.31c}$.

---

39. **REASONING** Since the crew is initially at rest relative to the escape pod, the length of 45 m is the proper length L_0 of the pod. The length of the escape pod as determined by an observer on earth can be obtained from the relation for length contraction given by Equation 28.2, $L = L_0\sqrt{1-\left(v_{PE}^2/c^2\right)}$. The quantity v_{PE} is the speed of the escape pod relative to the earth, which can be found from the velocity-addition formula, Equation 28.8. The following are the relative velocities, assuming that the direction away from the earth is the positive direction:

v_{PE} = velocity of the escape **P**od relative to **E**arth.

v_{PR} = velocity of escape **P**od relative to the **R**ocket = −0.55c. This velocity is negative because the rocket is moving away from the earth (in the positive direction), and the escape pod is moving in an opposite direction (the negative direction) relative to the rocket.

v_{RE} = velocity of **R**ocket relative to **E**arth = +0.75c

These velocities are related by the velocity-addition formula, Equation 28.8.

SOLUTION The relative velocity of the escape pod relative to the earth is

$$v_{PE} = \frac{v_{PR}+v_{RE}}{1+\dfrac{v_{PR}v_{RE}}{c^2}} = \frac{-0.55c+0.75c}{1+\dfrac{(-0.55c)(+0.75c)}{c^2}} = +0.34c$$

The speed of the pod relative to the earth is the magnitude of this result, or 0.34c. The length of the pod as determined by an observer on earth is

$$L = L_0\sqrt{1-\frac{v_{PE}^2}{c^2}} = (45\text{ m})\sqrt{1-\frac{(0.34c)^2}{c^2}} = \boxed{42\text{ m}}$$

---

43. **REASONING AND SOLUTION** The diameter D of the planet, as measured by a moving spacecraft, is given in terms of the proper diameter D_0 by Equation 28.2. Taking the ratio of the diameter D_A of the planet measured by spaceship A to the diameter D_B measured by spaceship B, we find

$$\frac{D_A}{D_B} = \frac{D_0\sqrt{1-\dfrac{v_A^2}{c^2}}}{D_0\sqrt{1-\dfrac{v_B^2}{c^2}}} = \frac{\sqrt{1-\dfrac{(0.60\,c)^2}{c^2}}}{\sqrt{1-\dfrac{(0.80\,c)^2}{c^2}}} = \boxed{1.3}$$

49. **REASONING** Only the sides of the rectangle that lie in the direction of motion will experience length contraction. In order to make the rectangle look like a square, each side must have a length of $L = 2.0$ m. Thus, we move along the long side, taking the proper length to be $L_0 = 3.0$ m. We can solve for the speed using Equation 28.2. Then, with this speed, we can use the relation for length contraction to find L for the short side as we move along it.

SOLUTION From Equation 28.2, $L = L_0\sqrt{1 - (v^2/c^2)}$, we find that

$$v = c\sqrt{1 - \left(\frac{L}{L_0}\right)^2} = c\sqrt{1 - \left(\frac{2.0 \text{ m}}{3.0 \text{ m}}\right)^2} = 0.75\,c$$

Moving at this speed along the short side, we take $L_0 = 2.0$ m and find L:

$$L = L_0\sqrt{1 - \left(\frac{v}{c}\right)^2} = (2.0 \text{ m})\sqrt{1 - \left(\frac{0.75\,c}{c}\right)^2} = 1.3 \text{ m}$$

The observed dimensions of the rectangle are, therefore, $\boxed{3.0 \text{ m} \times 1.3 \text{ m}}$, since the long side is not contracted due to motion along the short side.

CHAPTER 29 | *PARTICLES AND WAVES*

3. **REASONING** According to Equation 29.3, the work function W_0 is related to the photon energy hf and the maximum kinetic energy KE_{max} by $W_0 = hf - KE_{max}$. This expression can be used to find the work function of the metal.

SOLUTION KE_{max} is 6.1 eV. The photon energy (in eV) is, according to Equation 29.2,

$$hf = \left(6.63 \times 10^{-34}\ J \cdot s\right)\left(3.00 \times 10^{15}\ Hz\right)\left(\frac{1\ eV}{1.60 \times 10^{-19}\ J}\right) = 12.4\ eV$$

The work function is, therefore,

$$W_0 = hf - KE_{max} = 12.4\ eV - 6.1\ ev = \boxed{6.3\ eV}$$

5. **REASONING** The energy of the photon is related to its frequency by Equation 29.2, $E = hf$. Equation 16.1, $v = f\lambda$, relates the frequency and the wavelength for any wave.

SOLUTION Combining Equations 29.2 and 16.1, and noting that the speed of a photon is c, the speed of light in a vacuum, we have

$$\lambda = \frac{c}{f} = \frac{c}{(E/h)} = \frac{hc}{E} = \frac{(6.63 \times 10^{-34}\ J \cdot s)(3.0 \times 10^8\ m/s)}{6.4 \times 10^{-19}\ J} = 3.1 \times 10^{-7}\ m = \boxed{310\ nm}$$

9. **REASONING AND SOLUTION** The number of photons per second, N, entering the owl's eye is $N = SA/E$, where S is the intensity of the beam, A is the area of the owl's pupil, and E is the energy of a single photon. Assuming that the owl's pupil is circular, $A = \pi r^2 = \pi\left(\frac{1}{2}d\right)^2$, where d is the diameter of the owl's pupil. Combining Equations 29.2 and 16.1, we have $E = hf = hc/\lambda$. Therefore,

$$N = \frac{SA\lambda}{hc} = \frac{(5.0 \times 10^{-13}\ W/m^2)\pi\left[\frac{1}{2}(8.5 \times 10^{-3}\ m)\right]^2(510 \times 10^{-9}\ m)}{(6.63 \times 10^{-34}\ J \cdot s)(3.0 \times 10^8\ m/s)} = \boxed{73\ photons/s}$$

13. *REASONING AND SOLUTION*

a. According to Equation 24.5b, the electric field can be found from $E = \sqrt{S/(\varepsilon_0 c)}$. The intensity S of the beam is

$$S = \frac{\text{Energy per unit time}}{A} = \frac{Nhf}{A} = \frac{Nh}{A}\left(\frac{c}{\lambda}\right)$$

$$= \frac{\left(1.30 \times 10^{18} \text{ photons/s}\right)\left(6.63 \times 10^{-34} \text{ J}\cdot\text{s}\right)}{\pi\left(1.00 \times 10^{-3} \text{ m}\right)^2}\left(\frac{3.00 \times 10^8 \text{ m/s}}{514.5 \times 10^{-9} \text{ m}}\right)$$

$$= 1.60 \times 10^5 \text{ W/m}^2$$

where N is the number of photons per second emitted. Then,

$$E = \sqrt{S/(\varepsilon_0 c)} = \boxed{7760 \text{ N/C}}$$

b. According to Equation 24.3, the average magnetic field is

$$B = E/c = \boxed{2.59 \times 10^{-5} \text{ T}}$$

---

15. *REASONING* The angle θ through which the X-rays are scattered is related to the difference between the wavelength λ' of the scattered X-rays and the wavelength λ of the incident X-rays by Equation 29.7 as

$$\lambda' - \lambda = \frac{h}{mc}(1 - \cos\theta)$$

where h is Planck's constant, m is the mass of the electron, and c is the speed of light in a vacuum. We can use this relation directly to find the angle, since all the other variables are known.

SOLUTION Solving Equation 29.7 for the angle θ, we obtain

$$\cos \theta = 1 - \frac{mc}{h}(\lambda' - \lambda)$$

$$= 1 - \frac{\left(9.11 \times 10^{-31} \text{ kg}\right)\left(3.00 \times 10^{8} \text{ m/s}\right)}{6.63 \times 10^{-34} \text{ J} \cdot \text{s}} \left(0.2703 \times 10^{-9} \text{ m} - 0.2685 \times 10^{-9} \text{ m}\right) = 0.26$$

$$\theta = \cos^{-1}(0.26) = \boxed{75^{\circ}}$$

---

21. **REASONING** The change in wavelength that occurs during Compton scattering is given by Equation 29.7:

$$\lambda' - \lambda = \frac{h}{mc}(1 - \cos \theta) \quad \text{or} \quad (\lambda' - \lambda)_{\max} = \frac{h}{mc}(1 - \cos 180^{\circ}) = \frac{2h}{mc}$$

$(\lambda' - \lambda)_{\max}$ is the maximum change in the wavelength, and to calculate it we need a value for the mass m of a nitrogen molecule. This value can be obtained from the mass per mole M of nitrogen (N_2) and Avogadro's number N_A, according to $m = M/N_A$ (see Section 14.1).

SOLUTION Using a value of $M = 0.0280$ kg/mol, we obtain the following result for the maximum change in the wavelength:

$$(\lambda' - \lambda)_{\max} = \frac{2h}{mc} = \frac{2h}{\left(\dfrac{M}{N_A}\right)c} = \frac{2\left(6.63 \times 10^{-34} \text{ J} \cdot \text{s}\right)}{\left(\dfrac{0.0280 \text{ kg/mol}}{6.02 \times 10^{23} \text{ mol}^{-1}}\right)\left(3.00 \times 10^{8} \text{ m/s}\right)}$$

$$= \boxed{9.50 \times 10^{-17} \text{ m}}$$

---

25. **REASONING AND SOLUTION** The de Broglie wavelength λ is given by Equation 29.8 as $\lambda = h/p$, where p is the magnitude of the momentum of the particle. The magnitude of the momentum is $p = mv$, where m is the mass and v is the speed of the particle. Using this expression in Equation 29.8, we find that $\lambda = h/(mv)$, or

$$v = \frac{h}{m\lambda} = \frac{6.63 \times 10^{-34} \text{ J} \cdot \text{s}}{\left(1.67 \times 10^{-27} \text{ kg}\right)\left(1.30 \times 10^{-14} \text{ m}\right)} = 3.05 \times 10^{7} \text{ m/s}$$

The kinetic energy of the proton is

$$KE = \tfrac{1}{2}mv^2 = \tfrac{1}{2}(1.67 \times 10^{-27} \text{ kg})(3.05 \times 10^7 \text{ m/s})^2 = \boxed{7.77 \times 10^{-13} \text{ J}}$$

---

31. **REASONING** The de Broglie wavelength λ is related to Planck's constant h and the magnitude p of the particle's momentum. The magnitude of the momentum can be related to the particle's kinetic energy. Thus, using the given wavelength and the fact that the kinetic energy doubles, we will be able to obtain the new wavelength.

SOLUTION The de Broglie wavelength is

$$\lambda = \frac{h}{p} \tag{29.8}$$

The kinetic energy and the magnitude of the momentum are

$$KE = \frac{1}{2}mv^2 \quad (6.2) \qquad p = mv \quad (7.2)$$

where m and v are the mass and speed of the particle. Substituting Equation 7.2 into Equation 6.2, we can relate the kinetic energy and momentum as follows:

$$KE = \frac{1}{2}mv^2 = \frac{m^2v^2}{2m} = \frac{p^2}{2m} \qquad \text{or} \qquad p = \sqrt{2m(KE)}$$

Substituting this result for p into Equation 29.8 gives

$$\lambda = \frac{h}{p} = \frac{h}{\sqrt{2m(KE)}}$$

Applying this expression for the final and initial wavelengths λ_f and λ_i, we obtain

$$\lambda_f = \frac{h}{\sqrt{2m(KE)_f}} \qquad \text{and} \qquad \lambda_i = \frac{h}{\sqrt{2m(KE)_i}}$$

Dividing the two equations and rearranging reveals that

$$\frac{\lambda_f}{\lambda_i} = \frac{\dfrac{h}{\sqrt{2m(KE)_f}}}{\dfrac{h}{\sqrt{2m(KE)_i}}} = \sqrt{\frac{(KE)_i}{(KE)_f}} \qquad \text{or} \qquad \lambda_f = \lambda_i \sqrt{\frac{(KE)_i}{(KE)_f}}$$

Using the given value for λ_i and the fact that $\mathrm{KE}_f = 2(\mathrm{KE}_i)$, we find

$$\lambda_f = \lambda_i \sqrt{\frac{(\mathrm{KE})_i}{(\mathrm{KE})_f}} = (2.7 \times 10^{-10} \text{ m}) \sqrt{\frac{\mathrm{KE}_i}{2(\mathrm{KE}_i)}} = \boxed{1.9 \times 10^{-10} \text{ m}}$$

---

35. **REASONING** The de Broglie wavelength λ of the electron is related to the magnitude p of its momentum by $\lambda = h/p$ (Equation 29.8), where h is Planck's constant. If the speed of the electron is much less than the speed of light, the magnitude of the electron's momentum is given by $p = mv$ (Equation 7.2). Thus, the de Broglie wavelength can be written as $\lambda = h/(mv)$.

When the electron is at rest, it has electric potential energy, but no kinetic energy. The electric potential energy EPE is given by EPE $= eV$ (Equation 19.3), where e is the magnitude of the charge on the electron and V is the potential difference. When the electron reaches its maximum speed, it has no potential energy, but its kinetic energy is $\frac{1}{2}mv^2$. The conservation of energy states that the final total energy of the electron equals the initial total energy:

$$\underbrace{\tfrac{1}{2}mv^2}_{\substack{\text{Final total}\\\text{energy}}} = \underbrace{eV}_{\substack{\text{Initial total}\\\text{energy}}}$$

Solving this equation for the final speed gives $v = \sqrt{2eV/m}$. Substituting this expression for v into $\lambda = h/(mv)$ gives $\lambda = h/\sqrt{2meV}$.

SOLUTION After accelerating through the potential difference, the electron has a de Broglie wavelength of

$$\lambda = \frac{h}{\sqrt{2meV}} = \frac{6.63 \times 10^{-34} \text{ J·s}}{\sqrt{2(9.11 \times 10^{-31} \text{ kg})(1.60 \times 10^{-19} \text{ C})(418 \text{ V})}} = \boxed{6.01 \times 10^{-11} \text{ m}}$$

---

37. **REASONING AND SOLUTION** According to the uncertainty principle, the minimum uncertainty in the momentum can be determined from $\Delta p_y \Delta y = h/(4\pi)$. Since $p_y = mv_y$, it follows that $\Delta p_y = m\Delta v_y$. Thus, the minimum uncertainty in the velocity of the oxygen molecule is given by

$$\Delta v_y = \frac{h}{4\pi m \Delta y} = \frac{6.63 \times 10^{-34} \text{ J} \cdot \text{s}}{4\pi \left(5.3 \times 10^{-26} \text{ kg}\right)\left(0.12 \times 10^{-3} \text{ m}\right)} = \boxed{8.3 \times 10^{-6} \text{ m/s}}$$

39. **REASONING** The uncertainty in the electron's position is $\Delta y = 3.0 \times 10^{-15}$ m. The minimum uncertainty Δp_y in the y component of the electron's momentum is given by the Heisenberg uncertainty principle as $\Delta p_y = h/(4\pi \Delta y)$ (Equation 29.10).

SOLUTION Setting $\Delta y = 3.0 \times 10^{-15}$ m in the relation $\Delta p_y = h/(4\pi \Delta y)$ gives

$$\Delta p_y = \frac{h}{4\pi \Delta y} = \frac{6.63 \times 10^{-34} \text{ J} \cdot \text{s}}{4\pi (3.0 \times 10^{-15} \text{ m})} = \boxed{1.8 \times 10^{-20} \text{ kg} \cdot \text{m/s}}$$

43. **REASONING** In order for the person to diffract to the same extent as the sound wave, the de Broglie wavelength of the person must be equal to the wavelength of the sound wave.

SOLUTION
a. Since the wavelengths are equal, we have that

$$\lambda_{sound} = \lambda_{person}$$

$$\lambda_{sound} = \frac{h}{m_{person} v_{person}}$$

Solving for v_{person}, and using the relation $\lambda_{sound} = v_{sound} / f_{sound}$ (Equation 16.1), we have

$$v_{person} = \frac{h}{m_{person} (v_{sound} / f_{sound})} = \frac{h f_{sound}}{m_{person} v_{sound}}$$

$$= \frac{(6.63 \times 10^{-34} \text{ J} \cdot \text{s})(128 \text{ Hz})}{(55.0 \text{ kg})(343 \text{ m/s})} = \boxed{4.50 \times 10^{-36} \text{ m/s}}$$

b. At the speed calculated in part (a), the time required for the person to move a distance of one meter is

$$t = \frac{x}{v} = \frac{1.0 \text{ m}}{4.50 \times 10^{-36} \text{ m/s}} \underbrace{\left(\frac{1.0 \text{ h}}{3600 \text{ s}}\right) \left(\frac{1 \text{ day}}{24.0 \text{ h}}\right) \left(\frac{1 \text{ year}}{365.25 \text{ days}}\right)}_{\substack{\text{Factors to convert} \\ \text{seconds to years}}} = \boxed{7.05 \times 10^{27} \text{ years}}$$

CHAPTER 30 | *THE NATURE OF THE ATOM*

1. **REASONING** Assuming that the hydrogen atom is a sphere of radius r_{atom}, its volume V_{atom} is given by $\frac{4}{3}\pi r_{atom}^3$. Similarly, if the radius of the nucleus is $r_{nucleus}$, the volume $V_{nucleus}$ is given by $\frac{4}{3}\pi r_{nucleus}^3$.

SOLUTION
a. According to the given data, the nuclear dimensions are much smaller than the orbital radius of the electron; therefore, we can treat the nucleus as a point about which the electron orbits. The electron is normally at a distance of about 5.3×10^{-11} m from the nucleus, so we can treat the atom as a sphere of radius $r_{atom} = 5.3 \times 10^{-11}$ m. The volume of the atom is

$$V_{atom} = \frac{4}{3}\pi r_{atom}^3 = \frac{4}{3}\pi \left(5.3 \times 10^{-11}\ \text{m}\right)^3 = \boxed{6.2 \times 10^{-31}\ \text{m}^3}$$

b. Similarly, since the nucleus has a radius of approximately $r_{nucleus} = 1 \times 10^{-15}$ m, its volume is

$$V_{nucleus} = \frac{4}{3}\pi r_{nucleus}^3 = \frac{4}{3}\pi \left(1 \times 10^{-15}\ \text{m}\right)^3 = \boxed{4 \times 10^{-45}\ \text{m}^3}$$

c. The percentage of the atomic volume occupied by the nucleus is

$$\frac{V_{nucleus}}{V_{atom}} \times 100\% = \frac{\frac{4}{3}\pi r_{nucleus}^3}{\frac{4}{3}\pi r_{atom}^3} \times 100\% = \left(\frac{1 \times 10^{-15}\ \text{m}}{5.3 \times 10^{-11}\ \text{m}}\right)^3 \times 100\% = \boxed{7 \times 10^{-13}\ \%}$$

5. **REASONING** The distance of closest approach can be obtained by setting the kinetic energy KE of the α particle equal to the electric potential energy EPE of the α particle. According to Equation 19.3, we have that EPE $= (2e)V$, where $2e$ is the charge on the α particle and V is the electric potential created by a gold nucleus. According to Equation 19.6, the electric potential of the gold nucleus (charge $= Ze = 79e$) is $V = k(79e)/r$, where r is the distance between the α particle and the gold nucleus. Therefore, we have that

$$\text{EPE} = (2e)V = (2e)\frac{k(79e)}{r} \tag{1}$$

In this expression, we note that $k = 8.99 \times 10^9\ \text{N} \cdot \text{m}^2/\text{C}^2$ and $e = 1.602 \times 10^{-19}$ C.

SOLUTION Solving Equation (1) for the distance r we obtain

$$r = \frac{(2e)k(79e)}{\text{EPE}} = \frac{(8.99 \times 10^9 \text{ N} \cdot \text{m}^2/\text{C}^2)2(79)(1.602 \times 10^{-19} \text{ C})^2}{5.0 \times 10^{-13} \text{ J}} = \boxed{7.3 \times 10^{-14} \text{ m}}$$

---

7. **REASONING AND SOLUTION**

a. The longest wavelength in the Pfund series occurs for the transition $n = 6$ to $n = 5$, so that according to Equation 30.14 with $Z = 1$, we have

$$\frac{1}{\lambda} = R\left(\frac{1}{5^2} - \frac{1}{n^2}\right) = \left(1.097 \times 10^7 \text{ m}^{-1}\right)\left(\frac{1}{5^2} - \frac{1}{6^2}\right) \qquad \text{or} \qquad \boxed{\lambda = 7458 \text{ nm}}$$

b. The shortest wavelength occurs when $1/n^2 = 0$, so that

$$\frac{1}{\lambda} = R\left(\frac{1}{5^2} - \frac{1}{n^2}\right) = \left(1.097 \times 10^7 \text{ m}^{-1}\right)\left(\frac{1}{5^2}\right) \qquad \text{or} \qquad \boxed{\lambda = 2279 \text{ nm}}$$

c. The lines in the Pfund series occur in the $\boxed{\text{infrared region}}$.

---

11. **REASONING** According to Equation 30.14, the wavelength λ emitted by the hydrogen atom when it makes a transition from the level with n_i to the level with n_f is given by

$$\frac{1}{\lambda} = \frac{2\pi^2 mk^2 e^4}{h^3 c} (Z^2)\left(\frac{1}{n_f^2} - \frac{1}{n_i^2}\right) \qquad \text{with} \quad n_i, n_f = 1, 2, 3, \ldots \quad \text{and} \quad n_i > n_f$$

where $2\pi^2 mk^2 e^4 /(h^3 c) = 1.097 \times 10^7 \text{ m}^{-1}$ and $Z = 1$ for hydrogen. Once the wavelength for the particular transition in question is determined, Equation 29.2 ($E = hf = hc/\lambda$) can be used to find the energy of the emitted photon.

SOLUTION In the Paschen series, $n_f = 3$. Using the above expression with $Z = 1$, $n_i = 7$ and $n_f = 3$, we find that

$$\frac{1}{\lambda} = \left(1.097 \times 10^7 \text{ m}^{-1}\right)\left(1^2\right)\left(\frac{1}{3^2} - \frac{1}{7^2}\right) \qquad \text{or} \qquad \lambda = 1.005 \times 10^{-6} \text{ m}$$

The photon energy is

$$E = \frac{hc}{\lambda} = \frac{\left(6.63 \times 10^{-34} \text{ J} \cdot \text{s}\right)\left(3.00 \times 10^8 \text{ m/s}\right)}{1.005 \times 10^{-6} \text{ m}} = \boxed{1.98 \times 10^{-19} \text{ J}}$$

17. **REASONING** A wavelength of 410.2 nm is emitted by the hydrogen atoms in a high-voltage discharge tube. This transition lies in the visible region (380–750 nm) of the hydrogen spectrum. Thus, we can conclude that the transition is in the Balmer series and, therefore, that $n_f = 2$. The value of n_i can be found using Equation 30.14, according to which the Balmer series transitions are given by

$$\frac{1}{\lambda} = R\left(1^2\right)\left(\frac{1}{2^2} - \frac{1}{n_i^2}\right) \qquad n = 3, 4, 5, \ldots$$

This expression may be solved for n_i for the energy transition that produces the given wavelength.

SOLUTION Solving for n_i, we find that

$$n_i = \frac{1}{\sqrt{\dfrac{1}{2^2} - \dfrac{1}{R\lambda}}} = \frac{1}{\sqrt{\dfrac{1}{2^2} - \dfrac{1}{(1.097 \times 10^7 \text{ m}^{-1})(410.2 \times 10^{-9} \text{ m})}}} = 6$$

Therefore, the initial and final states are identified by $\boxed{n_i = 6 \text{ and } n_f = 2}$.

---

19. **REASONING** Singly ionized helium, He^+, is a hydrogen-like species with $Z = 2$. The wavelengths of the series of lines produced when the electron makes a transition from higher energy levels into the $n_f = 4$ level are given by Equation 30.14 with $Z = 2$ and $n_f = 4$:

$$\frac{1}{\lambda} = \left(1.097 \times 10^7 \text{ m}^{-1}\right)(2^2)\left(\frac{1}{4^2} - \frac{1}{n_i^2}\right)$$

SOLUTION Solving this expression for n_i gives

$$n_i = \left[\frac{1}{4^2} - \frac{1}{4\lambda(1.097 \times 10^7 \text{ m}^{-1})}\right]^{-1/2}$$

Evaluating this expression at the limits of the range for λ, we find that $n_i = 19.88$ for $\lambda = 380$ nm, and $n_i = 5.58$ for $\lambda = 750$ nm. Therefore, the values of n_i for energy levels from which the electron makes the transitions that yield wavelengths in the range between 380 nm and 750 nm are $\boxed{6 \le n_i \le 19}$.

27. **REASONING** The values that ℓ can have depend on the value of n, and only the following integers are allowed: $\ell = 0, 1, 2, \ldots (n-1)$. The values that m_ℓ can have depend on the value of ℓ, with only the following positive and negative integers being permitted: $m_\ell = -\ell, \ldots -2, -1, 0, +1, +2, \ldots +\ell$.

SOLUTION Thus, when $n = 6$, the possible values of ℓ are 0, 1, 2, 3, 4, 5. Now when $m_\ell = 2$, the possible values of ℓ are 2, 3, 4, 5, $\ldots$ These two series of integers overlap for the integers 2, 3, 4, and 5. Therefore, the possible values for the orbital quantum number ℓ that this electron could have are $\boxed{\ell = 2, 3, 4, 5}$.

---

29. **REASONING AND SOLUTION**
a. For the angular momentum, Bohr's value is given by Equation 30.8, with $n = 1$,

$$L_n = \frac{nh}{2\pi} = \boxed{\frac{h}{2\pi}}$$

According to quantum theory, the angular momentum is given by Equation 30.15. For $n = 1$, $\ell = 0$

$$L = \sqrt{\ell(\ell+1)}\left(\frac{h}{2\pi}\right) = \sqrt{0(0+1)}\left(\frac{h}{2\pi}\right) = \boxed{0 \text{ J} \cdot \text{s}}$$

b. For $n = 3$; Bohr theory gives

$$L_n = \frac{nh}{2\pi} = \boxed{\frac{3h}{2\pi}}$$

while quantum mechanics gives

$[n = 3, \ell = 0]$ $L = \sqrt{\ell(\ell+1)}\left(\frac{h}{2\pi}\right) = \sqrt{0(0+1)}\left(\frac{h}{2\pi}\right) = \boxed{0 \text{ J} \cdot \text{s}}$

$[n = 3, \ell = 1]$ $L = \sqrt{\ell(\ell+1)}\left(\frac{h}{2\pi}\right) = \sqrt{1(1+1)}\left(\frac{h}{2\pi}\right) = \boxed{\frac{\sqrt{2}h}{2\pi}}$

$[n = 3, \ell = 2]$ $L = \sqrt{\ell(\ell+1)}\left(\frac{h}{2\pi}\right) = \sqrt{2(2+1)}\left(\frac{h}{2\pi}\right) = \boxed{\frac{\sqrt{6}h}{2\pi}}$

37. ***REASONING*** This problem is similar to Example 11 in the text. We use Equation 30.14 with the initial value of n being $n_i = 2$, and the final value being $n_f = 1$. As in Example 11, we use a value of Z that is one less than the atomic number of the atom in question (in this case, a value of $Z = 41$ rather than 42); this accounts approximately for the shielding effect of the single K-shell electron in canceling out the attraction of one nuclear proton.

SOLUTION Using Equation 30.14, we obtain

$$\frac{1}{\lambda} = (1.097 \times 10^7 \text{ m}^{-1})(41)^2 \left(\frac{1}{1^2} - \frac{1}{2^2} \right) \qquad \text{or} \qquad \boxed{\lambda = 7.230 \times 10^{-11} \text{ m}}$$

---

43. ***REASONING*** In the spectrum of X-rays produced by the tube, the cutoff wavelength λ_0 and the voltage V of the tube are related according to Equation 30.17, $V = hc / (e\lambda_0)$. Since the voltage is increased from zero until the K_α X-ray just appears in the spectrum, it follows that $\lambda_0 = \lambda_\alpha$ and $V = hc / (e\lambda_\alpha)$. Using Equation 30.14 for $1/\lambda_\alpha$, we find that

$$V = \frac{hc}{e\lambda_\alpha} = \frac{hcR(Z-1)^2}{e} \left(\frac{1}{1^2} - \frac{1}{2^2} \right)$$

In this expression we have replaced Z with $Z-1$, in order to account for shielding, as explained in Example 11 in the text.

SOLUTION The desired voltage is, then,

$$V = \frac{\left(6.63 \times 10^{-34} \text{ J} \cdot \text{s} \right)\left(3.00 \times 10^8 \text{ m/s} \right)\left(1.097 \times 10^7 \text{ m}^{-1} \right)(47-1)^2}{\left(1.60 \times 10^{-19} \text{ C} \right)} \left(\frac{1}{1^2} - \frac{1}{2^2} \right) = \boxed{21\,600 \text{ V}}$$

---

45. ***REASONING*** The number of photons emitted by the laser will be equal to the total energy carried in the beam divided by the energy per photon.

SOLUTION The total energy carried in the beam is, from the definition of power,

$$E_{\text{total}} = Pt = (1.5 \text{ W})(0.050 \text{ s}) = 0.075 \text{ J}$$

The energy of a single photon is given by Equations 29.2 and 16.1 as

$$E_{\text{photon}} = hf = \frac{hc}{\lambda} = \frac{\left(6.63 \times 10^{-34} \text{ J} \cdot \text{s} \right)\left(3.00 \times 10^8 \text{ m/s} \right)}{514 \times 10^{-9} \text{ m}} = 3.87 \times 10^{-19} \text{ J}$$

where we have used the fact that 514 nm $= 514 \times 10^{-9}$ m. Therefore, the number of photons emitted by the laser is

$$\frac{E_{\text{total}}}{E_{\text{photon}}} = \frac{0.075 \text{ J}}{3.87 \times 10^{-19} \text{ J/photon}} = \boxed{1.9 \times 10^{17} \text{ photons}}$$

---

53. **REASONING** In the theory of quantum mechanics, there is a selection rule that restricts the initial and final values of the orbital quantum number ℓ. The selection rule states that when an electron makes a transition between energy levels, the value of ℓ may not remain the same or increase or decrease by more than one. In other words, the rule requires that $\Delta\ell = \pm 1$.

SOLUTION
a. For the transition 2s → 1s, the electron makes a transition from the 2s state $(n = 2, \ \ell = 0)$ to the 1s state $(n = 1, \ \ell = 0)$. Since the value of ℓ is the same in both states, $\Delta\ell = 0$, and we can conclude that this energy level transition is $\boxed{\text{not allowed}}$.

b. For the transition 2p → 1s, the electron makes a transition from the 2p state $(n = 2, \ \ell = 1)$ to the 1s state $(n = 1, \ \ell = 0)$. The value of ℓ changes so that $\Delta\ell = 0 - 1 = -1$, and we can conclude that this energy level transition is $\boxed{\text{allowed}}$.

c. For the transition 4p → 2p, the electron makes a transition from the 4p state $(n = 4, \ \ell = 1)$ to the 2p state $(n = 2, \ \ell = 1)$. Since the value of ℓ is the same in both states, $\Delta\ell = 0$, and we can conclude that this energy level transition is $\boxed{\text{not allowed}}$.

d. For the transition 4s → 2p, the electron makes a transition from the 4s state $(n = 4, \ \ell = 0)$ to the 2p state $(n = 2, \ \ell = 1)$. The value of ℓ changes so that $\Delta\ell = 1 - 0 = +1$, and we can conclude that this energy level transition is $\boxed{\text{allowed}}$.

e. For the transition 3d → 3s, the electron makes a transition from the 3d state $(n = 3, \ \ell = 2)$ to the 3s state $(n = 3, \ \ell = 0)$. The value of ℓ changes so that $\Delta\ell = 0 - 2 = -2$, and we can conclude that this energy level transition is $\boxed{\text{not allowed}}$.

55. **REASONING** According to the Bohr model, the energy E_n (in eV) of the electron in an orbit is given by Equation 30.13: $E_n = -13.6(Z^2/n^2)$. In order to find the principal quantum number of the state in which the electron in a doubly ionized lithium atom Li^{2+} has the same total energy as a ground state electron in a hydrogen atom, we equate the right hand sides of Equation 30.13 for the hydrogen atom and the lithium ion. This gives

$$-(13.6)\left(\frac{Z^2}{n^2}\right)_H = -(13.6)\left(\frac{Z^2}{n^2}\right)_{Li} \qquad \text{or} \qquad n_{Li}^2 = \left(\frac{n^2}{Z^2}\right)_H Z_{Li}^2$$

This expression can be evaluated to find the desired principal quantum number.

SOLUTION For hydrogen, $Z = 1$, and $n = 1$ for the ground state. For lithium Li^{2+}, $Z = 3$. Therefore,

$$n_{Li}^2 = \left(\frac{n^2}{Z^2}\right)_H Z_{Li}^2 = \left(\frac{1^2}{1^2}\right)(3^2) \qquad \text{or} \qquad \boxed{n_{Li} = 3}$$

---

59. **REASONING AND SOLUTION** For the Paschen series, $n_f = 3$. The range of wavelengths occurs for values of $n_i = 4$ to $n_i = \infty$. Using Equation 30.14, we find that the shortest wavelength occurs for $n_i = \infty$ and is given by

$$\frac{1}{\lambda} = \left(1.097 \times 10^7 \text{ m}^{-1}\right)(1)^2\left(\frac{1}{n_f^2} - \frac{1}{n_i^2}\right) = \left(1.097 \times 10^7 \text{ m}^{-1}\right)\left(\frac{1}{3^2}\right) \quad \text{or} \quad \underbrace{\lambda = 8.204 \times 10^{-7} \text{ m}}_{\substack{\text{Shortest wavelength in} \\ \text{Paschen series}}}$$

The longest wavelength in the Paschen series occurs for $n_i = 4$ and is given by

$$\frac{1}{\lambda} = \left(1.097 \times 10^7 \text{ m}^{-1}\right)\left(\frac{1}{3^2} - \frac{1}{4^2}\right) \quad \text{or} \quad \underbrace{\lambda = 1.875 \times 10^{-6} \text{ m}}_{\substack{\text{Longest wavelength in} \\ \text{Paschen series}}}$$

For the Brackett series, $n_f = 4$. The range of wavelengths occurs for values of $n_i = 5$ to $n_i = \infty$. Using Equation 30.14, we find that the shortest wavelength occurs for $n_i = \infty$ and is given by

$$\frac{1}{\lambda} = \left(1.097 \times 10^7 \text{ m}^{-1}\right)(1)^2\left(\frac{1}{n_f^2} - \frac{1}{n_i^2}\right) = \left(1.097 \times 10^7 \text{ m}^{-1}\right)\left(\frac{1}{4^2}\right) \quad \text{or} \quad \underbrace{\lambda = 1.459 \times 10^{-6} \text{ m}}_{\substack{\text{Shortest wavelength in} \\ \text{Brackett series}}}$$

The longest wavelength in the Brackett series occurs for $n_i = 5$ and is given by

$$\frac{1}{\lambda} = \left(1.097 \times 10^7 \text{ m}^{-1}\right)\left(\frac{1}{4^2} - \frac{1}{5^2}\right) \qquad \text{or} \qquad \underline{\lambda = 4.051 \times 10^{-6} \text{ m}}$$

Longest wavelength in
Brackett series

Since the longest wavelength in the Paschen series falls within the Brackett series, the wavelengths of the two series overlap.

---

61. **_REASONING AND SOLUTION_**

a. To find an expression for v_n, we use Equation 30.8, $L_n = mv_n r_n = nh/(2\pi)$, and substitute for r_n from Equation 30.9:

$$mv_n\left(\frac{h^2 n^2}{4\pi^2 mke^2 Z}\right) = \frac{nh}{2\pi} \qquad \text{or} \qquad \boxed{v_n = \frac{2\pi ke^2 Z}{nh}}$$

b. For the hydrogen atom ($Z = 1$) in the $n = 1$ orbit,

$$v_n = \frac{2\pi \; (8.99 \times 10^9 \text{ N} \cdot \text{m}^2/\text{C}^2)(1.602 \times 10^{-19} \text{ C})^2 (1)}{(1)(6.626 \times 10^{-34} \text{ J} \cdot \text{s})} = \boxed{2.19 \times 10^6 \text{ m/s}}$$

c. For the $n = 2$ orbit of hydrogen,

$$v_n = \frac{2\pi \; (8.99 \times 10^9 \text{ N} \cdot \text{m}^2/\text{C}^2)(1.602 \times 10^{-19} \text{ C})^2 (1)}{(2)(6.626 \times 10^{-34} \text{ J} \cdot \text{s})} = \boxed{1.09 \times 10^6 \text{ m/s}}$$

d. The speeds found in parts (b) and (c) are well below the speed of light, 3.0×10^8 m/s, and $\boxed{\text{are consistent with ignoring relativistic effects}}$.

CHAPTER 31 | *NUCLEAR PHYSICS AND RADIOACTIVITY*

1. **REASONING** For an element whose chemical symbol is X, the symbol for the nucleus is $_Z^A X$, where A represents the total number of protons and neutrons (the nucleon number) and Z represents the number of protons in the nucleus (the atomic number). The number of neutrons N is related to A and Z by Equation 31.1: $A = Z + N$.

   SOLUTION For the nucleus $_{82}^{208} Pb$, we have $Z = 82$ and $A = 208$.

   a. The net electrical charge of the nucleus is equal to the total number of protons multiplied by the charge on a single proton. Since the $_{82}^{208} Pb$ nucleus contains 82 protons, the net electrical charge of the $_{82}^{208} Pb$ nucleus is

$$q_{net} = (82)(+1.60 \times 10^{-19} \ C) = \boxed{+1.31 \times 10^{-17} \ C}$$

   b. The number of neutrons is $N = A - Z = 208 - 82 = \boxed{126}$.

   c. By inspection, the number of nucleons is $A = \boxed{208}$.

   d. The approximate radius of the nucleus can be found from Equation 31.2, namely

$$r = (1.2 \times 10^{-15} \ m) A^{1/3} = (1.2 \times 10^{-15} \ m)(208)^{1/3} = \boxed{7.1 \times 10^{-15} \ m}$$

   e. The nuclear density is the mass per unit volume of the nucleus. The total mass of the nucleus can be found by multiplying the mass $m_{nucleon}$ of a single nucleon by the total number A of nucleons in the nucleus. Treating the nucleus as a sphere of radius r, the nuclear density is

$$\rho = \frac{m_{total}}{V} = \frac{m_{nucleon} A}{\frac{4}{3} \pi r^3} = \frac{m_{nucleon} A}{\frac{4}{3} \pi \left[(1.2 \times 10^{-15} \ m) A^{1/3} \right]^3} = \frac{m_{nucleon}}{\frac{4}{3} \pi (1.2 \times 10^{-15} \ m)^3}$$

Therefore,

$$\rho = \frac{1.67 \times 10^{-27} \ kg}{\frac{4}{3} \pi (1.2 \times 10^{-15} \ m)^3} = \boxed{2.3 \times 10^{17} \ kg / m^3}$$

5. **REASONING AND SOLUTION** Equation 31.2 gives for the radius of the $^{48}_{22}$Ti nucleus that

$$r = (1.2 \times 10^{-15} \text{ m})A^{1/3} = (1.2 \times 10^{-15} \text{ m})(48)^{1/3} = \boxed{4.4 \times 10^{-15} \text{ m}}$$

9. **REASONING** According to Equation 31.2, the radius of a nucleus in meters is $r = \left(1.2 \times 10^{-15} \text{ m}\right) A^{1/3}$, where A is the nucleon number. If we treat the neutron star as a uniform sphere, its density (Equation 11.1) can be written as

$$\rho = \frac{M}{V} = \frac{M}{\frac{4}{3}\pi r^3}$$

Solving for the radius r, we obtain,

$$r = \sqrt[3]{\frac{M}{\frac{4}{3}\pi\rho}}$$

This expression can be used to find the radius of a neutron star of mass M and density ρ.

SOLUTION As discussed in Conceptual Example 1, nuclear densities have the same approximate value in all atoms. If we consider a uniform spherical nucleus, then the density of nuclear matter is approximately given by

$$\rho = \frac{M}{V} \approx \frac{A \times (\text{mass of a nucleon})}{\frac{4}{3}\pi r^3} = \frac{A \times (\text{mass of a nucleon})}{\frac{4}{3}\pi\left[(1.2 \times 10^{-15} \text{ m})\,A^{1/3}\right]^3}$$

$$= \frac{1.67 \times 10^{-27} \text{ kg}}{\frac{4}{3}\pi(1.2 \times 10^{-15} \text{ m})^3} = 2.3 \times 10^{17} \text{ kg} / \text{m}^3$$

The mass of the sun is 1.99×10^{30} kg (see inside of the front cover of the text). Substituting values into the expression for r determined above, we find

$$r = \sqrt[3]{\frac{(0.40)(1.99 \times 10^{30} \text{ kg})}{\frac{4}{3}\pi(2.3 \times 10^{17} \text{ kg} / \text{m}^3)}} = \boxed{9.4 \times 10^3 \text{ m}}$$

11. **REASONING** To obtain the binding energy, we will calculate the mass defect and then use the fact that 1 u is equivalent to 931.5 MeV. The atomic mass given for $^7_3$Li includes the 3 electrons in the neutral atom. Therefore, when computing the mass defect, we must account for these electrons. We do so by using the atomic mass of 1.007 825 u for the hydrogen atom $^1_1$H, which also includes the single electron, instead of the atomic mass of a proton.

SOLUTION Noting that the number of neutrons is $7 - 3 = 4$, we obtain the mass defect Δm as follows:

$$\Delta m = \underbrace{3(1.007\ 825\ \text{u})}_{\substack{\text{3 hydrogen atoms} \\ \text{(protons plus electrons)}}} + \underbrace{4(1.008\ 665\ \text{u})}_{\text{4 neutrons}} - \underbrace{7.016\ 003\ \text{u}}_{\substack{\text{Intact lithium atom} \\ \text{(including 3 electrons)}}} = 4.2132 \times 10^{-2}\ \text{u}$$

Since 1 u is equivalent to 931.5 MeV, the binding energy is

$$\text{Binding energy} = \left(4.2132 \times 10^{-2}\ \text{u}\right)\left(\frac{931.5\ \text{MeV}}{1\ \text{u}}\right) = \boxed{39.25\ \text{MeV}}$$

---

17. *REASONING* Since we know the difference in binding energies for the two isotopes, we can determine the corresponding mass defect. Also knowing that the isotope with the larger binding energy contains one more neutron than the other isotope gives us enough information to calculate the atomic mass difference between the two isotopes.

SOLUTION The mass defect corresponding to a binding energy difference of 5.03 MeV is

$$(5.03\ \text{MeV})\left(\frac{1\ \text{u}}{931.5\ \text{MeV}}\right) = 0.005\ 40\ \text{u}$$

Since the isotope with the larger binding energy has one more neutron ($m = 1.008\ 665\ \text{u}$) than the other isotope, the difference in atomic mass between the two isotopes is

$$1.008\ 665\ \text{u} - 0.005\ 40\ \text{u} = \boxed{1.003\ 27\ \text{u}}$$

---

19. *REASONING AND SOLUTION* The general form for β^+ decay is

$$\underbrace{{}^{A}_{Z}\text{P}}_{\substack{\text{Parent} \\ \text{nucleus}}} \rightarrow \underbrace{{}^{A}_{Z-1}\text{D}}_{\substack{\text{Daughter} \\ \text{nucleus}}} + \underbrace{{}^{0}_{+1}\text{e}}_{\substack{\beta^+ \text{ particle} \\ \text{(positron)}}}$$

a. Therefore, the β^+ decay process for ${}^{18}_{9}\text{F}$ is $\boxed{{}^{18}_{9}\text{F} \rightarrow {}^{18}_{8}\text{O} + {}^{0}_{+1}\text{e}}$.

b. Similarly, the β^+ decay process for ${}^{15}_{8}\text{O}$ is $\boxed{{}^{15}_{8}\text{O} \rightarrow {}^{15}_{7}\text{N} + {}^{0}_{+1}\text{e}}$.

21. **REASONING AND SOLUTION** The general form for β^- decay is

$$\underset{\substack{\text{Parent} \\ \text{nucleus}}}{\underbrace{{}^{A}_{Z}\text{P}}} \rightarrow \underset{\substack{\text{Daughter} \\ \text{nucleus}}}{\underbrace{{}^{A}_{Z+1}\text{D}}} + \underset{\substack{\beta^- \text{ particle} \\ \text{(electron)}}}{\underbrace{{}^{0}_{-1}\text{e}}}$$

Therefore, the β^- decay process for ${}^{35}_{16}\text{S}$ is $\boxed{{}^{35}_{16}\text{S} \rightarrow {}^{35}_{17}\text{Cl} + {}^{0}_{-1}\text{e}}$.

---

31. **REASONING** Energy is released during the β decay. To find the energy released, we determine how much the mass has decreased because of the decay and then calculate the equivalent energy. The reaction and masses are shown below:

$$\underset{21.994\,434\text{ u}}{\underbrace{{}^{22}_{11}\text{Na}}} \rightarrow \underset{21.991\,383\text{ u}}{\underbrace{{}^{22}_{10}\text{Ne}}} + \underset{5.485\,799 \times 10^{-4}\text{ u}}{\underbrace{{}^{0}_{+1}\text{e}}}$$

SOLUTION The decrease in mass is

$$21.994\,434\text{ u} - \left(21.991\,383\text{ u} + 5.485\,799 \times 10^{-4}\text{ u} + 5.485\,799 \times 10^{-4}\text{ u}\right) = 0.001\,954\text{ u}$$

where the extra electron mass takes into account the fact that the atomic mass for sodium includes the mass of 11 electrons, whereas the atomic mass for neon includes the mass of only 10 electrons.

Since 1 u is equivalent to 931.5 MeV, the released energy is

$$\left(0.001\,954\text{ u}\right)\left(\frac{931.5\text{ MeV}}{1\text{ u}}\right) = \boxed{1.82\text{ MeV}}$$

---

33. **REASONING AND SOLUTION** The number of radioactive nuclei that remains in a sample after a time t is given by Equation 31.5, $N = N_0 e^{-\lambda t}$, where λ is the decay constant. From Equation 31.6, we know that the decay constant is related to the half-life by $T_{1/2} = 0.693 / \lambda$; therefore, $\lambda = 0.693 / T_{1/2}$ and we can write

$$\frac{N}{N_0} = e^{-(0.693/T_{1/2})t} \qquad \text{or} \qquad \frac{t}{T_{1/2}} = -\frac{1}{0.693}\ln\left(\frac{N}{N_0}\right)$$

When the number of radioactive nuclei decreases to one-millionth of the initial number, $N / N_0 = 1.00 \times 10^{-6}$; therefore, the number of half-lives is

$$\frac{t}{T_{1/2}} = -\frac{1}{0.693} \ln(1.00 \times 10^{-6}) = \boxed{19.9}$$

---

37. ***REASONING AND SOLUTION*** According to Equation 31.5, $N = N_0 e^{-\lambda t}$, the decay constant is

$$\lambda = -\frac{1}{t} \ln\left(\frac{N}{N_0}\right) = -\frac{1}{20 \text{ days}} \ln\left(\frac{8.14 \times 10^{14}}{4.60 \times 10^{15}}\right) = 0.0866 \text{ days}^{-1}$$

The half-life is, from Equation 31.6,

$$T_{1/2} = \frac{0.693}{\lambda} = \frac{0.693}{0.0866 \text{ days}^{-1}} = \boxed{8.00 \text{ days}}$$

---

39. ***REASONING*** We can find the decay constant from Equation 31.5, $N = N_0 e^{-\lambda t}$. If we multiply both sides by the decay constant λ, we have

$$\lambda N = \lambda N_0 e^{-\lambda t} \qquad \text{or} \qquad A = A_0 e^{-\lambda t}$$

where A_0 is the initial activity and A is the activity after a time t. Once the decay constant is known, we can use the same expression to determine the activity after a total of six days.

SOLUTION Solving the expression above for the decay constant λ, we have

$$\lambda = -\frac{1}{t} \ln\left(\frac{A}{A_0}\right) = -\frac{1}{2 \text{ days}} \ln\left(\frac{285 \text{ disintegrations/min}}{398 \text{ disintegrations/min}}\right) = 0.167 \text{ days}^{-1}$$

Then the activity four days after the second day is

$$A = (285 \text{ disintegrations/min}) e^{-(0.167 \text{ days}^{-1})(4.00 \text{ days})} = \boxed{146 \text{ disintegrations/min}}$$

---

45. ***REASONING*** According to Equation 31.5, the number of nuclei remaining after a time t is $N = N_0 e^{-\lambda t}$. Using this expression, we find the ratio N_A / N_B as follows:

$$\frac{N_A}{N_B} = \frac{N_{0A} e^{-\lambda_A t}}{N_{0B} e^{-\lambda_B t}} = e^{-(\lambda_A - \lambda_B)t}$$

where we have used the fact that initially the numbers of the two types of nuclei are equal $(N_{0A} = N_{0B})$. Taking the natural logarithm of both sides of the equation above shows that

$$\ln(N_A / N_B) = -(\lambda_A - \lambda_B)t \qquad \text{or} \qquad \lambda_A - \lambda_B = \frac{-\ln(N_A / N_B)}{t}$$

SOLUTION Since $N_A / N_B = 3.00$ when $t = 3.00$ days, it follows that

$$\lambda_A - \lambda_B = \frac{-\ln(3.00)}{3.00 \text{ days}} = -0.366 \text{ days}^{-1}$$

But we need to find the half-life of species B, so we use Equation 31.6, which indicates that $\lambda = 0.693 / T_{1/2}$. With this expression for λ, the result for $\lambda_A - \lambda_B$ becomes

$$0.693\left(\frac{1}{T_{1/2}^A} - \frac{1}{T_{1/2}^B}\right) = -0.366 \text{ days}^{-1}$$

Since $T_{1/2}^B = 1.50$ days, the result above can be solved to show that $\boxed{T_{1/2}^A = 7.23 \text{ days}}$.

---

47. **REASONING AND SOLUTION** The answer can be obtained directly from Equation 31.5, combined with Equation 31.6:

$$\frac{N}{N_0} = e^{-\lambda t} = e^{-(0.693)t/T_{1/2}} = e^{-(0.693)(41\,000 \text{ yr})/(5730 \text{ yr})} = 0.0070$$

The percent of atoms remaining is $\boxed{0.70 \%}$.

---

51. **REASONING** According to Equation 31.5, $N = N_0 e^{-\lambda t}$. If we multiply both sides by the decay constant λ, we have

$$\lambda N = \lambda N_0 e^{-\lambda t} \qquad \text{or} \qquad A = A_0 e^{-\lambda t}$$

where A_0 is the initial activity and A is the activity after a time t. The decay constant λ is related to the half-life through Equation 31.6: $\lambda = 0.693 / T_{1/2}$. We can find the age of the fossils by solving for the time t. The maximum error can be found by evaluating the limits of the accuracy as given in the problem statement.

SOLUTION The age of the fossils is

$$t = -\frac{T_{1/2}}{0.693} \ln\left(\frac{A}{A_0}\right) = -\frac{5730 \text{ yr}}{0.693} \ln\left(\frac{0.100 \text{ Bq}}{0.23 \text{ Bq}}\right) = \boxed{6900 \text{ yr}}$$

The maximum error can be found as follows:

When there is an error of +10.0 %, $A = 0.100 \text{ Bq} + 0.0100 \text{ Bq} = 0.110 \text{ Bq}$, and we have

$$t = -\frac{5730 \text{ yr}}{0.693} \ln\left(\frac{0.110 \text{ Bq}}{0.23 \text{ Bq}}\right) = 6100 \text{ yr}$$

Similarly, when there is an error of −10.0 %, $A = 0.100 \text{ Bq} - 0.0100 \text{ Bq} = 0.090 \text{ Bq}$, and we have

$$t = -\frac{5730 \text{ yr}}{0.693} \ln\left(\frac{0.090 \text{ Bq}}{0.23 \text{ Bq}}\right) = 7800 \text{ yr}$$

The maximum error in the age of the fossils is 7800 yr − 6900 yr = $\boxed{900 \text{ yr}}$.

---

59. **REASONING AND SOLUTION** As shown in Figure 31.18, if the first dynode produces 3 electrons, the second produces 9 electrons (3^2), the third produces 27 electrons (3^3), so the Nth produces 3^N electrons. The number of electrons that leaves the 14th dynode and strikes the 15th dynode is

$$3^{14} = \boxed{4\,782\,969 \text{ electrons}}$$

CHAPTER 32 | *IONIZING RADIATION, NUCLEAR ENERGY, AND ELEMENTARY PARTICLES*

1. **REASONING** The biologically equivalent dose (in rems) is the product of the absorbed dose (in rads) and the relative biological effectiveness (RBE), according to Equation 32.4. We can apply this equation to each type of radiation. Since the biologically equivalent doses of the neutrons and α particles are equal, we can solve for the unknown RBE.

SOLUTION Applying Equation 32.4 to each type of particle and using the fact that the biological equivalent doses are equal, we find that

$$\underbrace{\left(\text{Absorbed dose}\right)_{\alpha} \text{RBE}_{\alpha}}_{\substack{\text{Biologically equivalent dose} \\ \text{of } \alpha \text{ particles}}} = \underbrace{\left(\text{Absorbed dose}\right)_{\text{neutrons}} \text{RBE}_{\text{neutrons}}}_{\substack{\text{Biologically equivalent dose} \\ \text{of neutrons}}}$$

Solving for RBE_{α} and noting that $\left(\text{Absorbed dose}\right)_{\text{neutrons}} = 6\left(\text{Absorbed dose}\right)_{\alpha}$, we have

$$\text{RBE}_{\alpha} = \frac{\left(\text{Absorbed dose}\right)_{\text{neutrons}}}{\left(\text{Absorbed dose}\right)_{\alpha}}\left(\text{RBE}_{\text{neutrons}}\right) = \frac{6\left(\text{Absorbed dose}\right)_{\alpha}}{\left(\text{Absorbed dose}\right)_{\alpha}}\left(2.0\right) = \boxed{12}$$

---

5. **REASONING AND SOLUTION** According to Equation 32.2, the absorbed dose (AD) is equal to the energy absorbed by the tumor divided by its mass:

$$\text{AD} = \frac{\text{Energy absorbed}}{\text{Mass}} = \frac{\left(25 \text{ s}\right)\left(1.6 \times 10^{10} \text{ s}^{-1}\right)\left(4.0 \times 10^{6} \text{ eV}\right)\left(\dfrac{1.60 \times 10^{-19} \text{ J}}{1 \text{ eV}}\right)}{0.015 \text{ kg}}$$

$$= 1.7 \times 10^{1} \text{ Gy} = 1.7 \times 10^{3} \text{ rad}$$

The biologically equivalent dose (BED) is equal to the product of the absorbed dose (AD) and the RBE (see Equation 32.4):

$$\text{BED} = \text{AD} \times \text{RBE} = (1.7 \times 10^{3} \text{ rad})(14) = \boxed{2.4 \times 10^{4} \text{ rem}} \tag{32.4}$$

11. *REASONING* The number of nuclei in the beam is equal to the energy absorbed by the tumor divided by the energy per nucleus (130 MeV). According to Equation 32.2, the energy (in joules) absorbed by the tumor is equal to the absorbed dose (expressed in grays) times the mass of the tumor. The absorbed dose (expressed in rads) is equal to the biologically equivalent dose divided by the RBE of the radiation (see Equation 32.4). We can use these concepts to determine the number of nuclei in the beam.

SOLUTION The number N of nuclei in the beam is equal to the energy E absorbed by the tumor divided by the energy per nucleus. Since the energy absorbed is equal to the absorbed dose (in Gy) times the mass m (see Equation 32.2) we have

$$N = \frac{E}{\text{Energy per nucleus}} = \frac{\left[\text{Absorbed dose (in Gy)}\right]m}{\text{Energy per nucleus}}$$

We can express the absorbed dose in terms of rad units, rather than Gy units, by noting that 1 rad = 0.01 Gy. Therefore,

$$\text{Absorbed dose (in Gy)} = \text{Absorbed dose (in rad)}\left(\frac{0.01\,\text{Gy}}{1\,\text{rad}}\right)$$

The number of nuclei can now be written as

$$N = \frac{E}{\text{Energy per nucleus}} = \frac{\left[\text{Absorbed dose (in rad)}\right]\left(\dfrac{0.01\,\text{Gy}}{1\,\text{rad}}\right)m}{\text{Energy per nucleus}}$$

We know from Equation 32.4 that the Absorbed dose (in rad) is equal to the biologically equivalent dose divided by the RBE, so that

$$N = \frac{E}{\text{Energy per nucleus}} = \frac{\left[\dfrac{\text{Biologically equivalent dose}}{\text{RBE}}\right]\left(\dfrac{0.01\,\text{Gy}}{1\,\text{rad}}\right)m}{\text{Energy per nucleus}}$$

$$= \frac{\left[\dfrac{180\,\text{rem}}{16}\right]\left(\dfrac{0.01\,\text{Gy}}{1\,\text{rad}}\right)(0.17\,\text{kg})}{\left(130\times10^{6}\,\text{eV}\right)\left(\dfrac{1.60\times10^{-19}\,\text{J}}{1\,\text{eV}}\right)} = \boxed{9.2\times10^{8}}$$

13. **REASONING** The reaction given in the problem statement is written in the shorthand form: $^{17}_{8}O\,(\gamma,\,\alpha\,n)\,^{12}_{6}C$. The first and last symbols represent the initial and final nuclei, respectively. The symbols inside the parentheses denote the incident particles or rays (left side of the comma) and the emitted particles or rays (right side of the comma).

SOLUTION In the shorthand form of the reaction, we note that the designation an refers to an α particle (which is a helium nucleus $^{4}_{2}He$) and a neutron ($^{1}_{0}n$). Thus, the reaction is

$$\boxed{\gamma + {}^{17}_{8}O \rightarrow {}^{12}_{6}C + {}^{4}_{2}He + {}^{1}_{0}n}$$

---

19. **REASONING** Energy is released from this reaction. Consequently, the combined mass of the daughter nucleus $^{12}_{6}C$ and the α particle $^{4}_{2}He$ is less than the combined mass of the parent nucleus $^{14}_{7}N$ and $^{2}_{1}H$. The mass defect is equivalent to the energy released. We proceed by determining the difference in mass in atomic mass units and then use the fact that 1 u is equivalent to 931.5 MeV (see Section 31.3).

SOLUTION The reaction and the atomic masses are as follows:

$$\underbrace{^{2}_{1}H}_{2.014\,102\,\text{u}} + \underbrace{^{14}_{7}N}_{14.003\,074\,\text{u}} \rightarrow \underbrace{^{12}_{6}C}_{12.000\,000\,\text{u}} + \underbrace{^{4}_{2}He}_{4.002\,603\,\text{u}}$$

The mass defect Δm for this reaction is

$$\Delta m = 2.014\,102\ \text{u} + 14.003\,074\ \text{u} - 12.000\,000\ \text{u} - 4.002\,603\ \text{u} = 0.014\,573\ \text{u}$$

Since 1 u = 931.5 MeV, the energy released is

$$(0.014\,573\ \text{u}) \left(\frac{931.5\ \text{MeV}}{1\ \text{u}} \right) = \boxed{13.6\ \text{MeV}}$$

---

21. **REASONING** The rest energy of the uranium nucleus can be found by taking the atomic mass of the $^{235}_{92}U$ atom, subtracting the mass of the 92 electrons, and then using the fact that 1 u is equivalent to 931.5 MeV (see Section 31.3). According to Table 31.1, the mass of an electron is $5.485\,799 \times 10^{-4}$ u. Once the rest energy of the uranium nucleus is found, the desired ratio can be calculated.

SOLUTION The mass of $^{235}_{92}U$ is 235.043 924 u. Therefore, subtracting the mass of the 92 electrons, we have

Mass of $^{235}_{92}$U nucleus = 235.043 924 u – 92(5.485 799 × 10⁻⁴u) = 234.993 455 u

The energy equivalent of this mass is

$$(234.993\ 455\ \text{u}) \left(\frac{931.5\ \text{MeV}}{1\ \text{u}} \right) = 2.189 \times 10^5\ \text{MeV}$$

Therefore, the ratio is

$$\frac{200\ \text{MeV}}{2.189 \times 10^5\ \text{MeV}} = \boxed{9.0 \times 10^{-4}}$$

---

27. **REASONING AND SOLUTION**
a. The number of nuclei in one gram of U-235 can be obtained as follows:

$$1\ \text{gram of U-235} = \left(\tfrac{1}{235}\ \text{mol} \right) \left(6.02 \times 10^{23}\ \text{nuclei/mol} \right) = 2.56 \times 10^{21}\ \text{nuclei}$$

Each nucleus yields 2.0×10^2 MeV of energy, so we have

$$E = \left(2.0 \times 10^8\ \text{eV} \right) \left(\frac{1.60 \times 10^{-19}\ \text{J}}{1\ \text{eV}} \right) \left(2.56 \times 10^{21} \right) = \boxed{8.2 \times 10^{10}\ \text{J}}$$

b. If 30.0 kWh of energy are used per day, the total energy use per year is

$$E_{\text{total}} = \left(30.0\ \text{kWh/d} \right) \left(\frac{3.60 \times 10^6\ \text{J}}{1\ \text{kWh}} \right) \left(\frac{365\ \text{d}}{1\ \text{yr}} \right) = 3.94 \times 10^{10}\ \text{J/yr}$$

The amount of U-235 needed in a year, then, is

$$m = \frac{3.94 \times 10^{10}\ \text{J}}{8.2 \times 10^{10}\ \text{J/g}} = \boxed{0.48\ \text{g}}$$

---

29. **REASONING** We first determine the total power generated (used and wasted) by the plant. Energy is power times the time, according to Equation 6.10, and given the energy, we can determine how many kilograms of $^{235}_{92}$U are fissioned to produce this energy.

SOLUTION Since the power plant produces energy at a rate of 8.0×10^8 W when operating at 25 % efficiency, the total power produced by the power plant is

$$\left(8.0 \times 10^8 \text{ W}\right) 4 = 3.2 \times 10^9 \text{ W}$$

The energy equivalent of one atomic mass unit is given in the text (see Section 31.3) as

$$1 \text{ u} = 1.4924 \times 10^{-10} \text{ J} = 931.5 \text{ MeV}$$

Since each fission produces 2.0×10^2 MeV of energy, the total mass of $^{235}_{92}\text{U}$ required to generate 3.2×10^9 W for a year (3.156×10^7 s) is

$$\overbrace{\left(3.2 \times 10^9 \text{ J/s}\right)\left(3.156 \times 10^7 \text{ s}\right)}^{\substack{\text{Power times time gives} \\ \text{energy in joules}}}$$

$$\times \underbrace{\left(\frac{931.5 \text{ MeV}}{1.4924 \times 10^{-10} \text{ J}}\right)}_{\substack{\text{Converts joules to} \\ \text{MeV. See Sec. 31.3.}}} \underbrace{\left(\frac{1.0 \ ^{235}_{92}\text{U nucleus}}{2.0 \times 10^2 \text{ MeV}}\right)}_{\substack{\text{Converts MeV to} \\ \text{number of nuclei}}} \underbrace{\left(\frac{0.235 \text{ kg}}{6.022 \times 10^{23} \ ^{235}_{92}\text{U nuclei}}\right)}_{\substack{\text{Converts number of nuclei to} \\ \text{kilograms}}} = \boxed{1200 \text{ kg}}$$

---

35. **REASONING** To find the energy released per reaction, we follow the usual procedure of determining how much the mass has decreased because of the fusion process. Once the energy released per reaction is determined, we can determine the amount of gasoline that must be burned to produce the same amount of energy.

SOLUTION The reaction and the masses are shown below:

$$\underbrace{3\ ^2_1\text{H}}_{3(\ 2.0141 \text{ u})} \rightarrow \underbrace{^4_2\text{He}}_{4.0026 \text{ u}} + \underbrace{^1_1\text{H}}_{1.0078 \text{ u}} + \underbrace{^1_0\text{n}}_{1.0087 \text{ u}}$$

The mass defect is, therefore,

$$3(2.0141 \text{ u}) - 4.0026 \text{ u} - 1.0078 \text{ u} - 1.0087 \text{ u} = 0.0232 \text{ u}$$

Since 1 u is equivalent to 931.5 MeV, the released energy is 21.6 MeV, or since it is shown in Section 31.3 that 931.5 MeV $= 1.4924 \times 10^{-10}$ J, the energy released per reaction is

$$(21.6 \text{ MeV}) \left(\frac{1.4924 \times 10^{-10} \text{ J}}{931.5 \text{ MeV}}\right) = 3.46 \times 10^{-12} \text{ J}$$

To find the total energy released by all the deuterium fuel, we need to know the number of deuterium nuclei present. Noting that 6.1×10^{-6} kg $= 6.1 \times 10^{-3}$ g, we find that the number of deuterium nuclei is

$$(6.1 \times 10^{-3} \text{ g}) \left(\frac{6.022 \times 10^{23} \text{ nuclei/mol}}{2.0141 \text{ g/mol}} \right) = 1.8 \times 10^{21} \text{ nuclei}$$

Since each reaction consumes three deuterium nuclei, the total energy released by the deuterium fuel is

$$\tfrac{1}{3} (3.46 \times 10^{-12} \text{ J/nuclei})(1.8 \times 10^{21} \text{ nuclei}) = 2.1 \times 10^9 \text{ J}$$

If one gallon of gasoline produces 2.1×10^9 J of energy, then the number of gallons of gasoline that would have to be burned to equal the energy released by all the deuterium fuel is

$$\left(2.1 \times 10^9 \text{ J}\right) \left(\frac{1.0 \text{ gal}}{2.1 \times 10^9 \text{ J}} \right) = \boxed{1.0 \text{ gal}}$$

---

43. **REASONING**   The momentum of a photon is given in the text as $p = E/c$ (see the discussion leading to Equation 29.6). This expression applies to any massless particle that travels at the speed of light. In particular, assuming that the neutrino has no mass and travels at the speed of light, it applies to the neutrino. Once the momentum of the neutrino is determined, the de Broglie wavelength can be calculated from Equation 29.6 ($p = h/\lambda$).

SOLUTION
a. The momentum of the neutrino is, therefore,

$$p = \frac{E}{c} = \left(\frac{35 \text{ MeV}}{3.00 \times 10^8 \text{ m/s}} \right) \left(\frac{1.4924 \times 10^{-10} \text{ J}}{931.5 \text{ MeV}} \right) = \boxed{1.9 \times 10^{-20} \text{ kg} \cdot \text{m/s}}$$

where we have used the fact that 1.4924×10^{-10} J = 931.5 MeV (see Section 31.3).

b. According to Equation 29.6, the de Broglie wavelength of the neutrino is

$$\lambda = \frac{h}{p} = \frac{6.63 \times 10^{-34} \text{ J} \cdot \text{s}}{1.9 \times 10^{-20} \text{ kg} \cdot \text{m/s}} = \boxed{3.5 \times 10^{-14} \text{ m}}$$

---

47. **REASONING**   The K^- particle has a charge of $-e$ and contains one quark and one antiquark. Therefore, the charge on the quark and the charge on the antiquark must add to give a total of $-e$. Any quarks or antiquarks that cannot possibly lead to this charge are the ones we seek.

SOLUTION

a. Any quark that has a charge of $+\frac{2}{3}e$ (u, c, or t) cannot be present in the K⁻ particle, because if it were, then the antiquark that is present would need to have a charge of $-\frac{5}{3}e$ to give a total charge of $-e$. Since there are no antiquarks with a charge of $-\frac{5}{3}e$, we conclude that

the K⁻ particle does not contain u, c, or t quarks

b. Any antiquark that has a charge of $+\frac{1}{3}e$ ($\overline{d}$, $\overline{s}$, or $\overline{b}$) cannot be present in the K⁻ particle, because if it were, then the quark that is present would need to have a charge of $-\frac{4}{3}e$ to give a total charge of $-e$. Since there are no quarks with a charge of $-\frac{4}{3}e$, we conclude that

the K⁻ particle does not contain $\overline{d}$, $\overline{s}$, or $\overline{b}$ antiquarks

Note: the K⁻ particle contains the s quark and the $\overline{u}$ antiquark.

---

51. **REASONING** To find the energy released per reaction, we follow the usual procedure of determining how much the mass has decreased because of the fusion process. Once the energy released per reaction is determined, we can determine the mass of lithium $^{6}_{3}$Li needed to produce 3.8×10^{10} J.

SOLUTION The reaction and the masses are shown below:

$$\underbrace{^{2}_{1}\text{H}}_{2.014\,\text{u}} + \underbrace{^{6}_{3}\text{Li}}_{6.015\,\text{u}} \rightarrow \underbrace{2\,^{4}_{2}\text{He}}_{2(4.003\,\text{u})}$$

The mass defect is, therefore, $2.014\,\text{u} + 6.015\,\text{u} - 2(4.003\,\text{u}) = 0.023\,\text{u}$. Since 1 u is equivalent to 931.5 MeV, the released energy is 21 MeV, or since the energy equivalent of one atomic mass unit is given in Section 31.3 as $1\,\text{u} = 1.4924 \times 10^{-10}\,\text{J} = 931.5\,\text{MeV}$,

$$(21\,\text{MeV})\left(\frac{1.4924 \times 10^{-10}\,\text{J}}{931.5\,\text{MeV}}\right) = 3.4 \times 10^{-12}\,\text{J}$$

In 1.0 kg of lithium $^{6}_{3}$Li, there are

$$(1.0\,\text{kg of }^{6}_{3}\text{Li})\left(\frac{1.0 \times 10^3\,\text{g}}{1.0\,\text{kg}}\right)\left(\frac{6.022 \times 10^{23}\,\text{nuclei/mol}}{6.015\,\text{g/mol}}\right) = 1.0 \times 10^{26}\,\text{nuclei}$$

Therefore, 1.0 kg of lithium ^6_3Li would produce an energy of

$$(3.4 \times 10^{-12} \text{ J/nuclei})(1.0 \times 10^{26} \text{ nuclei}) = 3.4 \times 10^{14} \text{ J}$$

If the energy needs of one household for a year is estimated to be 3.8×10^{10} J, then the amount of lithium required is

$$\frac{3.8 \times 10^{10} \text{ J}}{3.4 \times 10^{14} \text{ J/kg}} = \boxed{1.1 \times 10^{-4} \text{ kg}}$$

NOTES

NOTES

NOTES

NOTES